THE PREHISTORY OF

Music is possessed by all human cultures, and archaeological evidence for musical activities pre-dates even the earliest known cave art. Music has been the subject of keen investigation across a great diversity of fields, from neuroscience and psychology to ethnography, archaeology, and its own dedicated field, musicology. Despite the great contributions that these studies have made towards understanding musical behaviours, much remains mysterious about this ubiquitous human phenomenon—not least, its origins.

In a ground-breaking study, this volume brings together evidence from these fields, and more, in investigating the evolutionary origins of our musical abilities, the nature of music, and the earliest archaeological evidence for musical activities amongst our ancestors. Seeking to understand the true relationship between our unique musical capabilities and the development of the remarkable social, emotional, and communicative abilities of our species, it is essential reading for anyone interested in music and human physical and cultural evolution.

Iain Morley is a palaeoanthropologist and archaeologist, and is Academic Coordinator of the School of Anthropology and Museum Ethnography, University of Oxford.

The Prehistory of Music

Human Evolution, Archaeology, and the Origins of Musicality

IAIN MORLEY

OXFORD
UNIVERSITY PRESS

OXFORD
UNIVERSITY PRESS

Great Clarendon Street, Oxford, OX2 6DP,
United Kingdom

Oxford University Press is a department of the University of Oxford.
It furthers the University's objective of excellence in research, scholarship,
and education by publishing worldwide. Oxford is a registered trade mark of
Oxford University Press in the UK and in certain other countries

© Iain Morley 2013

The moral rights of the author have been asserted

First published 2013
First published in paperback 2018

All rights reserved. No part of this publication may be reproduced, stored in
a retrieval system, or transmitted, in any form or by any means, without the
prior permission in writing of Oxford University Press, or as expressly permitted
by law, by licence or under terms agreed with the appropriate reprographics
rights organization. Enquiries concerning reproduction outside the scope of the
above should be sent to the Rights Department, Oxford University Press, at the
address above

You must not circulate this work in any other form
and you must impose this same condition on any acquirer

Published in the United States of America by Oxford University Press
198 Madison Avenue, New York, NY 10016, United States of America

British Library Cataloguing in Publication Data
Data available

Library of Congress Cataloging in Publication Data
Data available

ISBN 978-0-19-923408-0 (Hbk.)
ISBN 978-0-19-882726-9 (Pbk.)

Links to third party websites are provided by Oxford in good faith and
for information only. Oxford disclaims any responsibility for the materials
contained in any third party website referenced in this work.

Dedicated to Evelyn, whose songs are inimitable, to Laura, who inspires them, and to Kristin Hersh, whose music is so much more than sound.

Preface

Interest in evolutionary perspectives on music has grown enormously since the turn of this century, when the research for this book began, both in terms of the evolutionary origins of music itself, and its relationship with the evolution of other human abilities. Relevant evidence has also burgeoned, and this book has grown considerably in the making, from its nascence in research for Master's and Doctoral theses. My PhD research at Cambridge from 2000 to 2003 was supervised by Preston Miracle in the Faculty of Archaeology and Anthropology and Ian Cross in the Faculty of Music, and I am greatly indebted to them for their advice and support at the time and since. The thesis was examined by Leslie Aiello and Paul Mellars and I am grateful to them for their comments and subsequent support. The research strategy pursued there, in terms of the types of evidence examined, was developed from that of my Master's thesis on 'The Origin and Evolution of the Human Capacity for Music', completed in 2000 as part of the MA in Cognitive Evolution at Reading University. I am grateful to Steven Mithen, director of the course, for pointing out that this was an important research area that had at that time received little dedicated study in Palaeolithic archaeology or studies of human cognitive evolution; he was very helpful in assisting with composing a funding proposal and suggesting foci for the research, and subsequently read that work closely.

When I began the research I took very seriously the stated requirement that it should constitute 'a substantial contribution to human knowledge'. In fact, the principal attraction of academic work in this area was that it *could* have the potential to have a fundamental impact upon the way in which we understand the human condition. Such ambitions can have a ring of immodesty and presumption: immodesty about the author's own abilities and the merits of their work, and presumption about its reception. It would certainly be pretentious to claim that this piece of work could single-handedly achieve those high ideals. I have been fortunate to land upon a field which still has major features to explore, even from the imperfect distance from which I view it, but there are now more minds of greater experience and dedicated expertise contributing to addressing the issues explored here; I am indebted to their work, which features throughout. I do nevertheless believe the subject matter at least to be fundamental to understanding humans, and can still hope that this might at least put one or two pieces together in the jigsaw of our understanding of how we are today and how we got here.

It is customary in forewords to acknowledge the shortcomings of the work; I do so here not out of custom but necessity. This is an imperfect study of the literature relevant to the question at its heart, not least because, as I have discovered, there is almost no limit to what is relevant to it in some way. There are doubtless many researchers who are carrying out work relevant to the questions of the origins of music whose research is not represented here, and to them I apologize. I look forward to their input. For those who are featured, I hope that I have accurately, even if not completely, represented their research and views.

Significant portions of this research were funded by scholarships from Trinity Hall, Cambridge, and the Arts and Humanities Research Board (now Council); the process of incorporating the latest research and preparing it for publication was funded in large part by a Hunt Fellowship of the Wenner-Gren Foundation of New York, to whom I am very grateful. I am also indebted to the Harold Hyam Wingate Foundation and Caroline Gale for their support and interest in the project at the outset.

Between 2000 and 2009 I was a member of the Faculty of Archaeology and Anthropology and the McDonald Institute for Archaeological Research at Cambridge, and my friends and colleagues there contributed to making it an excellent intellectual home, as did those in Trinity Hall and Darwin College. I am particularly grateful to Colin Renfrew and Graeme Barker for the opportunities and inspiration that they provided. In addition, much valuable archaeological excavation experience was gained through work with Cambridge Archaeological Unit and on the Departmental Palaeolithic projects in Croatia, Moravia, and Libya, and I am grateful to their directors for my involvement (Chris Evans, Preston Miracle, Martin Jones, Graeme Barker). Work on the book was completed whilst a Fellow of Keble College, Oxford, and then as a member of the School of Anthropology and Museum Ethnography, and a Fellow of St Hugh's College, Oxford.

Thanks go to all the people who have recommended papers to me over the years and who have shown such polite interest and enthusiasm for the subject matter, including (but not limited to) the members of the Science & Music Group (Cambridge), the Palaeolithic-Mesolithic Group (Cambridge), the Palaeolithic-Quaternary Group (Oxford), Nicholas Bannan, Elizabeth Blake, Uniit Carruyo, Fred Coolidge, Michel Dauvois, Francesco d'Errico, Becky Farbstein, Tahl Holtzman, Graeme Lawson, Stephen Malloch, Hugh Mellor, Chris Scarre, Jay Schulkin, Krish Seetah, Mila Simões de Abreu, Alice Stevenson, Colwyn Trevarthen, Ivan Turk, and Simon Wyatt. Earlier research by Paula Scothern provided very important material at the start.

My own musical (and other) experience has been very positively influenced by John Alderton, Alasdair Hood, Graeme Hood, Hugh Kleinberg, Roy Gill, Mike Morley, Roman Roth, and my parents. Clearly such contributions cannot be rated too highly.

I am also very grateful to Hilary O'Shea at OUP, for her interest in and commissioning of this project and for her patience over its long development, to Taryn Das Neves, Rowena Anketell, Cathryn Steele, Helen Hill, Annie Rose, Dora Kemp, and Albert Stewart for their excellent work seeing it through to publication, and to the anonymous reviewers of the proposal and manuscript.

Profound thanks go to Laura Preston (Laura Morley) for, amongst other things, her generosity in drawing all of the figures as well as providing very useful comments on drafts of the chapters. Finally, I am forever grateful to all of my family who, I hope, know all the things for which I'd like to thank them.

Contents

List of Figures	xv
List of Tables	xvi

1. **Conceiving Music in Prehistory** — 1
   - Introduction — 1
   - Conceptualizing music and prehistory — 3
   - The organization of the book — 9

2. **Implications of Music in Hunter-Gatherer Societies** — 11
   - Introduction — 11
   - Native Americans of the plains (Blackfoot and Sioux) — 15
   - African Pygmies of the equatorial forest (Aka and Mbuti) — 19
   - Australian Aborigines of the Western Desert (Pintupi) — 22
   - The Eskimo of south-west Alaska (Yupik) and Canada (Inuit) — 24
   - Conclusions — 29
     - *Some common features in the uses and nature of music in four hunter-gatherer societies* — 29
     - *Methods and materials of construction of instruments: implications for the archaeological record* — 30

3. **Palaeolithic Music Archaeology 1: Pipes** — 32
   - Introduction — 32
     - *Introduction to the Upper Palaeolithic* — 33
     - *Introduction to Palaeolithic pipes* — 34
   - The earliest reputed pipes — 35
     - *Mousterian musicianship?* — 38
   - Upper Palaeolithic pipes — 41
     - *The Swabian Alb (Geissenklösterle, Hohle Fels, Vogelherd)* — 42
     - *Isturitz* — 51
     - *Other sites* — 88
   - Representations of instruments — 89
   - The use of bone for instrument manufacture in the Middle and Upper Palaeolithic — 90
     - *Raw material availability?* — 91
     - *Neanderthal use of avian fauna for subsistence and as a raw material* — 92
     - *Use of avian fauna and technological limitations* — 93
     - *Use of avian fauna due to environmental stress* — 94
   - Cultural revolution? — 96

Contents

4. Palaeolithic Music Archaeology 2: Other Sound-Producers ... 99
   Introduction ... 99
   Other aerophones ... 100
      Phalangeal whistles ... 100
      Bullroarers (free aerophones) ... 105
   Percussive instruments ... 109
      Rasps (scraped idiophones) ... 109
      Struck percussion ... 114
   Caves and lithophones ... 115
   Music and dance in later prehistory ... 121
   Archaeology conclusions ... 124

5. The Palaeoanthropology of Vocalization 1: Vocal Anatomy ... 130
   Introduction ... 130
   The vocal apparatus and fossil evidence for its evolution ... 131
      The larynx and basicranial flexion ... 135
      The hyoid bone and mandible ... 144
      The hypoglossal canal and tongue ... 146
      Vertebral innervation, intercostal musculature, and breathing control ... 148
   Some previous explanations for increased tonal range ... 153
   Conclusions ... 158

6. The Palaeoanthropology of Vocalization 2: The Brain and Hearing ... 161
   Introduction ... 161
   Evidence for the evolution of vocal control in the brain ... 161
      Fossil endocasts ... 161
      Neurology of vocal production in primates and humans ... 164
         Voluntary control of emotional content ... 165
         The role of the periaqueductal grey matter (PAG) ... 166
         The FoxP2 genetic mutations ... 168
   The ear, sound perception, and evolution ... 169
   Conclusions ... 174

7. Neurological Relationships Between Music and Speech ... 177
   Introduction ... 177
   Hemispheric organization: language in the left brain, music in the right? ... 180
   Identifying functional neuroanatomy: brain scanning and neuropathology ... 181
   Speech and melody production ... 183
   Processing of tonal information in music and speech ... 185
   Tonal and rhythmic information processing ... 190
   Does the brain have a neurological modular specialization dedicated uniquely to music? ... 194
   Conclusions ... 197

Contents

8. Vocal Versatility and Complexity in an Evolutionary Context	201
Introduction	201
Evidence for an inherited capacity for the perception of melody and rhythm	202
Early vocal behaviours in primate infants	202
Infant-directed speech, music, and vocalization	204
Proto-music/language: rationales for a shared ancestry	214
Social vocalization in primates	220
Evolutionary rationales for complexity of vocalization: proto-music, proto-language, and social vocalization	221
Conclusions	225
9. Vocal Control and Corporeal Control—Vocalization, Gesture, Rhythm, Movement, and Emotion	228
Introduction	228
Vocal content and manual gesture	229
Gesture and vocalization in infants	231
Gesture, vocalization, and meaning	233
Rhythm, corporeal movement, and emotion	237
Entrainment	243
Conclusions	250
10. Emotion and Communication in Music	255
Introduction	255
Intrinsic and extrinsic emotional content of music	256
Ecological context, social context, and the human in music	259
The human in music	261
Autism, Asperger and Williams syndromes	265
The role of the social context in which music is experienced	267
Physiological, neurological, and neurochemical correlates with the experience of emotion in music	268
Conclusions	273
11. Rationales for Music in Evolution	275
Introduction	275
Non-adaptive models of musical origins	277
Some adaptive rationales for the use of music	279
Music and group cohesion	280
Music and dance as a coalition signalling system	283
Music and sexual selection	285
Music and group selection	289
Music's multiple meanings and cognitive development	291
Music and cognitive evolution	294
Cognitive modularity and symbolic thought	294
Mimesis, culture, and cognition	297
Enhanced Working Memory	301
Conclusions	303

12. Conclusions	307
The nature of music	308
Conceiving the foundations of music	315
A timeline for the emergence of musicality	319
Appendix	326
Table 1: Inventory of Palaeolithic reputed pipes and flutes	327
Table 2: Inventory of Palaeolithic objects originally reputed to be pipes and flutes but since deemed unlikely	369
Table 3: Inventory of Palaeolithic reputed phalangeal whistles	378
References	389
Index	427

List of Figures

1.1. Chronology of major known hominin species	4
2.1. Examples of sound-producers used by some Native American hunter-gatherers	18
3.1. Divje babe I bone	40
3.2. Hohle Fels Flute 1 (vulture), Aurignacian	46
3.3. Bird radius and ulna	47
3.4. Geissenklösterle Flute 1 (swan), Aurignacian	48
3.5. Geissenklösterle Flute 3 (mammoth ivory), Aurignacian	50
3.6. Plan of Isturitz cave	52
3.7. Isturitz flute (vulture), Aurignacian	85
3.8. Isturitz flute (vulture), Gravettian	86
4.1. Ungulate leg indicating phalanx and location of piercing	101
4.2. Phalangeal whistles	101
4.3. Possible bullroarers from La Roche, Lespugue, Laugerie Basse, and Badegoule	107
4.4. Possible rasps from Pekárna, Abri Lafaye Bruniquel, and Mas d'Azil	111
4.5. Bison horn held by the Venus of Laussel	113
5.1. Supralaryngeal tract and soundspace	134
5.2. Adult human hyoid bone and larynx location	136
5.3. Comparison of key vocalization anatomy in chimpanzees and adult humans	137
5.4. Basicranial flexion and styloid process	137
5.5. Mylohyoid groove	146
5.6. Hypoglossal canal	147
5.7. Comparison of the ribcage shapes of humans and chimpanzees	149
6.1. Location of Broca's area	162
6.2. Location of anterior cingulate cortex and periaqueductal grey matter (PAG)	165
6.3. Labyrinthine capsule of the inner ear – quadrupedal and bipedal orientations	172
7.1. Diagram showing locations of parts of the brain referred to in text. (A: *Left hemisphere*; B: *Right hemisphere*; C: *Section*.)	180

List of Tables

3.1. Pipe finds from the Ach Valley, Germany	43
3.2. Stratigraphy of the two chambers at Isturitz, France, showing the correspondence between the layers excavated by Passemard and the Saint-Périers	54
3.3. Inventory of Isturitz pipes discussed by Buisson (1990) in chronological order	55
3.4. Isturitz pipes listed as 'lost' by Scothern (1992)	77
3.5. Two more objects from Isturitz described by Lawson and d'Errico (2002)	82
Appendix Table 1: Inventory of Palaeolithic reputed pipes and flutes	327
Appendix Table 2: Inventory of Palaeolithic objects originally reputed to be pipes and flutes but since deemed unlikely	369
Appendix Table 3: Inventory of Palaeolithic reputed phalangeal whistles	378

1

Conceiving Music in Prehistory

> Since music is the only language with the contradictory attributes of being at once intelligible and untranslatable, the musical creator is a being comparable to the gods, and music itself the supreme mystery of the science of man.
>
> Claude Lévi-Strauss (1970, p. 18)[1]

INTRODUCTION

The anthropologist Claude Lévi-Strauss was not alone in regarding music to be a 'supreme mystery'. A century earlier, Charles Darwin had made the observation that since 'neither the enjoyment nor the capacity of producing musical notes are faculties of the least use to man in reference to his daily habits of life, they must be ranked among the most mysterious with which he is endowed' (1871, pp. 369–570). Andersson and Ulvaeus (1978) wonder how music started, and who it was who found out that a melody can capture a heart like nothing else. Indeed, they ask what we would be without a song or a dance. Musical experience of some form is an important part of the life of virtually all of us, across the globe. It pervades our daily lives, our celebrations and mourning, our leisure and socializing; it is used to help solicit our purchasing power, our votes, and our romantic attentions. How can this be? It is clearly more than just a leisure activity. Why do we respond emotionally to music? Are music and language abilities really connected? When did the capacities for musical behaviours first develop? What might musical behaviours have been like when our ancestors first carried them out? Perhaps one of the most mysterious issues concerns why they would have done so. Were there selective reasons for these developments within human ancestors? This book aims to investigate this supreme mystery, and how it is that it makes us what we are.

[1] Also quoted in Storr (1992), p. xi.

Investigation into the evolution of (other aspects of) human culture and behaviour has a long history (for an excellent discussion and critique of the different approaches that have been taken since the advent of Darwinism, see Laland and Brown, 2011), and studies of the evolutionary development of the human brain and cognition have burgeoned in the last couple of decades. This research has spanned a number of disciplines, including archaeology, biological anthropology, psychology, philosophy, and linguistics. Many areas of cognition have received attention in the context of this research, such as cultural and sexual behaviours (e.g. Dunbar et al., 1999), intelligence and cognitive processing systems (e.g. Barkow et al., 1992; Byrne, 1995; Carter, 1998; Deacon, 1997; Dennett, 1998; Donald, 1991, 2001; Coolidge and Wynn, 2009), and (the focus of a great deal of interest) the evolution of human language abilities (e.g. Aiello and Dunbar, 1993; Pinker, 1997; Jablonski and Aiello, 1998; Dunbar, 1998; and many others). Work in cognitive archaeology has paralleled this research, drawing upon the fossil and archaeological record to make inferences about the development and nature of intelligence and cognition in human ancestors (e.g. Ingold and Gibson, 1993; Mellars and Gibson, 1996; Mithen, 1996; Renfrew and Scarre, 1998; Renfrew and Morley, 2009; de Beaune, Coolidge, and Wynn, 2009; Coolidge and Wynn, 2009).

Within this body of research, however, the evolution of musical abilities has been relatively neglected, especially in the latter half of the twentieth century. Not so long ago, the question as to whether musical behaviours had any biological basis, as opposed to being solely a cultural phenomenon, was considered to be highly contentious. Although it had not always been so, the idea that music was nothing more than a satisfying but evolutionarily superfluous phenomenon—'auditory cheesecake' (Pinker, 1997)—had become one that would have been echoed by much of the scientific community. Perhaps ironically, Pinker's description has become one of the most oft-quoted passages in literature relating to music's functions and origins, and is discussed further in Chapter 11, but those who are interested in a succinct and effective summary of difficulties with this view are pointed to Trehub (2003); the evidence presented in this book will, it is hoped, contribute to viewing music in a different way. Whilst researchers in particular areas of musical cognition have increasingly turned to discussing the evolutionary implications of the work in their own fields (e.g. Peretz and Zatorre, 2003; Patel, 2008), it is only recently that any unifying approach to the issue has been attempted, bringing together the results of numerous cognitive and behavioural fields and situating them in an evolutionary context (Cross, 1999b, 2003b; Wallin, Merker, and Brown, 2000; Morley, 2002a, 2003; Mithen, 2005; Cross and Morley, 2009; Malloch and Trevarthen, 2009a; Schulkin, 2013).

Neuropsychological and developmental studies suggest that the human musical ability has a deep evolutionary history (e.g. Wallin, 1991; Carter, 1998) but this contrasts with the evidence (discovered to date) from the archaeological record, in which musical instruments and evidence for the possible use of acoustics do not appear until around 40,000 years ago (e.g. Scothern, 1992; Lawson et al., 1998; Lawson and d'Errico, 2002; D'Errico et al., 2003; Conard et al., 2009; Morley, 2002a, 2003, 2006, 2011; Higham et al., 2012). This is at least 200,000 years after the first anatomically modern humans appeared and a great deal longer after the emergence of various other important cognitive capabilities. What is music's true place in our evolutionary past, and what can this in turn tell us about music? Was Darwin right to observe that neither the enjoyment nor the capacities to produce musical notes are of the least use in the daily life of humans?

It is hoped that this book can make a contribution to the further development of cognitive archaeology and to our understanding of human musical ability. It examines the evidence for the emergence of the capacities underlying musical behaviours, their interrelationship, development, and ultimate manifestation in the Palaeolithic period. The aim is to understand the development of these capacities, potential selective reasons for their development within human ancestors, functional and cognitive links between musical abilities and other abilities (including the earliest language abilities), evolutionary rationales for human emotional responses to music, and the possible nature of and reasons for the early use of music. To these ends, a multidisciplinary approach is taken, bringing together evidence from a broad diversity of fields, including palaeoanthropology, archaeology, ethnomusicology, neuroscience, developmental and social psychology, and evolutionary biology.

CONCEPTUALIZING MUSIC AND PREHISTORY

This book is entitled 'The Prehistory of Music', so we ought to be clear at the outset what is meant by each of those terms. Prehistory is relatively easy. Prehistory in its strict sense refers to the period prior to written records; this does not have a specific start date, and the end of prehistory and 'start' of history is very different in different parts of the world. In this book, dealing as it does in human prehistory, it is taken to refer to the period since our last common ancestor with chimpanzees, around 6 million years ago; this book does consider evidence from right up to the present day, from both humans and our primate relatives, but this is mostly regarding its relevance for our evolutionary context. The major prehistoric hominin species of the last 6 million years are shown in Figure 1.1, many of which are discussed in this book. In a broader sense, the term prehistory is also often taken to refer to the ancestral forms of things, and it is meant here in this way too.

Defining music is, notoriously, a trickier proposition. Mithen (2006) has said that attempts to define 'music' can be rather obscure and that terminological

4 *The Prehistory of Music*

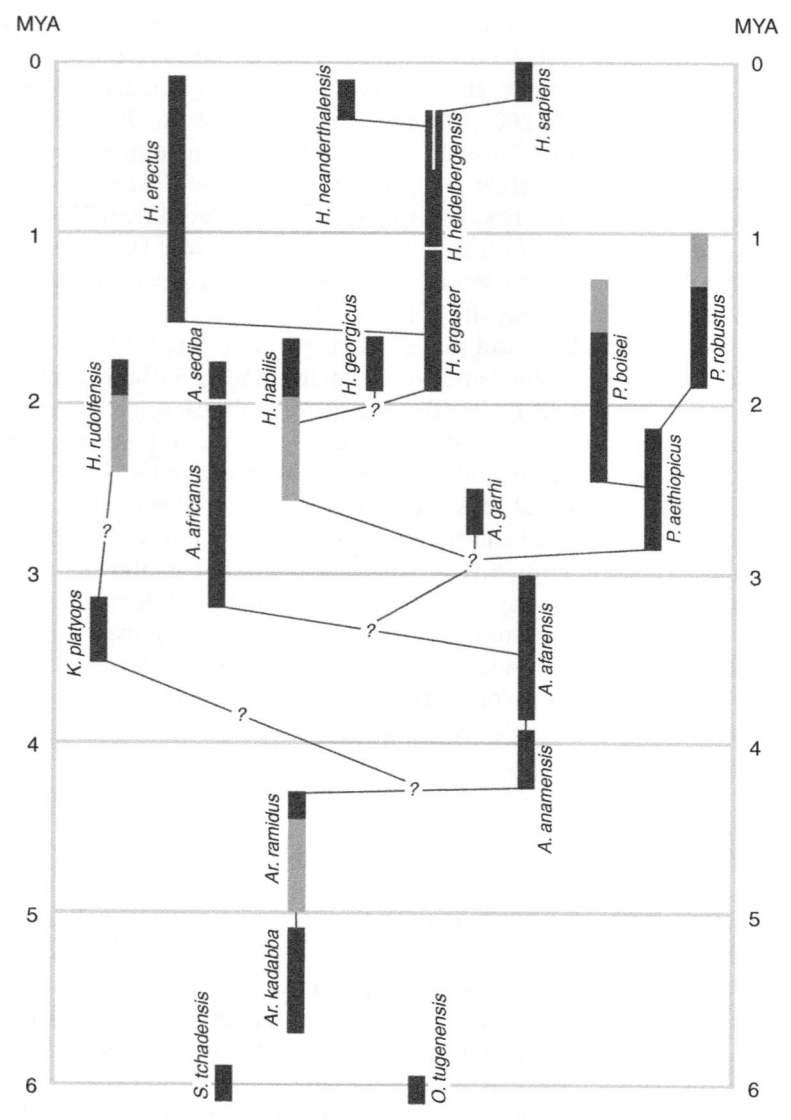

Fig. 1.1. Major known hominin species since our last common ancestor with chimpanzees. Genus names are abbreviated: H. = Homo; A. = Australopithecus; P. = Paranthropus (these species are sometimes also known as Australopithecus); K. = Kenyanthropus; Ar. = Ardipithecus; S. = Sahelanthropus; O. = Orrorin. It has been suggested that K. platyops, H. rudolfensis and H. habilis should be re-classified as Australopithecus. Question marks signify hypothesized connections between species via as-yet unknown hominins. Grey sections of bars represent possible temporal extent beyond that confidently known at present. MYA = Millions of Years Ago (Modified from Klein 2009, fig. 4.49, p. 244.)

debates regarding the definitions of such concepts as music, language, and symbolism deal with culturally specific products, so are not productive regarding studies of their evolutionary origins. Accordingly his book (2005) on 'The origins of music, language, mind and body' does not enter such debate, though he does subsequently (2006) state that we need to make an explicit distinction between 'a "natural biologically based musicality" and music as a culturally constructed phenomenon which builds upon that biological basis' (p. 109). This is indeed an important distinction to draw, and we are interested in both, in trying to understand how the latter has emerged from the former, the biological foundations. Indeed, one of our main objectives in seeking to understand the prehistoric origins of biologically based musicality is to better understand what music really *is* and how it can achieve the effects that it does in the many contexts in which it is used—our conception of music must itself be informed by our investigation into the phenomenon and, in the end, the investigation should allow us to better formulate a definition. This means that any *definition* of music that we start with might risk the circularity of us having too narrow a focus and investigating only what we already think we understand it to be. But clearly we need to be able to articulate, in broad terms, what the focus of our investigation must be; in order to understand what the biological foundations are that we need to investigate we must have a clear idea of what core features the cultural phenomenon of music possesses.

In literature dealing with music psychology and anthropology it is widely asserted that all cultures and societies have music (e.g. Clynes, 1982; Storr, 1992; Blacking, 1995; Brown et al., 2000), but definitions of what constitutes music are few and far between. It is perhaps indicative of a sense that musical knowledge is somewhat intuitive that few authors consider it necessary to define the term. In studies of music cognition and psychology, music tends to be dealt with either holistically in terms of (Western) composed pieces or, in contrast, as constituent elements, such as discrete pitches or transitions; neither requires a definition of music to be specified. Developmental studies are also generally concerned with the emergence of *elements* of music perception and production, or of production and perception of Western music, again bypassing the necessity of defining the entity as a whole. Palaeolithic music archaeology has generally simply equated music with instrumentation (in contrast, d'Errico et al., 2003, are noteworthy in eschewing this view).

Dictionary definitions, such as 'the art of combining sounds of voices or instruments so as to achieve beauty of form and expression of emotion' (*Concise Oxford English Dictionary*; Sykes, 1983) and 'the art or science of arranging sounds in notes and rhythms to give a desired pattern or effect' (*New Dictionary of Music*; Jacobs, 1972) may seem to describe what is familiar to us in music, but are not adequate to encompass the diversity of uses and effects of music across all human cultures (Cross and Morley, 2009). Much of the problem stems from the difficulty of identifying universality in different cultures' conceptions of music; for example, some cultures contain no single word that relates to what we recognize as music, whilst in many cultures no clear distinction is made (or is indeed possible) between music and dance as separate activities, both being viewed as facets of the same activity (Cross and Morley, 2009). Examples include *nkwa* of the Igbo of Nigeria (Waterman, 1991), the Sanskrit *sangīta,* and Thai Buddhist *wai khruu* (Bohlman, 2002). Indeed, the trend in ethnomusicology between the 1940s and 1970s was to frown upon attempts to identify musical universals, a pursuit seen as devaluing the cultural diversity of behaviour. This view has gradually been accompanied by a return of an interest in some quarters in identifying definitive features of musical activities across cultures (Nettl, 2000).

Is 'music' really an infinitely variable cultural product (albeit of common biological capabilities) that is impossible to tie down to a single conception? The fact that it is possible to recognize a diverse range of activities across all cultures as 'musical' would suggest that there is, at least, an identifiable common set of underlying characteristics that occur in various combinations in different cultures. In order to come up with a conception of music that gets beyond the culturally specific elements, we will need to identify universal features of the activities, shared by all cultures. Bruno Nettl (2000) reports that 'All societies have vocal music.... All societies have at least some music that conforms to a meter or contains a pulse.... All societies have some music that uses only three or four pitches, usually combining major seconds and minor thirds' (Nettl, 2000, p. 468). Rhythm may be produced by striking objects or the body (e.g. foot-stamping, clapping) or may be produced by rhythmic vocalization. Melody may be produced instrumentally or with the voice.

Trehub (2003) outlines some common features in scale structures across cultures: there is typically a division of the octave into five to seven discrete pitches; this may be related to human cognitive factors (Burns, 1999). These discrete pitches tend to be separated unequally across the scale, in some cases separated by a tone, in others by a semitone (Sloboda, 1985). It appears that the human perceptual system is set up to better process unequal scale steps than equal ones (Butler, 1989; Shepard, 1982), from infancy (Trehub et al., 1999). The 'perfect fifth' interval is also ubiquitous and is processed more effectively than other intervals by all ages (Schellenberg and Trehub, 1994, 1996a, 1996b). Finally, the perception of consonance and harmony, and of dissonance, are also direct products of universal properties of the human auditory system (Tramo et al., 2003), with aversive emotional reactions to dissonance being innate (Trehub, 2003; Gosselin et al., 2006). Trehub (2003) observes that musics of the world capitalize on processing predispositions such as these and, as a consequence, musical experience in general tends to intensify these initial predispositions.

In summary, it would appear that musical behaviours amongst all humans involve the encoding of sounds into pitches (usually between three and seven) which are unequally separated across the scale, including the perfect fifth, favouring consonance and harmony over dissonance, and organizing sequences of sounds so that they have a deliberate temporal relationship to each other. It is clear that although the cultural manifestations of these behaviours are very varied, frequently incorporating other features too, the variation builds upon a limited number of core common elements, which are derived from underlying biological predispositions. The ways in which these elements and their relations are used varies from culture to culture but the mechanisms required to make use of them, and to process that use, are universal, varying only in their application relative to the conventions of the culture. To understand the efficacy of the behaviours we will need to understand the nature of the biological predispositions.

So what can we say, at the outset, about the capabilities that such musical activities require? A single musical event (e.g. a tone or a chord or a beat) is not, on its own, music. It only becomes musical in the context of its relation to other, similar, elements—and the consistencies and differences between them. Peretz (2003) proposes that there are two principal 'anchorage points' of brain specialization for music: 'the encoding of pitch along musical scales and the ascribing of a regular beat to incoming events' (p. 201). Carrying out musical activities requires the production of finely controlled pitches and, usually, changes between them. Production itself obviously involves refined perception of pitches and pitch changes, as does the experience of musical stimuli. The pitch information that is extracted from the stimulus is also processed by other mechanisms that take a role in attributing significance—or meaning—to auditory stimuli.

The production of musical stimuli also involves the production of sounds with a particular rhythm—i.e. with a deliberately controlled temporal relationship between the sounds—and often a regular tempo, or pulse. This, like tonal vocalization, requires fine control of sequential muscular movements. Like vocalization the successful performance of this also requires the effective extraction of information from the signal, and moderation of the activity in response to that. This may also involve the achievement of synchronization with other people producing rhythmic auditory information at the same time. Again, as with other auditory stimuli, other mechanisms must take a role in attributing significance to the stimulus.

Reviewing many diverse characteristics of activities considered to be *musical* in different cultures, including exceptions to several traits often quoted as universal, Brandt et al. (2012) ultimately define music as 'creative play with sound' (p. 3). To Brandt et al. the abilities to produce and perceive timbre, pitch, tonal contour, dynamic stress, and rhythm are used in such diverse ways in different cultures that a conception of music should be based not on the specific nature of those characteristics within the musical context but the process by which they are produced and perceived. The implication is that any sound produced and/or perceived by humans in a creative, playful way is music. The relationship with creative play is a potentially important one, and the voluntary, creative, and subjective experience of these stimuli is clearly very significant; however, there may be other characteristics of musical stimuli that it is important to consider, and it is important that we also keep in mind the underlying capabilities that are required.

Some authors, whilst not seeking specifically to define music, have highlighted critical properties of music. Besson and Schön (2003) state that 'Music is the acoustic result of action' (p. 271). Not all acoustic results of action are viewed as music, however (for example, hammering a nail into a wall—although Brandt et al. (2012), might contend that it *could* be), but this does make the fundamental point that music is an embodied activity. The production of sound of any kind unavoidably involves use of the body, and the production of planned and controlled sequences of sound, rhythmic and melodic, involves precise control and sequencing of bodily action. This is clearly a fundamental capability underlying musical ability, and we should expect it to have important implications for our perception of music too. Turner and Ioannides (2009) also address this element of music when they refer to music 'as a structured and intentional succession of movement-produced sounds' (p. 148), and they include the important criteria that the sounds are structured and intended. The temporal structure is particularly emphasized by Cross (2003b), who says that 'Music embodies, entrains and transposably intentionalises time in sound and action' (p. 24). This description encompasses both the corporeal and auditory elements of musical performance and perception, and its potential to have multiple meanings—it means different things to different people in different contexts (referred to as 'transposable intentionality'). It may be noted that dance also embodies, entrains, and transposably intentionalizes time in action, and usually also in sound (either intentionally or as a by-product); this is important, and the commonalities—and interdependences—of music and dance will become clearer over the coming chapters.

So, on the basis of the above properties of music we can say that musical activities rely on the ability to voluntarily produce sequences of sounds moderated for intensity and/or pitch and/or contour, generated by metrically organized muscular movements (often coordinated (entrained) with an internally or externally perceived pulse), plus the ability to process and extract information from such sounds. Dancing also clearly involves the voluntary generation of finely-controlled metrically-organized muscular movements, often coordinated (entrained) with an internally or externally perceived pulse. The investigation into the prehistory of these behaviours must be concerned with the prehistory of these abilities.

THE ORGANIZATION OF THE BOOK

This book draws upon a great diversity of evidence from different disciplines in its aim of building up a picture of the interrelationship and co-evolution of the various capacities that underlie musical behaviours. Each of the chapters draws upon evidence from a number of different fields of investigation, for example palaeoanthropological studies, archaeology, neurological studies, primate studies, and studies of infant capabilities, amongst others. Although some chapters focus more on one type of evidence than others, the reader will not find one chapter which deals with neurological evidence, and one which deals with developmental evidence, for example. It makes more sense to approach this material on the basis of the implications of these different types of evidence for particular issues, rather than by 'type of evidence'. Each chapter draws upon evidence from related areas of investigation, and the findings in each are relevant to the findings in the others. These chapters could have been presented in a different order; the uniting theme is the implications of these various sources of evidence, and their proposed interpretations, for the development of the investigated capacities in an evolutionary context.

An important source of information about the possible roles that musical practices can fulfil is, of course, the ways in which they are used today across the wide variety of human cultures. Chapter 2 looks at the nature and roles of musical behaviours amongst four very different types of hunting and gathering societies from around the world, in order to gain a broader conception of what constitutes music than can be gained from familiar, 'Western', music, which is in many respects shaped by relatively recent changes in our technology and economic organization. This allows the identification of some more cross-cultural characteristics of musical activities and their uses, as well as refining our expectations regarding what evidence for these we might expect to see in the archaeological record.

The earliest archaeological evidence for musical activities constitutes the focus of the next two chapters. Chapter 3 examines the earliest evidence for pipes ('flutes'), associated with some of the first *Homo sapiens* in Europe, dating to around 40,000 years ago, as well as some earlier reputed examples. This includes detailed discussion of the examples from the Ach Valley sites in Germany and Isturitz in the French Pyrenees, as well as later examples from throughout the Upper Palaeolithic period. Chapter 4 goes on to look at evidence for other possible sound-producing artefacts, including rasps, bullroarers, and caves themselves, and concludes with a discussion of the implications of all of the Palaeolithic archaeological evidence for our understanding of the appearance of recognizable musical behaviours in humans.

The preceding chapters investigate the evidence for recognizable musical behaviours amongst *Homo sapiens*; the subsequent chapters go on to look at evidence for the evolutionary development of the capacities that make musical behaviours possible, and how they shape the effects that musical activities have upon us. Chapters 5 and 6 focus principally on the evolution of the physiological and neurological capacities that support tonal vocalization, a key component of musicality. Chapter 5 examines the evidence for the evolution of the physiological features that support vocal capabilities, and rationales that have been offered for this development. Chapter 6 looks at fossil and neurological evidence for the development of control of vocalization, and for the evolution of the ear and sound perception. Chapter 7 looks in particular at neurological evidence for relationships between musical and speech production and perception, including both tonal and rhythmic content, and what these can tell us about the emergence of these capabilities in the human brain.

Chapters 8, 9, and 10 move on from looking at the physiological (including neurophysiological) relationships between musical and speech vocalization to looking at relationships between them in terms of their functionality. Chapter 8 discusses the elements of melody and rhythm production that are innate, biological capabilities, present in primate and human infants, and the roles of tonal vocalizations in social interaction in humans and other higher primates. Chapter 9 looks at the relationships between vocal control and gesture, rhythm, movement, and emotion, including the role of entrainment—the ability to synchronize with a pulse—in human musical experience. Chapter 10 is concerned specifically with emotion and communication in music, including types of emotional content in music, the role of the context in which music is experienced, and the physiological reactions that occur when experiencing emotion in response to musical stimuli.

All of the chapters outlined so far deal with evidence for the evolution of specific aspects of musicality. Chapter 11 goes on to examine ways in which musical behaviours as a whole might have selective benefits, and the differences between selection for the capacities for the behaviours and selection for musical behaviours as a whole in the context of their cultural practice.

Each of these different elements of the investigation into the prehistory of music has a direct bearing on the significance of each of the others. Chapter 12 brings together the implications of all these sources of evidence in order to build an understanding of the nature of musicality, the origins of music, and its place in human evolution.

2

Implications of Music in Hunter-Gatherer Societies

> You see in your mind's eye the power in everything, even in a blade of grass. That's why people become singers. The world is too beautiful for even our language to describe.
>
> Willy Selam of the Chinook, quoted in Layton (2006), p. 93

INTRODUCTION

An important source of information about the roles that musical behaviours might fulfil is, of course, their uses today. Whilst examination of musical practices in Western society can provide us with many insights into the nature of those behaviours and the capabilities that underlie them, we should note that, although varied, the forms of music that predominate today in the Western world, and the practices associated with them, do not encompass the variety of musical behaviours practised by humans today, or in the past.

To develop a fuller picture of the form and some of the roles of musical behaviours we must look more widely, at musical practices in other types of societies. As discussed in Chapter 1, musical behaviours amongst all humans involve the encoding of pitches into between three and seven pitches which are unequally separated across the scale, including the perfect fifth, favouring consonance and harmony over dissonance, and organizing sequences of such pitches to a regular pulse. We have seen too that these are part of a continuum with dance, both involving voluntary metrically-organized expressive movement. These are common features of the musics, but are there also commonalities in the nature of the practices and their roles? Examination of the ways that musical behaviours are used in societies that subsist by hunting and gathering may highlight elements that have more in common with traditional musical practices than those of Western society do, because of recent changes to the form of our use and experience of music wrought by changes in our technology and economic organization. The organization of settled agricultural societies is very atypical of that which was predominant throughout all but the last 5,000–10,000 years of human history. In the rest of that time, as well as during the last 10,000 years in many locations, subsistence by a combination of gathering and hunting, coupled usually with a degree of mobility, was the norm—and the most effective and successful means of subsistence in those circumstances.

It is very important to note, however, that modern and recent-historical hunter-gatherer societies are *not* prehistoric populations preserved into the present day, but societies for whom hunting and gathering continues to represent the most effective and efficient means of subsistence in their given environment. The intention of looking at recent examples of hunter-gatherer music is not for them to be used as direct models for the interpretation of the archaeological record of musical activities in prehistoric hunter-gatherer societies, but to examine and illustrate a wider diversity of the musical behaviours that exist, and demonstrate that what constitutes these behaviours can be rather different from our expectations based on our own experience of music—and for this knowledge to ultimately inform a more inclusive interpretation of prehistoric evidence. Complementary to the aim of illustrating the diversity of musical behaviours is the aim of identifying shared features of musical behaviours across these cultures. We have already outlined some common structural features shared between different cultures' musical behaviours (Chapter 1); in spite of the diversity of musical practices that build upon those foundations, are there also shared elements to the uses of music?

A further aim of this chapter is to examine the nature and the uses of instruments, such as are used, how they are manufactured, and what resources are employed in their construction. This aspect of the evidence may indeed have implications for interpretation of the archaeological record. Western music of the last few centuries has been dominated by instrumental sound production; we must ask whether a conception of musical practice of this form is really representative of typical musical practice more widely across human populations. To what extent are instruments really important in other cultures' musical practices? Are the resources used to produce sound-making devices subject to complex modification or are they used in their natural state? Are the instruments melodic, percussive, or both? It may be possible to create very effective sound-producers without the technological sophistication required for modern instruments or even for other tools. The answers to these questions could have important implications for the capability of prehistoric and pre-modern humans to undertake similar behaviours, and the likelihood of the preservation of such artefacts archaeologically.

As a consequence of these analyses it should be possible to identify a diversity of types of instrumentation, including some which may have been available to be used by early humans, which we might expect to be able to find or, on the contrary, which have been lost to the archaeological record. Through looking at the uses of music within societies which hunt and gather, it may also be possible to identify potential benefits to be had from the use of music within such a social and subsistence organization.

Despite a particular interest in the nature and uses of music in 'traditional' and hunter-gatherer societies in the late nineteenth and early twentieth century (see Toner, 2007, for an interesting overview of the history of cross-cultural music research), the majority of the anthropological literature of the last forty years deals quite cursorily with the topic of music, if at all. This is probably as a consequence of the parallel emergence of the separate discipline of *ethnomusicology*, which developed from late-nineteenth- and early-twentieth-century *comparative musicology*. Whilst initially motivated, as its original name suggests, by musicological analysis and cross-cultural comparison, as well as early psychological studies (Toner, 2007), ethnomusicology has subsequently focused more on analysis of the music itself and the methodology of its study than the traditional uses of the music and its instrumentation. *Anthropological* aspects of the use of music, if studied, have tended to focus not on the traditional forms, but on the influence and adoption of other forms as a consequence of contact with outside influences.

The early anthropological sources, although at least taking an interest in its nature and importance, often deal with music in a rather ethnocentric way, with many references to 'savages' and 'primitives', up until at least the nineteen-fifties. More recent authors who do address the topic have tended to concentrate on the uses of music for political communication and as an indication of social change today, rather than on the traditional uses of music within the cultures (e.g. Barac, 1999; Broughton et al., 1994). For example, the *Cambridge Encyclopedia of Hunters and Gatherers* (Lee and Daly, 1999), despite including excellent sections dealing with the languages, traditions, lifestyles, and habitats of most of the world's hunters and gatherers, dedicates only six of its 500 pages specifically to music in those societies, four and a half of those concerned with modern use of music (Barac, 1999). Many other volumes pertaining to hunter-gatherers contain no mention of music, instruments, or singing at all (e.g. Bettinger, 1991; Schrive, 1984, to name just two). Fortunately, the work of ethnomusicologists such as John Blacking, Bruno Nettl, David McAllester, David Locke, and Jerome Lewis does make detailed study of the instrumentation and purposes of the music of the cultures they studied. Considering that their studies suggest that music impinges on a great variety of aspects of life in these cultures, it is all the more surprising that there has been such a dearth of reporting and discussion elsewhere.

This chapter describes the types of instruments used traditionally by four groups of modern hunter-gatherers from around the world, and the nature of the musical activities in which they are engaged. These groups are Native Americans of the Plains (Blackfoot and Sioux populations), the Aka and

Mbuti African Pygmies, the Pintupi-speaking Australian Aborigines, and the Arctic-dwelling Yupik (south-west Alaska) and Inuit (Canada). These groups occupy four very different types of environments: rolling temperate grasslands, wet rainforest, arid desert, and Arctic tundra respectively; also, with the exception of Asia, between them there is a representative from each of the continents still inhabited by hunter-gatherers. In each case, the ways in which music is used within those cultures, the use of instruments, and their materials of manufacture are examined with a view to illuminating possible natures and roles of musical behaviours within the context of hunter-gatherer subsistence and social organization.

Whilst the examples examined are by necessity selective, and there is a huge diversity of musical practice in other cultures not represented here, those discussed have been chosen with particular characteristics in mind. First, these populations all subsist by hunting and gathering, which is the subsistence strategy that predominated throughout human history. This means that we can be confident that particular traits of their musical practices are not the product of a settled and/or agricultural mode of subsistence (and all else that entails). Second, they each occupy very different ecological niches, and are united solely by their subsistence method and their humanity. Third, these populations (with the probable exception of the Plains and Arctic Native Americans from each other) have been geographically separated from each other by tens of thousands of years. For the latter two reasons, elements of their use of music that they have in common with each other are unlikely to be shared as a consequence of either their ecology or cross-cultural contact, but more likely a product of other things that they have in common, namely, their subsistence method, a shared heritage, or a convergence[1] in practice on the basis of properties of human cognition and its responses to music.

Given the diversity of habitats and the temporal and spatial separation of the hunter-gatherer peoples considered, and that they are united only in their humanity and subsistence strategy, similarities in their musical behaviours could suggest information about the underlying foundations of these behaviours. As noted, close parallels between the groups would imply either convergent development of these behaviours or a shared cultural heritage. The former (convergence) would suggest that there are important driving forces towards those common behaviours, either through human biology or through their efficacy in those subsistence circumstances, these being the principal common factors between the groups. The latter situation (shared heritage) would indicate some very ancient traditions in musical behaviour, as these groups are probably separated by at least 50,000 years. For a detailed discussion

---

[1] Convergence, or *equifinality*, describes the situation whereby different species or genera (or, by extension, communities) develop the same physical or behavioural adaptation under different circumstances.

of a particular possible case of this, concerning equatorial rainforest Pygmies and desert bushmen in Africa, see Grauer (2006) and critical replies, and Grauer (2007), which also discusses Native Americans.

Bearing in mind the issues and limitations (and potential uses) associated with making analogies between past societies and the present ethnographic record, the aims of the present chapter are quite specific. It does not seek to explain Palaeolithic musical evidence by drawing direct parallels with the present or recent history. Instead, it seeks to draw attention to the diversities and similarities of musical behaviours and instrumentation evident, and to highlight the implications of these for a richer conception of how and why such behaviours may have been manifested in human populations than could be gained from knowledge of familiar Western musical practices alone.

Although modern hunter-gatherers are not to be considered to be a direct analogy for Palaeolithic humans, some legitimate parallels may be drawn, in terms of shared constraints. Their subsistence strategies, in a broad sense, are similar, and have an influence on the ways in which they must organize their lives and interact with the world in order to survive. Furthermore, the tools and raw materials which these peoples have available to them for the creation of instruments often resemble those available to early humans. For example, before contact with Westerners, the Plains Indians (Taylor, 1991), Pygmies (Ichikawa, 1999), and Australian Aborigines (Morton, 1999) made no use of metal in the manufacture of their tools and artefacts, using only wood, stone, and occasionally bone.

NATIVE AMERICANS OF THE PLAINS (BLACKFOOT AND SIOUX)

The first group of hunter-gatherers to be looked at are the Native Americans of the Plains of central North America. Although there were at least thirty-two tribes occupying this region in the year 1800, some of the best documentation relates to the Sioux and the Blackfoot tribes. These tribes both lived in areas of relatively high humidity, producing rolling grasslands as their major habitat, the Blackfoot in the northern plains, and the Sioux in the east (Taylor, 1991). This Plains environment may in fact be more reminiscent of the habitat occupied by ancestral hunter-gatherers in Africa than those habitats that are home to hunter-gatherers in Africa today. This is because hunter-gatherers in Africa today occupy areas either of very low rainfall (e.g. the !Kung San of the Kalahari Desert) or of very high rainfall (e.g. the rainforest-dwelling Pygmies), with all of the temperate grassland now occupied by settled agriculturalists (Foley, 1992).

The Blackfoot were traditionally nomadic hunters of antelope and bison (buffalo) and, until the introduction of horses by Europeans in the eighteenth century, hunted and travelled on foot. Interestingly, they used a hunting method which was also used by Middle and Upper Palaeolithic hunters, both Neanderthal and modern humans (Chase, 1989). The men would drive a herd of the animals into a v-shaped drive leading over a cliff edge or ditch until the animals fell to their deaths, ready to be collected and processed (Taylor, 1991). Particularly relevant is the use of song in this procedure amongst the Blackfoot. The herd was initially enticed towards the drive area by a young man singing a spiritually potent song in the manner of a bleating calf (Kehoe, 1999).

All able-bodied members of the group would help process such a kill. Blackfoot women also harvested plant foods from the surrounding environment (and some limited cultivation), such as berries, bulbs, and turnips (Kehoe, 1999). They followed the buffalo herds' annual movements, living in bands of ten to twenty tipis, with around eight persons in each, moving to grasslands in the spring, meeting with other groups in the summer, and then to sheltered river valleys in the autumn (Epp, 1988).

The summer meeting between groups involved the resolution of disputes, policies, and trade, as well as the performance of the 'Sun Dance', to bring prosperity and health. Groups had leaders, and certain families were seen as privileged, such that they did not have to participate in daily menial work, but were instead concerned with leadership activities. However, leadership was subject to the views of the members of the group (including children), and individual autonomy was also highly valued (Kehoe, 1999). Although the nomadic nature of their existence is curtailed by the reservation system today, many of the other activities of the society, including music, are still practised (Nettl, 1992).

The music of the Plains Indians resembles that of the majority of the rest of the Native Americans, in that it is almost always *monophonic* (contains only one melody proceeding at any one time), and the melody is nearly always vocal rather than instrumental (Nettl, 1992; McAllester, 1996). Such instrumentation as there is consists predominantly of percussion in the form of drums or rattles, which are used to accompany the vocal melody. This limited variety of types of instruments is in fact typical of the whole of the North American continent, where the selection of musical instruments used by the native inhabitants is surprisingly small, considering the widespread use of music and variety of peoples.

Although almost all conform to basic types of idiophone (instruments whose bodies vibrate to produce their sound), the instruments within this category are very varied. There are rattles made from dried gourds (which rattle due to the dry seeds or stones inside), tree bark, and spiders' nests, as well as deer hooves and turtle shells, for example (McAllester, 1996). The deer-hoof rattles consist of around twenty doe hooves suspended from a stick. Drums tend to be either frame drums or barrel drums, made from wood and skin, some consisting only of a piece of rawhide suspended from a stake (Nettl, 1992). Sometimes simply a plank of wood is beaten (Nettl, 1956) and idiophones made from a notched stick are also used (Nettl, 1992). Much less common, but also used, are some aerophones such as bullroarers and whistles/flutes. The latter often have no finger holes, and are made of wood or bird-bone; others have up to six finger holes (Nettl, 1956). It is noteworthy that all of the instruments used are manufactured from organic matter, occurring naturally in the Plains Indians' environment, and that many require little, if any, modification before use. Figure 2.1 illustrates some examples of Native American instruments, from a selection of groups.

It may seem that the music is rather simplistic, but amongst the Plains Indians the value of the music is not measured in terms of its complexity. Instead, it is its ability to integrate ceremonial and social events, to integrate society in general and represent it to outsiders, and to evoke supernatural influence that is important (Nettl, 1992). The Plains Indians traditionally believed that music came to people through supernatural input in dreams, so little credit for agency in composition (in the contemporary Western sense) was given to individuals (Nettl, 1956), although considerable credit could be given for being the recipient of such supernatural input. The use of music to evoke supernatural power is particularly prevalent amongst the Blackfoot Indians, who have specific songs for each act in a ceremony. For example, in a ceremony to influence the weather, a bundle of objects is opened and the correct song must be sung for each, in the right order, to 'activate' each object. Some of these ceremonial bundles can contain over 160 items (Nettl, 1992).

Such religious activities are very frequently accompanied by dancing as well as the music, and this is the most common use of music amongst the Plains Indians. The second most common use is to accompany *social* dancing. War dances and puberty rites are other important social and ceremonial occasions accompanied by music and dance, and the aforementioned deer-hoof rattles accompany the latter in particular (Nettl, 1956). In all these instances, the men perform the majority of the dancing and singing. In the religious activities involving the opening of bundles of items, the performance is limited to the medicine man performing the ritual, but in the more social activities, the performance involves many members of the group. This includes the women, who join in certain of the men's songs, and walk around the periphery of the dancing area, rather than dancing themselves (Nettl, 1992).

Both the vocal and the rhythmic elements of the Blackfoot and Sioux music consist of a great deal of repetition, with very subtle variations on a theme (McAllester, 1996). The vocal technique used in the Sioux 'Grass Dance' (a war dance) is of particular interest. This uses *vocables* (non-lexical,

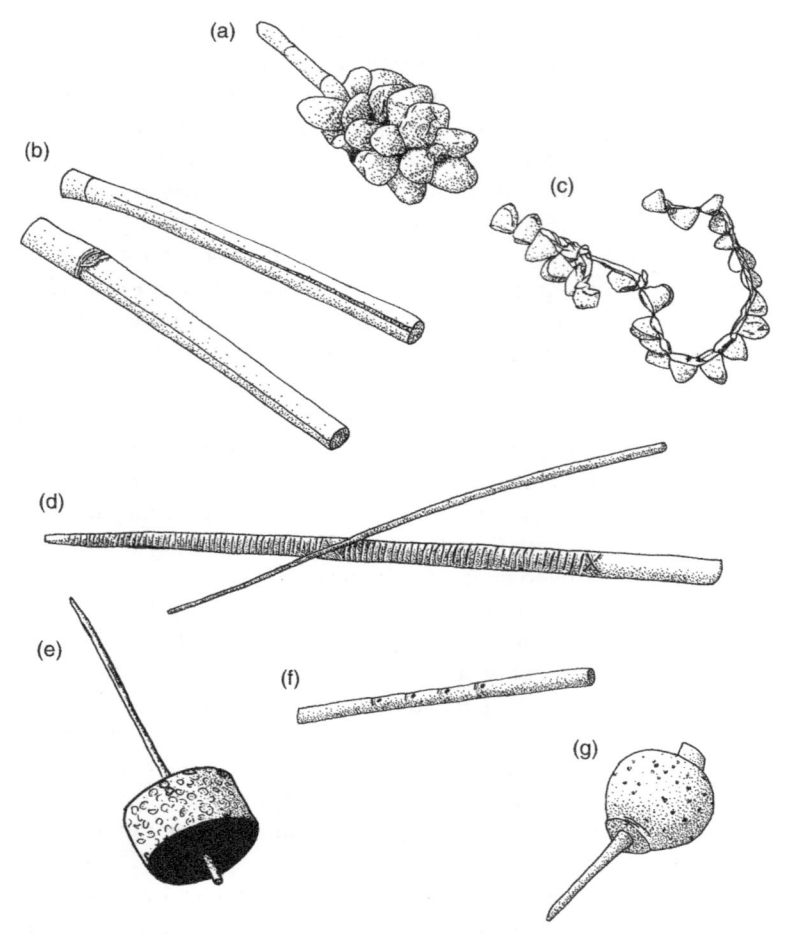

Fig. 2.1. A selection of Native American musical instruments (redrawn from Taylor, 1991). A wide variety of sound-producers is used, the majority percussive. None of those illustrated would be likely to be preserved archaeologically.

(a) Rattle made of cocoons with small stones inserted
(b) Pair of split-stick rattles
(c) Cocoons strung on a cord, worn as leg rattles
(d) Wooden rasp and scraping stick (scraped idiophone)
(e) Wooden rattle
(f) Elder-wood end-blown flute
(g) Gourd rattle

meaningless syllables) with high emotive input rather than translatable words as the basis of the song (McAllester, 1996). This is, in fact, a common feature of virtually all Plains Indian songs, with any words occupying only a tiny portion of the melody (Nettl, 1992). For example, the aforementioned Blackfoot Sun Dance ceremony contains only the words 'Sun says to sing', and a medicine bundle ceremony contains only the words 'It is spring, let others see you'; all the rest of the melody, although verbal, consists of vocables (Nettl, 1989).

Although this vocable music is clearly used in close association with symbolic activities, and a particular song can relate very specifically to a particular activity, the music itself is said to have no symbolic content. According to Nettl: 'native informants are able to say almost nothing on the symbolic aspect of their [non-lexical] music' (Nettl, 1956, p. 25). It would seem that the main purpose of this type of vocalization is to contribute to the emotional responses evoked by the music (McAllester, 1996), and has no (conscious) symbolism behind it. This is in contrast to lyrical music, or to instrumental pieces which deliberately attempt to evoke the various traits of their subjects, be they *Mars, the war-bringer, Winter,* or the sounds of the Australian bush, as is the case with the Pintupi described in a later section. The only exception to this amongst the Plains Indians would appear to be the imitation of a bleating calf used in the hunting tactic described at the beginning of this section.

AFRICAN PYGMIES OF THE EQUATORIAL FOREST (AKA AND MBUTI)

The second group of hunter-gatherers to be discussed are the Pygmies of the Ituri forest of equatorial Africa, in particular the Aka (or BaAka) and Mbuti (or BaMbuti) communities. These two groups are particularly worthy of note as they still receive the majority of their subsistence from wild foods, which they forage and hunt. This is in contrast to the majority of the other Pygmy groups (such as the Efe and Baka) who now tend to live near to villages of farmers, and provide services in exchange for a diet of more cultivated food (Hitchcock, 1999).

Both groups live in dense humid forest near the equator, in central Africa. Their principal habitation is family domes made from bent branches and thatched with leaves, organized into communities of thirty to a hundred. These communities are nomadic, and tend to move on every month or two, following their quarry. The communities are egalitarian and cooperative, with no formalized hierarchy, although individuals may be acknowledged as experts in particular skills. Ownership of personal possessions is minimal, largely as a product of the nomadic lifestyle (Turino, 1992), and they do not make any use of metal or ceramic in manufacturing their tools and containers (Ichikawa, 1999), which tend to be made of wood or bark (Bahuchet, 1999).

Hunting is carried out using nets, spears, and poisoned darts (Bahuchet, 1999; Ichikawa, 1999). Amongst the Aka, subsistence from hunting accounts for about two-thirds of their diet. The other third is made up of foraged plants, fungi, and animals, such as wild yam, leaves, nuts, snails, tortoises, weevils, beetles, and caterpillars. This activity is predominantly carried out by women, whereas the men collect honey and do most of the hunting of large game. The game is mainly elephants, antelope, river hogs, gorillas, monkeys, and chimpanzees. Monkeys are hunted with bows and poisoned darts; large game is hunted with spears. During the dry season, when many Aka communities band together, communal hunting often occurs, with all the men, women, and children getting together to 'beat' duiker into rings of nets (Bahuchet, 1999). This activity is also carried out by the Mbuti, although hunting among the Mbuti accounts for less of their subsistence than among the Aka. The Mbuti also rely on the foraging of wild vegetables, honey, nuts, and invertebrates as a portion of their diets, although in recent years more of their diet has been provided by nearby farmers (Ichikawa, 1999).

The music of the Aka and Mbuti is, like most aspects of the culture, a communal activity, without specialist musicians. Like that of the Plains Indians, it is predominantly vocal, with little instrumentation; this is one of several features which the Pygmy and Plains Indian music share. Aka and Mbuti music is also considered to be of supernatural origin, told to individuals by their ancestors in their sleep, and is used as a way of communicating with the spirits of their surrounding environment (Ichikawa, 1999). They believe the way to communicate with the divine is through sound alone; as a result song texts are minimal, often consisting of only one line such as 'the forest is good' among the Aka (Turino, 1992) or 'we are the children of the forest' among the Mbuti (Turnbull, 1962). The rest of the vocal element of the music consists, as with Plains Indians, of vocables (Locke, 1996).

Some music is performed individually, such as lullabies, but communal singing for ceremonial activities is most important (Turino, 1992). Such music is often related to specific subsistence activities, with particular songs being performed for each activity; thus, there are songs for net-hunting, elephant hunting, and honey collecting. For example, the *mabo* song and dance of the Aka is a net-hunting dance, and is used for the ritual purpose of the administration of hunting medicine. It is also used as a method for instruction in dance, including between neighbouring settlements (Kisliuk, 1991, cited in Locke, 1996), in this respect acting both as a pedagogical aid and as a diplomatic mechanism. The *mobandi* ritual dance is related to honey collecting, and is unusual in that it is the only one that is seasonal rather than being carried out before or after the hunt. It is performed after the rains when the *mbaso* tree flowers, and is a purification ritual. The Aka believe that any misfortune to occur in hunting and gathering is the result of human misconduct, which jeopardizes the help of the spirits in the activity. To avoid this risk, during the *mobandi* all the members of the group gently hit themselves with branches to transfer any evil forces out of themselves and onto the branch (Bahuchet, 1999).

Songs also mark formal 'rites of passage' ceremonies such as circumcision, girls' puberty, marriages, and funerals (Ichikawa, 1999). On these occasions they are usually performed by women, unlike the hunting-related songs, which are performed by men (Turino, 1992). It is only in these ceremonies that gender or age distinctions are made in the performance of music; in other performances, anyone may participate who wishes to (Turino, 1992).

Instrumentation consists mainly of percussive accompaniment to the vocal tunes using claps, drumming on skins which cover the end of cone-shaped logs, and rapping sticks on the drum body (Locke, 1996). There are also a few rattles, as well as some end-blown flute pipes made from cane (Turino, 1992). Like all of the Pygmies' traditional artefacts, these instruments are made from organic materials occurring in their environment. In these respects, the instrumentation is similar to that of the Plains Indians described in the previous section. However, despite its similarities, the music of the Pygmies differs from that of the Plains Indians in that it is *polyphonic*; a performance consists of several different melodies sung simultaneously by different groups, which 'interlock' with each other to form a multilayered piece of music (Turino, 1992).

The majority of the music itself appears to lack any direct symbolic content (in terms of lexical meaning or mimicry), although it is clearly associated with activities that do have symbolic content and associations (see also Lewis, 2009). The melodies do not attempt to emulate or evoke any other thing, but follow particular structures for different activities and dances. The exception to this is the *molimo* music, which is performed specifically to wake the forest if hunting is bad. This ceremony is performed at night and can last several months, and is rooted in the belief that if hunting is poor, it is because the forest is sleeping. It uses the molimo trumpet, a single end-blown tube, which is considered to mimic the sounds of the forest and answer the men's singing (Turino, 1992). As such, the molimo sound does not have an abstract symbolic meaning, in the sense of encoding information, but does have an iconic association with the forest.[2]

---

[2] In the sense of **Peirce's semiotics of signs**. An information-carrying 'sign' (which may be visual or auditory, for example, such as a representation, a sound, or a word) may be classed as either an icon, an index, or a symbol, depending upon what the relationship is between the sign and its referent (the thing to which it refers).

An *Icon* is a sign which refers to the object that it denotes merely by virtue of characters of its own—a relationship by *similarity*. e.g. a picture of an elephant, an imitation of the sound of a bird.

An *Index* is a sign which refers to the object that it denotes by virtue of being really affected by that object—a *causal* relationship. e.g. a footprint, the actual sound of a bird.

A *Symbol* is a sign which is constituted a sign merely or mainly by the fact that it is used and understood as such—it is a *conventional* relationship. e.g. A 'No Entry' sign (red circle with a horizontal white bar), a word. (Based on a summary in White, H., 1978.)

The use of symbols is considered to be a critical step in the development of modern human behaviour, being essential for the development of modern-type language, in which the majority of words are *symbols*. It is worth noting that some signs may simultaneously act as more than one type. For example, the sound of thunder is an index of the electrical discharge (lightning) that produced it, but at the same time may have significant symbolic meaning, as an omen, for example, depending upon who perceives it. The case of the molimo trumpet sound may be

Fundamentally, musical and dancing activity forms the focus of group social activity on a large scale, and is the main occasion where social relationships manifest themselves (Bundo, 2002). It is notable that these roles would in the past have been fulfilled by both hunting and musical activities, but that in recent years as the importance of large-group hunting activities has declined the role is most commonly fulfilled by musical activities.

AUSTRALIAN ABORIGINES OF THE WESTERN DESERT (PINTUPI)

The Aboriginal population of Australia falls into two main language groups, with peoples being divided more by geographical features than by 'tribal' groupings (Peterson, 1999). Communities exist within these categories, but widespread systems of exchange across the continent, of goods, ceremonies, social practices, and some intermarriage, have led to an indigenous way of life which is fairly homogeneous throughout the continent. Archaeological evidence suggests that systems of long-distance exchange were in place at least 30,000 years ago (Mulvaney, 1999); amongst historical populations at least, these exchanges were as much for the benefit of social relationships as for practical purposes (Peterson, 1999).

Of particular note amongst the Aboriginal groups are the Pintupi-speaking people of the Western Desert, because they were the last Aboriginal population to be in any way incorporated into the settler society, during the 1960s (Myers, 1999). Consequently, they were (until then) less severely influenced by Western lifestyle than many of the other groups, and recent records exist of their traditional way of life. During the 1970s they started to create a few autonomous satellite communities around Papunya, and by the early 1980s, most had returned to their traditional territories of the Gibson Desert (Myers, 1999). The traditions outlined here are related in the past tense, as they describe the features of the Pintupi before their resettlement. However, since members of the culture have begun to return to their traditional territories today, hopefully they will perpetuate many of these cultural elements and subsistence methods, as have other Aboriginal Australian groups in recent years.

The Gibson Desert is an area of sandy dunes, plains, and hills, with sparse vegetation. Rainfall is minimal, mostly falling only during January and February. The main fauna are lizards, feral cats, kangaroos, wallabies, and emu. These animals were traditionally hunted by the men using stone axes, spears,

---

argued not to be solely indexical, but to be more akin to an onomatopoeic word such as 'miaow'—being a conventionalized—and thus symbolic—version of an index. The importance of symbolism to music is discussed further in Ch. 8.

and spear-throwers. Women collected seeds, fruit, vegetables, honey-ants, and grubs using digging-sticks and containers made from tree bark. These foods gathered by the women constituted the majority of the Pintupi diet (60–80 per cent) for most of the year (Gould, 1969); the majority of protein was provided by lizards, caught by both men and women (Myers, 1999). Hunting generally occurred alone or in small groups, and large game was shared amongst all the families of a residential group (ten to thirty members) (Myers, 1999). As with the Plains Indians and the Pygmies, the dry season (dri*est* season, in the case of the Pintupi) was the time when many separate communities of the same group would gather together for social and ceremonial activities; amongst the Pintupi, up to ten such groups would gather in these instances. Traditional shelters for the family groups were made of brush and branches, and would house a nuclear family. Groups of unmarried men still going through initiation and widowed women sometimes had their own habitations (Myers, 1999).

Ownership of the land was seen as being dictated by the content of songs, stories, and rituals pertaining to the Dreaming (see the next paragraph), with each group being guardians of a particular set of this information (Breen, 1994). This 'ownership', however, did not relate strongly to the actual group boundaries, which were quite flexible (Myers, 1999). Social organization was largely non-hierarchical, without designated leaders, but additional respect was given on the basis of age or the successful performance of rituals. Gender relations were predominantly egalitarian (Myers, 1999).

Like that of the Arrernte (Morton, 1999), Cape York (Martin, 1999), and Ngarrindjeri (Tonkinson, 1999) Aborigines, music amongst the Pintupi (Myers, 1999) was tied up closely with story and ritual describing their hereditary land. These combined renditions acted out events from the *Tjukurrpa*, or 'The Dreaming'. During Tjukurrpa, ancestor spirits travelled the land leaving their marks on the landscape and the essence of people and animals on the land. Pintupi songs incorporate these mythologies into the performances, and describe the world around them (Myers, 1999). The precise preservation of these songs was seen by them as fundamental to their continued survival and the constant renewal of nature, so was considered extremely powerful and valuable. As they had no written records and notation, this oral tradition was the only way in which this information could be passed on and preserved, and the musical form had important mnemonic benefits to this end (Jones, 1983).

Unlike the music of the Plains Indians and the Pygmies described so far, the music of the Pintupi was lyrically dominated; in fact, the perpetuation and communication of lyrical information was central to the purpose of much of the singing. However, the music does resemble that of the Plains Indians and Pygmies (and other Aboriginal groups) in that it was predominantly vocal, and featured little instrumentation (Breen, 1994). In fact, much of the music relied on percussion produced without instruments at all, such as clapping hands, slapping the body, and stamping feet on the ground. When instrumentation was used, it was the same as that used by the majority of other Aboriginal groups (Nettl, 1992). Percussion instrumentation consisted traditionally of two eucalyptus sticks bashed together (Nettl, 1992) called *bilma* (Breen, 1994).

The only non-percussive instrument used was the *didjeridu*, a long pipe made from the limb of a eucalyptus tree, hollowed out by termites (Breen, 1994). This instrument, now familiar to many Westerners, is played like a trumpet, creating a low single-tone droning sound, the 'colour' of which can be varied by changing the shape of the mouth (Nettl, 1992). This originated amongst the Aboriginals of the north of the continent, but is now used universally. Originally, it was only played by men specially selected by elder members of the group, and to learn it, players were encouraged to listen to all of the sounds of the bush, in order to be able to imitate the sounds of nature (Breen, 1994).

The music of the Pintupi had, as other Aboriginal music still does, explicit symbolic associations. This symbolism was seen as very important, describing the world and history of the people, and its preservation was seen as fundamental to the well-being of the group. The sound of the only non-percussive instrumentation that there is, the didjeridu, also had symbolic meaning; players aimed to mimic the sounds of the surrounding environment, adding to the symbolism of the song it accompanied (Breen, 1994).

THE ESKIMO OF SOUTH-WEST ALASKA (YUPIK) AND CANADA (INUIT)

The preceding sections have considered hunter-gatherers of the temperate plains, equatorial jungle, and the desert; this section looks at hunter-gatherers living in subarctic tundra. This environment is a demanding one in which to subsist, with a limited range of resources and difficult environmental conditions.

The peoples inhabiting the area to the south of the Bering Strait and Norton Sound in south-west Alaska are known collectively as the Yupik. The majority of the research specifically regarding the music of the Yupik has been carried out by Johnston, who provides the one comprehensive description of their musical traditions (Johnston, 1989). As such, much of the following discussion draws upon this work. The climate in this area is milder than that further north, where the Inupiaq Eskimo live, and consists mainly of low marshy tracts close to sea level. This area is largely treeless, although the Eskimo do move upstream to penetrate the forested inland areas too (Nelson, 1899/1983). Although there are no fully formed trees in this area, there is a variety of other vegetation. This particularly grows in sheltered areas on hillsides and along the courses of river valleys, and includes willows and alders, as well as the grasses, sphagnum, and flowering plants that occur on the rest of the landscape (Nelson, 1899/1983).

The population of Yupik numbers around 17,000 today, and they subsist in the traditional manner on sea-mammal, salmon-fishing, and reindeer, as well as wild berries. These are gathered, mainly by women and older children, and are stored until needed in winter in underground chambers (Johnston, 1989). This population exists in groups of up to 100 people, who associate themselves with particular tracts of land. This occupancy is not normally considered as ownership of the land or its resources, but merely as the community being the 'traditional occupants' of that particular tract (Riches, 1995). The social groups are best described as having a structure that is flexible, egalitarian, and individualistic (Gardner, 1991), with little ranking within the group, or between groups. Leadership tends to be attributed on the basis of exhibiting exemplary behaviour in some aspect of life, and the security of the leaders' position rests purely with the fact that others have *chosen* to follow them (Riches, 1995). The provision of food is a reciprocal system, with all members of a 'task-group' (up to thirty people) contributing to the total, and those contributing the most receiving some prestige as a consequence (Riches, 1984, quoted in Riches, 1995).

The music of the Yupik is predominantly vocal, with only two different instruments used, if any. All melody is carried by the vocal parts, whilst the instruments provide percussion. The most common instrument is a simple frame drum, made from a round wooden frame covered with a sea-mammal membrane, and each village tends to have a specialist maker of traditional drums. The other instrument sometimes used is a rattle; this may be made, for example, from a hoop strung with puffin-bills which rattle against each other when shaken (Johnston, 1989).

The Yupik themselves classify their songs and musical styles into thirteen categories; these are described in some detail by Johnston (1989), and his findings are summarized below. The first six of these are for adults: dance songs, shamans' songs, hunting songs, teasing songs, travelling songs, and berry-picking songs. The remaining seven are for children (either to be performed by children, or by adults for children): story songs, juggling-game songs, jump-rope game songs, ghost-game songs, bird identification songs, fish identification songs, and *inqum* 'cooing' songs (Johnston, 1989).

Dance songs tend to feature drumming. This is performed by four to eight people simultaneously who drum in synchrony, whilst chanting the words of the song. Drummers do not participate in dancing, and dancers do not sing either, although the audience may sing along. The drummer-singers are almost always male, although there is no formal rule against women drumming (Johnston, 1989). There are three main areas of subject matter for Yupik dance songs: the relating of past adventures of cultural/community heroes, description of comic incidents and portrayal of animal- or bird-life. Since the songs often form a repository of knowledge of past adventures and lore for the community, the singers are usually those who have most of interest to relate. The dance motions reflect this storytelling nature of the song, and mimic activities of daily life. Men's dance actions tend to resemble activities such as sledding and harpooning whilst the women's dance motions mimic activities such as carcass cutting and feather plucking.

The melody of the songs tends to consist only of two or three notes, with the note changing with each syllable of the song. As well as the narrative words there is a lot of use made of vocable non-linguistic syllables. Many of these communal songs place an emphasis on comic occurrences and events, and can have a strong bonding and socially lubricating effect on the community. As Johnston (1989) points out, this can be very important in communities of people confined in small quarters in some of the harshest environmental conditions in the world.

Another way in which song is used as a social lubricant is in the so-called teasing songs. These are used as a form of legal mechanism and as a way of avoiding direct conflict between two protagonists. The offended individual would compose a witty, chiding song aimed at the offender, which they would perform in front of the community. This would embarrass the latter, and hopefully also act as a deterrent for future repetitions of the offending action (Johnston, 1989).

Music and song are not only used as aids to social interactions, but also to influence the world around the Yupik. The shamans of the Yupik were traditionally the intermediaries between the hunters and the gods of hunting, who control the success of the hunt. It was thought that the shaman's soul could leave his body and travel to the moon, where the hunting gods resided, in order to commune with them; this process was aided by the shaman's drumming as he sang. These songs are still often used at important festivals, and composed for the event itself (Johnston, 1989). The hunters themselves also use music to influence the unpredictable environment in which they have to operate, using song to affect the hunt. The hunting songs are performed during the hunt itself and are thought to have the power to stop wounded game from escaping.

Songs may also be used to hold important knowledge, whilst being a source of entertainment; this is the case with the travelling songs and the berry-picking songs. Yupik frequently travel long distances for subsistence purposes and for the maintenance of alliances between different communities. Whilst they travel, they sing songs which often relate adventure narratives; the importance and significance of the Yupiks' awareness of their environment is reflected in the fact that these songs frequently include highly detailed descriptions of prevailing weather conditions as the journey is described.

Environmental information is also contained within the berry-picking songs performed by the women and children as they travel to pick the cloudberries and blueberries for storage for the winter. These songs are often concerned with the tundra and its features, particularly descriptive names of features which aid in their location. In this respect, these songs closely parallel the content and use of the songs of the Pintupi (and other) Australian Aborigines (see earlier sections), specifically used as a mnemonic aid describing the surrounding environment.

The Yupik songs performed for or by children also seem to have educational purposes: as an aid to developing coordination skills (the juggling-game, string-figure-game, and the jump-rope songs), as an educational medium concerning the natural world (bird-identification and fish-identification songs), encouraging creativity and interaction (the ghost-game songs), and as examples of triumph over adversity (the story songs).

The jump-rope songs take a question-answer form, with the rope-holders asking a question and the rope-jumper answering whilst avoiding getting caught in the skipping rope. The juggling-game songs are performed both by adults and children, and are reported in communities throughout Greenland, Canada, Alaska, and Siberia; it is thus likely that they are an extremely old form of song. The performer sings the song, which tends to have satirical or rather indelicate subject matter, whilst juggling stones in one hand. The string-figure game is a variation of cat's-cradle, with a loop of string held between the hands and feet or forehead of the performer. As they perform the song, the individual winds extra twists and loops into the string in time with the rhythm and syllables of the song. This game develops dexterity and spatial-directional skills whilst the juggling game develops hand–eye coordination skills (Johnston, 1989).

The bird and fish identification songs teach important information about the appearance and behaviour of birds, small animals, and fish. They are sung by elders to children and contain information about markings, nesting, courting, migrations, and the way these relate to the environment and the weather. They also aid in the hunting and catching of these creatures for subsistence, which children are encouraged to do at a young age. The Yupik also aim in the context of these songs to mimic the sounds and behaviours of the animals they find in their surroundings. A seal hunter may, for example, entice curious seals by accurately mimicking seal noises. Fish that are caught are often sung to in the hope that they will enjoy being caught and return the next year in their reincarnated form to be caught again (Johnston, 1989).

Like the jump-rope songs, the ghost-game songs are bipartate, consisting of question and answer between the participants; however, in this instance they are improvised replies to the question (usually concerning the whereabouts of an undesirable spirit). The children are thus required to be spontaneously creative; any participant who fails to come up with an answer becomes the ghost and has to chase and catch the other children (Johnston, 1989). It is thus a game of physical activity too.

It can be seen that there is a very great variety of song and music amongst the Yupik of the subarctic tundra, and that they also serve a variety of purposes, on an individual and community level. Music and song, for the Yupik, can be informative and educational, a social lubricant, an aid to hunting, a way of learning physical and mental skills, or purely a pastime.

Very different, but worthy of note are the throat-games of the Inuit, known as Katajjait. These games are found in eastern and central Canada, northern Quebec, and South Baffin Land, performed by Igloolik, Caribou, and Nelsilik Inuit (Nattiez, 1983; Baghemil, 1988). They are generally played by two women who face each other at close range (Baghemil, 1988), sometimes even holding each other's shoulders (Nattiez, 1983). Both participants produce sequences of short repeated vocalizations. These are strung together into longer phrases and syncopated into complex sound structures that are a mixture of droning and complex gutteral sounds (Baghemil, 1988). The game is a sort of endurance test, in that it ends when one participant runs out of energy, inspiration, or starts to laugh (Baghemil, 1988). The game is performed playfully, but the performance is valued on the basis of endurance, virtuosity, and aesthetics of sound—participants have to win with 'beautiful sounds' (Nattiez, 1983).

The sounds themselves take the form of morphemes and vocables (Beaudry, 1978), and form repeated motifs 'made up of a morpheme, a rhythm, an intonation contour and a pattern of voiced and voiceless, inspirated and expirated sounds' (Nattiez, 1983). There tend to be no more than three or four tone levels used for the utterances (Baghemil, 1988). This small number of elements is constantly recombined in novel ways during the course of the competition. The sounds produced may have subject matter in the sense of using real words, which may form a narrative, but most often there is no narrative at all. More frequently the sounds imitate animal noises such as goose-cries, ducks, walruses, panting dogs, or mosquitos. Alternatively they may just be abstract sounds made for aesthetic effect (Baghemil, 1988). Baghemil (1988) reports that the Inuit themselves consider the Katajjait to be the language spoken by the Tunnituaruit, or 'flying heads', mythical creatures that are half-woman, half-bird. The dead are also thought to communicate with this language, especially as manifested as the aurora borealis. Nattiez (1983), however, reports contrarily that the superstitious/spiritual language element is only rarely linked to the games and that there does not seem to be any consistent 'deep meaning' or symbolism to the games; they are usually performed for their own sake as entertainment. This is also consistent with what Beaudry (1978) was told by the performers she observed. The degree of symbolism and perceived content may be something that has varied historically, or an aspect that has only been revealed to observers in more recent times. According to Nattiez (1983), the games used to be performed by up to four people at once, at any time of day, month, or year. The performers, although usually women, could also be men or boys. The games seemed to be multifunctional, and could be performed as entertainment, celebration, or just to keep a baby quiet, for example (Nattiez, 1983).

These throat games are usually studied as a form of music, but Baghemil (1988) argues that they should be better analysed as a form of language. They seem in some ways to straddle both music and language in the vocalizations they use. Baghemil observes that the voicing and airstream mechanisms used in the games have more in common with several world languages, and only rarely occur in song. Interestingly, it also seems to be a system independent of the Inuktitut language—the features of vocalization that occur in the games have not been transferred across from the language as they would be in traditional song (Baghemil, 1988).

CONCLUSIONS

Some common features in the uses and nature of music in four hunter-gatherer societies

It can be seen from the preceding sections that the musical activities of these four hunter-gatherer peoples have a number of features in common, despite being separated across three continents, and occupying four very different environments:

All four societies meet up with fellow communities during their most difficult subsistence season, during which time there is increased performance of ceremonial and communal music and dance. In all four cultures, the ceremonial and social use of music is very important, is communal, and is almost always accompanied by rhythmic dancing. In all four cultures music is also performed purely as a communal activity for pleasure.

All four have music which is predominantly vocal, and accompanied mainly by percussion instruments. The use of melodic instruments is minimal and, when used, invariably consists of end-blown pipes. These are usually single-toned instruments. All the instruments, whether percussive or melodic, are made from naturally occurring organic materials.

Amongst the Blackfoot and Sioux Plains Indians and the Aka and Mbuti Pygmies, song has minimal lyrical content, vocalizations instead consisting of vocables, emotive sounds with no obvious inherent symbolism (they have no lexical meaning). In these two cultures, music accompanies rites of passage and rituals relating to subsistence and hunting. The Pintupi are a contrast in this respect, in that their music is almost exclusively lyrical, and is a repository for community knowledge and mythologies. The Inuit and Yupik music contains both types of content. Much consists of non-lyrical vocables and animal sounds, or sounds made just for aesthetic purposes. The greater proportion relates stories, descriptions of events, environments, journeys, and subsistence sources, so constitute an important repository of knowledge for both adults and children. Some songs and music are inextricably related to games designed to develop important skills of subsistence and coordination. It is worth noting here that with the exception of the Yupik, in all the examples examined here the literature does not mention the role and nature of music as related to children. Whether this is a consequence of an absence of such musics in these other cultures, or whether it is a consequence of a lacuna in research is not clear.

All four peoples believe themselves to have come from the land, to be akin with the other fauna of their environment, and use music to try to influence the world around them. Music and dance can have important uses in engendering group cohesion, altering mood, as an aid to the teaching of dance, and can facilitate group interactions and communality, within and between groups. In the majority of these instances the music itself has no inherent symbolism, but it can be used to accompany symbolic activities. As with the Yupik, amongst the Pintupi, song is a repository for knowledge of tribal history and legend, and is a mnemonic aid. The mnemonic benefits of a musical structure to knowledge are also mentioned by John Sloboda:

> Songs and rhythmically organised poems and sayings form the major repository of knowledge in non-literate cultures. This seems to be because such organised sequences are much easier to remember than the type of prose which literate societies use in books. (Sloboda, 1985, p. 267)

Methods and materials of construction of instruments: implications for the archaeological record

There appears to have been in the past a general conception when considering Palaeolithic music that the earliest instruments that we find archaeologically are likely to indicate the first musical behaviours. This is due, at least in part, to our perception that music is a predominantly instrumental melodic activity, as is indeed the case in the art music of the last few hundred years of Western history. The evidence considered in this chapter gives the lie to these assumptions, for two reasons.

First, the music of the hunter-gatherers described here is largely non-instrumental, with melody frequently being carried by vocalization alone and much of the percussion being provided by clapping, slapping, and foot-stamping. Instrumentation, when it is used, is almost exclusively percussive. Second, the instrumentation produced is constructed from readily occurring natural resources with relatively little modification; most use skins, wood, cane, or other vegetable matter such as gourds. Wood is used more readily than bone for scrapers, flutes, drums, and drumsticks; the use of bone is rare among the Plains Indians and there is no mention of its use at all amongst the Pintupi and Pygmies. Although ivory is used by the Yupik and Inuit, there it does not tend to be used for instrumentation at all. Apart from bone, none of these instruments would leave an archaeological trace, as vegetable or animal matter degrades rapidly under most natural conditions. Had they been used by early humans, there would be no evidence of it in the archaeological record.

As for the technological capability of earlier hominins to produce such artefacts, the Schöningen spears and other wooden artefacts (dated to 420,000–380,000 years ago) provide ample evidence of the ability of *Homo heidelbergensis* to work wood very finely (Thieme, 2005, 1997; Dennell, 1997). The ability to cut skins would have been a necessity from the earliest scavenging or hunting, for the processing of carcasses. Furthermore, many of the instruments described in this chapter require very little modification before use anyway. A naturally dried gourd will rattle as the seeds inside move freely, cane and bird bones form hollow pipes in their natural state (only the ends need be removed), and a tree trunk can be hollowed by termites, as is the case with the didjeridu. Much of the traditional percussion used by the hunter-gatherers described consists simply of two pieces of wood, or a piece of wood and a tree stump being struck together. It is clear that the earliest archaeological evidence for musical instruments should not be assumed to represent the first incidence of musical behaviours in humans.

The preceding evidence illustrates that there should be no expectation that we would see evidence for instrumentation as soon as humans participated in musical behaviour, or that the first incidence of instrumentation indicates the first incidence of musical behaviour. Secondly, it also highlights fundamental similarities in the nature and roles of music between these diverse groups of hunter-gatherers; this could imply either convergent development of these behaviours in the different environments in which we find these peoples today, or a shared cultural heritage (or both). The former (convergence) would suggest that there are important driving forces towards those common behaviours, either as a consequence of subsistence method or of human biology, these being the principal common factors between the groups. The latter situation (shared heritage) would indicate a very ancient tradition of musical behaviour, as these groups are probably separated by at least 50,000 years.

3

Palaeolithic Music Archaeology 1: Pipes

The eye of the flute is the doorway to thought
Fernando Librado, aka Kitsepawit of the Chumash, b. 1804.

Librado (ed. Hudson) (1977)

INTRODUCTION

Chapters 3 and 4 examine the archaeological evidence for the use of instruments and music in the Middle and Upper Palaeolithic (*c.*250,000–12,000 years ago), and the possible nature of music at this time. Whilst a major objective of this book is to address the questions of the origins of the *capacities* for musical behaviours in the human lineage, its overriding theme is the development of a picture of the emergence of the practice of musical behaviours throughout the Palaeolithic, and we are very fortunate that in spite of the huge time spans involved, archaeological evidence for musical instruments has actually been preserved from these distant times.

In fact, considerations of the prehistory of music have often taken as their starting point *only* the direct archaeological evidence from the Upper Palaeolithic, and treated this as synonymous with the origin of music. The evidence considered in Chapters 5–11 of this book deals with the deeper prehistory of these human capabilities, their evolution and relation to other human capacities. It is in those chapters that we deal with the *origins* of musicality—the emergence of the underlying capabilities which actually make possible the carrying-out of those activities that we would recognize as musical. But it is in the later Palaeolithic archaeological record that we find the first *direct* evidence of musical activities being carried out, in the form of musical instruments. It should be noted at the outset that musical behaviours and instrumental use are not synonymous; the former can occur without the latter, and musical capabilities are likely to have pre-dated the occurrence of instruments in the archaeological record by many years. Nevertheless, the identification of musical instruments from the Palaeolithic constitutes the first concrete evidence of recognizable instrumental behaviours in modern humans, and has the potential to demonstrate the importance of musical behaviours in human societies.

This analysis is complemented by a comprehensive database of all reputed pipes/flutes from the Middle and Upper Palaeolithic, their locations, dates, and status as possible sound-producers (Appendix Tables 1 and 2).

Introduction to the Upper Palaeolithic

The overwhelming majority of Palaeolithic archaeological finds reputed to be musical instruments come from Europe, and from the period known as the Upper Palaeolithic—or, rather, from contexts associated with Upper Palaeolithic technology—which date from around 45,000 years ago to around 12,000 years ago. There are several different technological repertoires that constitute the Upper Palaeolithic in different times and places within Europe, but the earliest of these appear to coincide with the arrival of anatomically modern humans (*Homo sapiens*) in Europe, and the latest of them with the warming of the world and end of the last Ice Age around 12,000 years ago. The hunter-gatherer lifestyle of Palaeolithic populations continued for several thousand years after the end of the last Ice Age (the various technologies associated with this period are called Epi-Palaeolithic and Mesolithic in different places), before this way of life was lost in Europe, replaced by a dominance of agricultural practices (Neolithic). Hunting and gathering subsistence strategies have remained highly successful, of course, in other parts of the world to the present day.

Neanderthals had occupied Europe, and some areas beyond the edges of what we today call Europe, for around 200,000 years before the appearance there of modern humans and Upper Palaeolithic technology (using technologies known as Middle Palaeolithic); their direct ancestors had occupied the same region for at least 300,000 years before that (using technologies known as Lower Palaeolithic). For the first few thousand years of the Upper Palaeolithic period Neanderthals and modern humans both occupied the European continent, but the last Neanderthals appear to have died out around 30,000 years ago. The extent to which Neanderthals developed or adopted Upper Palaeolithic-type technologies and behaviours prior to and during this period is a source of great debate. This extends to musical artefacts and behaviours; a few objects argued to represent sound-producers have been found in Mousterian (Middle Palaeolithic) contexts, and are relevant both to debates concerning the origins of musical behaviours and debates about Neanderthals' behavioural repertoires and capabilities.

The first of these two chapters examines artefacts interpreted as pipes, or flutes, looking in detail at the so-called Neanderthal flute from Slovenia, and the early Upper Palaeolithic pipes from the Ach Valley (Germany) and Isturitz (France). Whilst pipes (or flutes) are the most prolific of the various objects interpreted as sound-producers from the Palaeolithic, probably partly because they are relatively easily recognized, other possible types of sound-producer have been uncovered too. The second of these chapters looks more widely at other possible forms of instrumentation and sound production represented in the archaeological record of the Palaeolithic, including whistles, bullroarers, rasps, and the use of acoustic features of the environment such as caves and other lithic objects.

Introduction to Palaeolithic pipes

Sections of animal bone (or worked ivory, horn, or tooth) with or without holes in their shafts, naturally hollow after the marrow has decayed or been removed, have been found in Upper and Middle Palaeolithic contexts since the nineteenth century (Scothern, 1992; Lawson et al., 1998). Many of these have been described as pipes, flutes, or whistles (the terms have often been used interchangeably), and the catalogue of these items is now quite large. As will be seen, the precise method of sound production for any given artefact is often a matter of some debate. Note that 'phalangeal whistles' (small whistles made not using a bone tube but a pierced toe bone, or phalanx) are a separate category, and are discussed in Chapter 4.

It is the former type (pipes, flutes, whistles) that have received by far the most attention in considerations of Palaeolithic music. Scothern (1992), as part of her doctoral thesis, carried out an inventory of these items, listing 106 sections of bone which had been variously described as pipes, flutes, or whistles, plus around ninety punctured phalanges interpreted as 'phalangeal whistles', all of which had come from Palaeolithic contexts. A further twenty-one were attributed to Mesolithic contexts. By those aware of it (it is available only in Cambridge University Library), this has generally been considered to be the most complete inventory of these items (Lawson et al., 1998). Until Scothern's work, the most complete catalogue of Palaeolithic pipes was that of Fages and Mourer-Chauviré (1983), who listed thirty (Dams, 1985; Lawson et al., 1998), and it is this list that is still most frequently cited by other authors discussing Palaeolithic instruments. The information from Fages and Mourer-Chauviré (1983) and Scothern (1992) has been corrected, expanded, revised, and updated for this publication to the extent that the complete database now includes around 144 objects (or parts of them) that have at various times been interpreted as pipes/flutes or whistles. These, and as much information about them as it has been possible to collate, are detailed in Appendix Tables 1 and 2.

This corpus has as its foundation the inventory prepared by Scothern (1992), and owes much to it; it is hoped that the version presented here provides a convenient and (to date) complete reference for further analysis. Each entry contains notes on its original find site, current location, period, age, a description of the object, and its status as a sound-producer, if such information is available. It also notes the principal sources in which each piece is featured. It is not intended that all this detail should be replicated in the body of the text here, as the tables provide the most concise and digestible form of this information, but some especially salient examples will be discussed here.

Together Appendix Tables 1 and 2 aim to provide a directory for all the items that have at one time or another been interpreted as potential sound-producing pipes/flutes/whistles. Table 1 presents items that have, at various times, been reputed to be possible flute-type sound-producers; it does not include objects that have, since original publication, and on the basis of further analysis, been subsequently deemed unlikely to be sound-producers[1]—these are listed in Table 2. Note that not all of the objects listed in Table 1 have been subjected to further analysis since original publication; some of these remain to be reanalysed and may yet require their status to be similarly revised. Many of the objects have, however, been subjected to detailed analysis, and where comments about their status are available, these are noted.

It is difficult to quote a definitive total number of artefacts, as a few items with separate museum catalogue numbers have since been reconstructed as single objects, and in a couple of cases, items that have been separately catalogued by previous authors may actually represent the same object (see comments in the tables). As it stands, the 'edited' version of the pipe inventory (Table 1) includes 104 entries in total. Of these four are mammoth ivory, thirteen are specified as from various other mammal species, and in thirty-three cases the bone source is unspecified; the remaining fifty-four are bird bone.

THE EARLIEST REPUTED PIPES

The majority of these objects come from contexts associated with Aurignacian (c.43,000–28,000 years ago), Gravettian (c.28,000–22,000 years ago), and Magdalenian (c.17,000–11,000 years ago) technological complexes in Europe, and there are a few from Solutrean (c.22,000–17,000 years ago) contexts. All of these are associated exclusively with modern *Homo sapiens*. There are also several from contexts associated with Mousterian technology (c.200,000–40,000 years

---

[1] Whilst excluding items deemed unlikely to be sound-producers, Appendix Table 1 does include items Scothern (1992) describes as 'not demonstrably a sound producer'.

ago), called Middle Palaeolithic in Europe and Middle Stone Age (MSA) in Africa. These are particularly interesting (and have been subject to greater controversy) because Mousterian technology is associated with Neanderthals in Europe (*Homo neanderthalensis*) and in some cases with the ancestors of behaviourally modern *Homo sapiens* in Africa and Asia. It pre-dates the time traditionally asserted for a so-called 'Upper Palaeolithic revolution' in Europe (e.g. Mellars, 1989; Stringer and Gamble, 1993), after which evidence for art and symbolic thought becomes prolific. The implications of these items, if they could be shown to be intentionally manufactured as sound-producers, would be great in relation to the cognitive abilities of early modern humans in Africa and Asia and the Neanderthals in Europe, prior to the arrival in Europe of anatomically modern humans.

There is an important distinction to be made between 'pipes' (discussed here) and 'phalangeal whistles' (discussed in Chapter 4). Pipes consist of a hollow(ed) bone (or, rarely, ivory), which may or may not have been perforated along its length. A pipe with no perforations would usually produce a single tone, and each perforation made in the tube would allow an additional tone to be made by covering or uncovering the holes, effectively varying the length, and thus wavelength, of the resonating column of air inside. The vibrating column of air can potentially be produced in a variety of ways: by blowing over the end of the tube, as in a modern flute, or against a notch in the top of the tube, or as in the modern Western recorder with a block directing the air against a notch (called 'block-and-duct'), or as in woodwind instruments (where a single or double reed creates the vibration), or with the vibration created by the player's lips ('lip-reed', as in modern brass instruments). It is important to note, however, that holes are not actually necessary to allow the production of a wide variety of tones from a single pipe. Furthermore a pipe instrument producing a variety of tones can also be produced by binding together several pipes of different lengths, as is the case with 'pan pipes'—a design used widely throughout the world and through history (see Schaeffner, 1936, and Grauer, 2007, for example). This fact actually raises important implications for the identification of such material archaeologically, in which its anthropogenic origin—and potential musical function—is often recognized precisely because of the presence of piercings. Would a non-pierced musical pipe ever be recognized as such?

Phalangeal whistles, on the other hand, consist of a single pierced phalanx (toe bone, usually from an ungulate), usually punctured with only one hole. A tone is produced by blowing across the top of this hole, in much the same way as one would if producing a tone by blowing over the top of a bottle (a 'vessel-flute'). This would produce a high-pitched sound, due to the small internal volume of the phalanx. Appendix Table 3 deals with the phalangeal whistles, discussed in detail in Chapter 4.

These items, of both types, have been subject to considerable debate over the years. Foremost is the issue of whether a given item can be confidently identified as an *intentional* sound-producing object; i.e. not merely an object which is capable of producing sound, but one which was manufactured with the intention of it being used to produce sound. Second, as noted earlier, is debate regarding the implications of such objects for the possibility of symbolic thought amongst their manufacturers, and whether they could have been produced only by *Homo sapiens*, or also by Neanderthals and the ancestors of each.

A high proportion of these objects detailed in Appendix Tables 1 and 2 were recovered from excavations in the latter part of the nineteenth century and earlier part of the twentieth century. The motivations, circumstances, and techniques of excavation were often rather different then from those of later excavations; in many cases the simple recovery of finds was the principal objective, and the potential value of recording the stratigraphical and spatial relationships between finds was less well understood, or deemed superfluous. As a consequence the provenance of these earlier finds is often unclear, with site reports often non-existent or lacking any mention of specific pieces; exact descriptions of many and of the materials used in their construction are frequently absent, and the basis for their interpretation as sound-producers seems often to have been merely upon their appearance and the absence of any directly contradictory evidence. Frequently the main publication detailing them was written years after their discovery, on the basis of examination in museum collections (e.g. Seewald, 1934; Megaw, 1960). In consequence, many important pieces lack detailed contextual data; furthermore, some of the (often widely cited) examples are of dubious authenticity as genuine sound-producers, and many others which are more likely to be genuine instruments are rarely, if ever, cited, let alone analysed.

In addition, the presence of one or more holes in a piece of bone cannot be taken, by itself, to be indicative of human intention in creating that object; there are many processes that can lead to the creation of holes in pieces of bone. The anthropic (man-made) origin of many perforated bone artefacts has been called into question by the work of d'Errico and Villa (1997) and d'Errico et al. (1998). For example, d'Errico and Villa concur with Davidson (1991) that the perforated segment of bone from the Haua Fteah cave in Libya (McBurney, 1967), associated with Middle Stone Age tools, is most likely the product of carnivore teeth on the basis of the morphology of the hole and the lack of unambiguous evidence of human workmanship. This was originally proposed as a '"flute" or multiple pitch whistle' (McBurney, 1967, p. 90) and has been reported as such since (e.g. Scothern, 1992, albeit amidst debate regarding its correct age).

Fortunately, the features which result from carnivore activity are possible to recognize if the objects in question are analysed closely. Generally the surface of the bone is pitted with tiny indentations (microconcavities) and the edges of the bone at the perforation are thin and sharp. Holes have cylindrical sections with straight edges and depressed margins, rather than the more conical or biconical (hourglass) shaped sections of holes deliberately produced with a borer or burin (d'Errico and Villa, 1997). It is clear that many of the original reports of bone flutes need to be viewed with some caution, and that many of the artefacts may require careful re-examination and reinterpretation using d'Errico and Villa's criteria.

In contrast, some of the more recent finds have been subject to far greater scrutiny and thorough contextual recording; included amongst these are the earliest currently known instruments associated with anatomically modern *Homo sapiens*, from sites in the Ach and Lone valleys in Germany (Hahn and Münzel, 1995; Turk and Kavur, 1997; Richter et al., 2000; Conard et al., 2004; Conard et al., 2009); these are discussed in detail later in the chapter.

Mousterian musicianship?

The issue of the existence of 'modern', instrumental, musical behaviour amongst Neanderthals and pre-modern humans has been a source of great debate. There are four objects from Mousterian/MSA contexts that have variously been interpreted as pipe or flute sound-producers. In addition to the example from the Haua Fteah cave in Libya (mentioned earlier), there is an example from Ilsenhöhle, Germany (Hülle, 1977), which is of doubtful stratigraphy, a hare bone from Kent's Cavern, England (Seewald, 1934; Megaw, 1960), which is unlikely to be a sound-producer according to Scothern's (1992) analysis, and the so-called 'Neanderthal flute' juvenile cave bear femur from Divje babe I cave in Slovenia (Turk, 1997; Kunej and Turk, 2000; numerous other papers; Morley, 2006, includes a full bibliography).

This latter piece (Divje babe I item no. 652) has received a great deal of publicity and academic attention due to its potential importance with regard to the origins of instrumental use, and as the oldest reputed musical object to date. It is notable that in general 'flute' finds elicit more attention and excitement than other reputed instruments (which include rasps, bullroarers, and whistles; see Chapter 4), perhaps because they *seem* to be easier to positively identify. When the object in question is reputed to be both the oldest and from contexts associated with a different species of hominin, it generates even more passionate debate. The literature generated regarding the Divje babe I object has been prolific, at times exhibiting research which is a paragon in its rigour, surpassing any analysis applied to comparable objects associated with modern humans, at other times apparently contradictory.

A thorough review of the literature regarding this object reveals that, whilst the correct interpretation of the object seems clear to some of the authors, their views are often not easily reconciled, and caveats have to be applied to each of the advocated explanations. The literature falls into two main groups: those who believe the features of the Divje babe I bone to be the product of carnivore activity (e.g. d'Errico et al., 1998; Chase and Nowell, 1998; Albrecht et al., 2001), and those who advocate its interpretation as a Mousterian flute (e.g. Fink, 1997, 2000; Otte, 2000), of whom some have considered the possibility of carnivore action, and have carried out experimental analysis of both possibilities (e.g. Turk et al., 1997; Bastiani and Turk, 1997; Kunej and Turk, 2000; Turk et al., 2001; Turk et al., 2005). Several reconstructions of this item as a flute have been attempted, for example, by Atema (Gray et al., 2001), Kunej and Turk (2000) and Fink (1997, 2000), which have shown that the object can indeed produce a variety of pitches.

The debates regarding the interpretation of the features of the bone, and the analyses of their possible origins (both through carnivore and human action) are extremely detailed and complex, and worthy of thorough consideration. There is not the space to do them justice here, but a comprehensive review of all of the analyses and arguments pertaining to this object, their relative merits and the problems associated with them, has been published in the *Oxford Journal of Archaeology* (Morley, 2006); rather than replicating this, a summary of the findings is discussed here.

The bone in question is the diaphysis of the left femur of a one- to two-year-old cave bear, found securely locked in context in breccia near a Mousterian hearth. Recent Electron Spin Resonance (ESR) dating of the layer (layer 8) has dated it to around 60,000 years BP (Turk et al., 2006). Layer 8 had previously been radiocarbon dated to 43,100 ± 700 years BP (Nelson, 1997); according to Turk (*pers. comm.*) the radiocarbon date is superseded by the more reliable ESR dates. The bone exhibits several features which have led to its interpretation as a flute. These are numbered here 1–5 (see Figure 3.1), following as closely as possible the precedents of previous literature. There are two holes in the posterior side (Features 1 and 2), and semicircular features at the proximal (3) and distal (4) ends of the posterior side, which have been suggested to be the remains of two further holes. In addition there is a single notch in the anterior side of the distal end (5), which has also been suggested to be the remains of a hole.

The principal arguments and problems presented by the advocates of the 'flute' interpretation are addressed quite effectively by the highly detailed analyses of the object carried out by its finder (and advocate), Turk, and colleagues. The depth and diversity of their analyses, as well as their generosity with access to the specimen, set a fine precedent for future studies of reputed musical artefacts. The strongest arguments for the object being a flute were initially *arguments from design*: namely, the features, appearance, and sound-reproductive properties of the object. In fact, many of the 'design features' of the object are easily explained by other means, and do not appear to have been

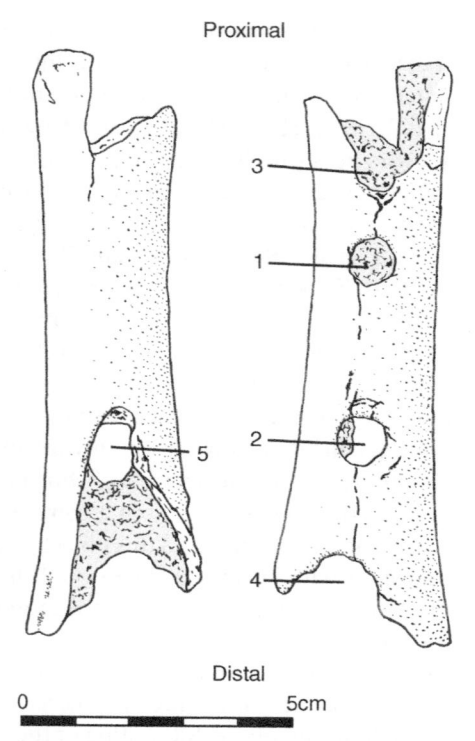

Fig. 3.1. Divje babe I object no. 652: the diaphysis of the left femur of a one- to two-year-old cave bear, from layer 8 of Divje babe cave, Slovenia. Key features are numbered 1–5. (Redrawn from Turk et al., 1997, 158, Fig. 11.1).

made contemporaneously with each other. In particular, the damage to the ends of the bone (including the three notches sometimes reputed to be the remains of complete holes) is typical of carnivore knawing damage, and this seems to have happened on different occasions. The bone only ever featured two complete holes.

These, however, have remained the focus of detailed analysis, and require some explanation. The two holes in the diaphysis of the femur are not typical of the damage inflicted by most possible carnivores, such as hyena, wolf, and cave lion. However, there is a suitable candidate for the punctures to the diaphysis of the bone, namely cave or brown bear, both of which are represented at the site. It has been illustrated that the canine teeth of brown bear and cave bear are the right shape to account for the holes in the Divje babe I bone, that they have the strength to create such holes (Turk et al., 2001), as well as that attacks on juveniles by mature males are well known in the behavioural repertoire of bears (Kurtén, 1976; Mattson et al., 1992; Swenson et al., 2001 and references therein), including on other bones at the site itself

(Turk et al., 2001). Furthermore, analyses have shown that this part of a fresh bone could be punctured in this way without cracking, and that in this particular bone, the punctured parts of the bone are in fact the thinnest (Turk et al., 2005). The odds of it puncturing once on this side have been shown experimentally to be 1 in 4 (Turk et al., 2001), meaning that the odds of it puncturing twice on the same side, even without the factor of its thinness at this point, would be about 1 in 16. Whilst there is the theoretical possibility of the two holes in the diaphysis being 'chipped out' using some of the types of tools present at the site, the direct evidence for human agency is, at best, highly ambiguous, and there is a lack of evidence of *other* possible human workmanship on the bone. The balance is further swung against hominin agency by the abundance of evidence of *other*, typical, carnivore activity on the bone, and similar damage on other bones at this and other sites.

The most reasonable conclusion, on the basis of the various experimental and microscopic analyses that have been carried out, would seem to be that, on the one hand, there is little reason to believe the object to have been made by humans with the available stone tools, but that, in contrast, there is every reason to believe that the two holes could have been made by a cave bear or brown bear with its teeth. It is most likely that the Divje babe I object as we see it today is the product of a number of stages of carnivore activity, and there is no need to invoke any hominin agency in the creation of the object. It should be noted that this conclusion makes no recourse to assumptions about Neanderthal aptitude or behavioural propensities—the same conclusion would be reached regardless of the hominin associated with the technology at the site.

UPPER PALAEOLITHIC PIPES

The reliability of the interpretation of such objects has often been assumed to be greater in the case of those of more recent origin, as a consequence of the existence of greater numbers of pieces and the existence of some individual specimens whose preservation is good enough that the human agency in their creation is unambiguous. It does have to be said though, that with the exception of the work of Lawson and d'Errico (2002) on the most complete Isturitz pipes, and the publications of the finds from Geissenklösterle, Hohle Fels, and Vogelherd (Hahn and Münzel 1995; Richter et al., 2000; Münzel et al., 2002; Conard et al., 2004; Conard et al., 2009), none of these objects from Upper Palaeolithic contexts have been subject to the same level of analysis as has the Mousterian object from Divje babe I. Scothern's (1992) analysis, however, went a good way towards addressing the objective analysis of many of the reputed artefacts. A full examination and description of each of these finds would warrant a book in itself; there is not space for that here, but the tables in the Appendix contain a full list of artefacts reputed, with varying likelihood, to have been musical instruments, coming from Palaeolithic contexts, and scholarly articles where they are discussed. The examples that have been subject to the greatest level of analysis come from two areas—Isturitz, in the French Basque region of the Pyrenees, and the Ach Valley in southern Germany, including the cave sites of Geissenklösterle (or Geißenklösterle), Hohle Fels, and Vogelherd. These two locations have produced both the best-preserved and oldest examples of Palaeolithic pipes found to date.

The Swabian Alb (Geissenklösterle, Hohle Fels, Vogelherd)

The Ach and Lone valley cave sites in the Swabian Jura of south-western Germany have produced many famous examples of early Upper Palaeolithic symbolic artefacts, including sculptured figurines of humans and animals, engraved artefacts, and other symbolic items, as well as hearths and thousands of stone and bone tools. Three of the caves in this area have also produced the most complete and oldest sound-producers currently known: Geissenklösterle, Hohle Fels, and Vogelherd. In fact all of the incontrovertibly-identified bone 'flutes' published in recent years have come from these sites. Together they have now produced four pipes (or parts of pipes) made from the wing bones of large birds, and four made from mammoth ivory. These were reconstructed from numerous fragments, but three of them, one from Hohle Fels and two from Geissenklösterle, make virtually complete specimens. The eight examples reported to date are summarized in Table 3.1.

Published in 2009, one of the most complete Upper Palaeolithic pipes found to date was discovered during excavation and wet-sieving at Hohle Fels (Conard et al., 2009). The virtually complete pipe, which was discovered in twelve pieces and reconstructed by Maria Malina, was made from the radius of *Gyps fulvus* (Griffon vulture) (see Figures 3.2 and 3.3). The pipe came from a context associated with the earliest Upper Palaeolithic occupation of the site, certainly more than 35,000 cal BP (see also Higham et al. 2012), and was associated with characteristic Aurignacian technology.

The pipe has many features in common with the Aurignacian and Gravettian examples from Isturitz in France (see later section). It was made using vulture wing bone, which shows evidence of having been smoothed by scraping—the surface features fine longitudinal striations. The surface close to one of the holes is also marked with a series of fine lines perpendicular to the length; these may constitute part of a measuring process for the spacing of the holes. Finally, the holes were made by thinning the surface of the bone, creating a cratered depression in which the finger can sit and make an airtight seal, and piercing a hole in the centre of this depression. The artefact has five holes preserved (one of which, at the distal end, is incomplete), and two

Table 3.1. Pipe finds from the Ach Valley, Germany

Cross-reference	Period	Origin	Described as	Material (if known)	Details of Age	Description	Status & Comments	Location (if known)	Main References
Swabian Alb 1	Aurignacian	Geissenklösterle, Germany	Flute	Swan radius	Upper Aurignacian; 43,150–39,370 cal BP Aurignacian II split bone points in same layer (Archaeological Horizon II)	Geissenklösterle Flute 1 3 holes preserved (diameter 5.3 × 3.4 mm, 3.5 × 3.0 mm, 2.8 × 2.4 mm), distance between holes 30–40 mm. Originally 18–19 cm long; now 12 cm, with the largest bore being 8–9 mm	Certainly sound-producer Reconstructed from 23 pieces excavated 1990, published 1995	Württembergisches Landesmuseum, Stuttgart	Hahn and Münzel, 1995; Turk and Kavur, 1997; Richter et al., 2000; Goldbeck, 2001; Münzel et al., 2002; d'Errico et al., 2003; Higham et al., 2012
Swabian Alb 2	Aurignacian	Geissenklösterle, Germany	Flute	Swan (radius?)	Upper Aurignacian; 43,150–39,370 cal BP Aurignacian II split bone points in same layer	Geissenklösterle Flute 2 1 hole remaining	Certainly sound-producer Reconstructed from 7 pieces excavated 1973, published 1995	Württembergisches Landesmuseum, Stuttgart	Hahn and Münzel, 1995; Turk and Kavur, 1997; Richter et al., 2000; Goldbeck, 2001; Münzel et al., 2002; d'Errico et al., 2003; Higham et al., 2012
Swabian Alb 3	Aurignacian	Geissenklösterle, Germany	Flute	Mammoth ivory	Upper Aurignacian; 43,150–39,370 cal BP	Geissenklösterle Flute 3 Length: 187 mm	Certainly a sound-producer. Complicated manufacture involving	University of Tübingen collections	Conard et al., 2004; Conard et al., 2009; Higham et al. 2012

(Continued)

Table 3.1. Continued

Cross-reference	Period	Origin	Described as	Material (if known)	Details of Age	Description	Status & Comments	Location (if known)	Main References
					Archaeological Horizon II	Numerous finely-carved notches along edges of the two halves to facilitate binding with resin	splitting ivory, hollowing, and then rebinding. Appears to replicate bird bone form. Reconstructed from 31 pieces excavated 1974–9, published 2004		
Swabian Alb 4	Aurignacian	Hohle Fels, Germany	Flute	Radius of *Gyps fulvus* (Griffon vulture)	Basal Aurignacian deposits of Layer Vb; initial Upper Palaeolithic occupation, c.40,000 ya; certainly >35,000 cal BP	Hohle Fels Flute 1 Length: 218 mm Diameter: 8 mm 5 holes preserved 2 V-shaped notches cut into proximal end Proximal end of radius = proximal end of flute. Length of unmodified radius c.340 mm	Reconstructed from 12 pieces excavated 2008, published 2009	University of Tübingen collections	Conard et al., 2009
Swabian Alb 5	Aurignacian	Hohle Fels, Germany	Flute fragment	Mammoth ivory	Basal Aurignacian, Feature 10, Base of Archaeological Horizon Va	Hohle Fels Flute 2 Length: 11.7 mm Width: 4.2 mm Thickness: 1.7 mm Portion of finger hole preserved	Excavated 2008, published 2009	University of Tübingen collections	Conard et al., 2009

Swabian Alb 6	Aurignacian	Hohle Fels, Germany	Flute fragment	Mammoth ivory	Basal Aurignacian, lowest Aurignacian unit of Archaeological Horizon Vb; initial Upper Palaeolithic occupation, c.40,000 ya; certainly >35,000 cal BP	Hohle Fels Flute 3 Length: 21.1 mm Width: 7.6 mm Thickness: 2.5 mm Incised lines on outer surface and 9 notches on edge. Striations on internal and external surfaces	Excavated 2008, published 2009 Greater thickness and dimensions suggest not part of same object as Flute 2	University of Tübingen collections	Conard et al., 2009
Swabian Alb 7	Aurignacian	Vogelherd, Germany	Flute fragment	Bird bone	From reworked contexts, but majority of finds from site are securely dated to Aurignacian; pre-date 30,000 years ago (Conard et al., 2009)	Vogelherd Flute 1 Length: 17.5 mm Width: 5.8 mm Thickness: 1.8 mm 1 partly preserved finger hole, 7 small notches along edge	Reconstructed from 3 fragments, excavated 2005, published 2006	University of Tübingen collections	Conard and Malina, 2006; Conard et al., 2009
Swabian Alb 8	Aurignacian	Vogelherd, Germany	Flute fragment	Mammoth ivory	From reworked contexts, but majority of finds from site are securely dated to Aurignacian; pre-date 30,000 ya (Conard et al., 2009)	Vogelherd Flute 2	Excavated 2008, published 2009	University of Tübingen collections	Conard et al., 2009

Inventory of sound-producing pipes found to date from the Ach Valley sites of Geissenklösterle, Hohle Fels, and Vogelherd. These currently constitute the earliest-known sound-producers associated with *Homo sapiens*, and date to close to the time of arrival of modern humans in Europe.

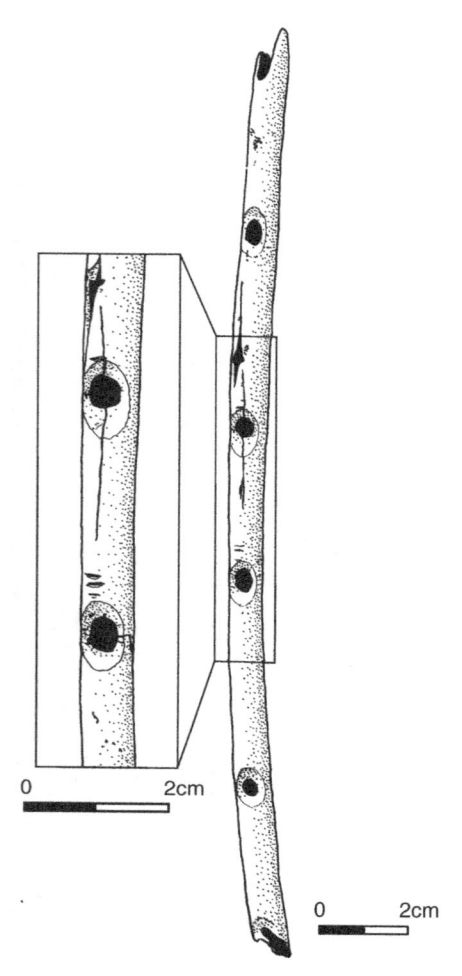

Fig. 3.2. Pipe from Hohle Fels cave, Ach Valley, Germany; vulture radius, dated to the earliest Aurignacian occupation of the site, c.40,000 years ago. (Redrawn from Conard et al. 2009, Fig. 1, p. 737.)

V-shaped notches carved into the proximal end (the proximal end of the bone is also the proximal end of the instrument[2]), which may be related to the method of sound production. Hohle Fels has also produced, from similarly dated contexts, fragments of two pipes made from mammoth ivory, which

---

[2] The proximal end of a bone is the end nearest to the body of the animal in life, the distal end is that furthest from the body. The proximal end of an instrument is the end closest to the player (into which they blow, in this case) and the distal end is furthest from the player. The proximal end of a pipe may therefore be either the proximal or distal end of the bone, depending upon how the bone is modified.

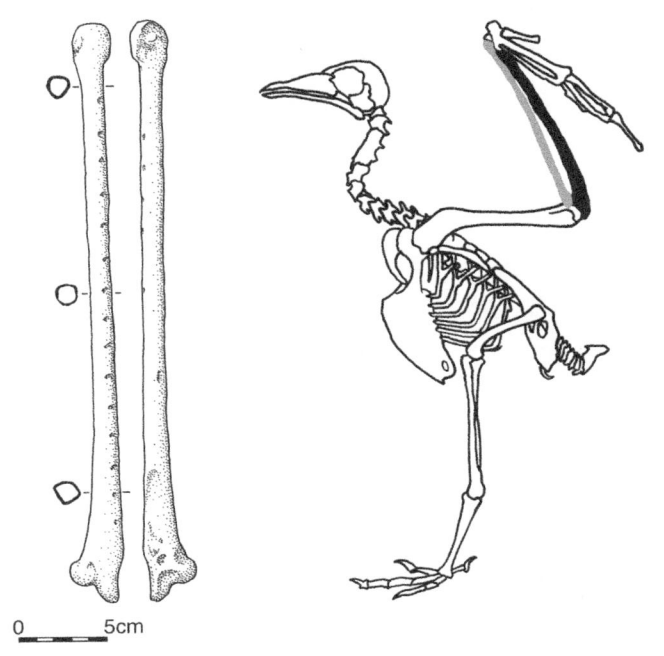

Fig. 3.3. Vulture ulna (left) and location of ulna (black) and radius (grey) bones in the wing of a raptor (vulture) (right). (Redrawn from Buisson 1990, Fig. 1, p. 423; vulture skeleton redrawn from illustration in Winston 1918.)

appear to be very similar to a mammoth ivory example found in Geissenklösterle cave nearby.

The three pipes recovered from Geissenklösterle appear to be part of a very similar tradition, and until the discovery of the Hohle Fels artefact, constituted the best-preserved examples known; recent redating (Higham et al. 2012) makes these the oldest-known musical artefacts in the world. Two pipes made from swan wing bones were published in 1995 (Hahn and Münzel, 1995), and are referred to as 'Flute 1' and 'Flute 2' (Münzel et al., 2002). Flute 1 (Figure 3.4) was excavated in 1990 and reconstructed from twenty-three fragments, and Flute 2 reconstructed from seven fragments originally excavated in 1973 (Conard et al., 2009). Flute 1 represents the well-preserved proximal portion of the radius of a swan, probably *Cygnus cygnus* (whooper swan), and Flute 2, whilst more fragmentary, is also of swan size (Münzel et al., 2002). Interestingly, no other swan bone remains were retrieved from the site, suggesting that the pipes were imported rather than being made there (Münzel et al., 2002); this is in contrast to vulture bones, which are present at various of the Swabian cave sites including Geissenklösterle and Hohle Fels, where the griffon vulture pipe was found (Conard et al., 2009), and also contrasts with the case at Isturitz (see later section).

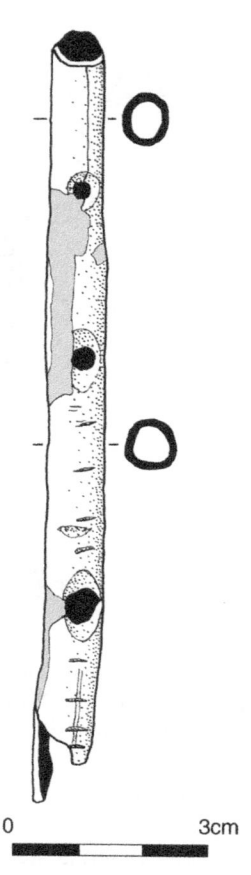

Fig. 3.4. Pipe from Geissenklösterle cave, Ach Valley, Germany ('Flute 1'); swan radius, dated to Aurignacian, c.40,000 years ago [AHII]. (Redrawn from Münzel et al. 2002, Fig. 5a, p. 114, and d'Errico et al. 2003, Fig. 11(b), p. 41.)

Faunal remains at the site suggest that it was occupied during the winter and spring, and for the majority of its occupation, from around 40,000 years ago to the last glacial maximum (LGM) 20,000 years ago, it would have had an arctic tundra steppe-type environment, and the valleys would have been relatively sheltered (Münzel et al., 2002).

As the better-preserved example, Flute 1 has been subject to the greatest amount of analysis, including construction of replicas to attempt to determine ways in which it might have been played (Münzel et al., 2002). Possible methods of playing these bone pipes are discussed later in this chapter. Like the Hohle Fels example (and those from Isturitz) the exterior of the bone has been smoothed by longitudinal scraping prior to the construction of the finger holes, and has also been incised with sequences of parallel lines perpendicular to its length. The holes have a conical cross-section, having been bored with a tool (Turk and Kavur, 1997), and are separated by 30–40 mm. The archaeological context ('Archaeological Horizon II', AHII) from which the fragments of Flute 1 were retrieved included tools of 'classic' Aurignacian type (Aurignacian II), and was originally dated to approximately 33,500 BP by radiocarbon (C14 AMS) dating and to about 37,000 BP by Thermoluminescence (TL) dating (Richter et al., 2000; Münzel et al., 2002). However, recent ultrafiltration C14 dating of the earliest Aurignacian Geissenklösterle layers by the Oxford Radiocarbon Laboratory has pushed that C14 date back to 43,150–39,370 cal BP (Higham et al. 2012). The artefact features three preserved finger holes, and could have been up to 180–200 mm long when complete, though 126.5 mm of it remains today.

Seeberger reconstructed Flute 1 using a radius bone of mute swan (*Cygnus olor*) (Münzel et al., 2002) and demonstrated that it can produce seven clear tones when played simply as a bevelled flute, blowing over the bevelled end of the bone as a whistle. Bevelled flutes are well known in Mediterranean Europe and North Africa, where their use is known to date back at least 5,000 years (Münzel et al., 2002), and possibly far longer. This is the most straightforward form of sound production with a bone pipe, requiring relatively little modification of the end and no additional components. It can also produce five tones simply by blowing into the end of the bone, with one of the finger holes thus acting as a sound window, though the sound thus produced is of a lower intensity, and Münzel et al. (2002) consider this method to have been less likely than that suggested by Seeberger. It is also possible to produce sound by producing the vibrating air column with the lips ('lip-reed', as in modern brass instruments) (Goldbeck, 2001), and this method has been suggested by d'Errico et al. (2003) to be most effective for some of the Isturitz examples (see later section). Lip-reed sound production creates a sound with a more humanlike pitch and timbre than bevelled-edge flutes, which are more bird-like (Lawson and d'Errico, 2002). Finally, it is possible to insert into the end of the flute a reed (as with a clarinet) or pair of reeds (as is the case with modern oboes and North African pipes) to create the vibrating air column inside the pipe. Unfortunately the organic material used for such a modification does not survive, but some pipes can be identified where this modification would, in theory, have been possible.

Perhaps the most remarkable find from Geissenklösterle so far, however, is the pipe made from mammoth ivory, known as 'Flute 3' (see Figure 3.5). This was also retrieved from Aurignacian II contexts (AHII, dated to 43,150–39,370) and reconstructed from thirty-one fragments excavated 1974–9, and published in 2004 (Conard et al., 2004, 2009).

Until the discovery of more ivory pipe fragments from nearby Hohle Fels and Vogelherd in 2008, Geissenklösterle Flute 3 was the only known ivory example from the entire Upper Palaeolithic. What is remarkable about this

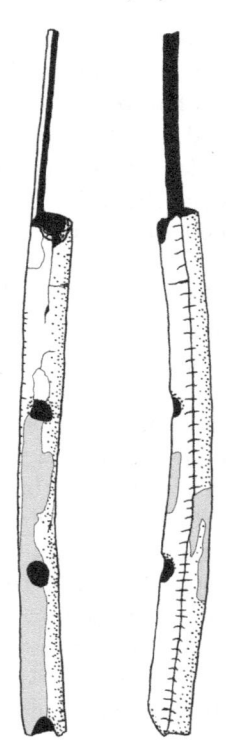

Fig. 3.5. Pipe from Geissenklösterle cave, Ach Valley, Germany ('Flute 3'); mammoth ivory, dated to Aurignacian, c.40,000 years ago [AHII]. Length: 187 mm. (Redrawn from Conard et al. 2004.)

material for pipe production (and all four known pieces appear to have been produced in the same way) is that it requires a huge amount of precision work in comparison to a bird bone pipe, in order to produce an object that is essentially identical to a bird bone version. Whereas the radius or ulna of a large bird such as swan or vulture is naturally hollow, the right size, and relatively easily worked, the reverse is the case for mammoth ivory. Ivory grows in layers (somewhat similar to tree rings), and in order to make it hollow these lamellae (layers) must be separated. To do this, a section of ivory must be sawn to the correct length, it must then be sawn in half along its length, the core lamellae must be removed, and then the two halves of the flute must be refitted and bound together with a bonding substance which must create an airtight seal in order for the pipe to be able to produce sound. The surfaces which have to reseal show evidence of having been roughened, presumably to provide a better bonding surface for the cementing material (probably resin) (Conard et al., 2009). This is a technically complex and challenging procedure. The fact that it was attempted at all is made all the more remarkable by the fact that the creator of the artefact appears to have gone out of their way to create an object that looks virtually identical to a bird bone pipe, in form, and finger holes, albeit slightly larger when a bird bone example could have been produced far more easily. This would suggest that the choice of raw material (mammoth ivory) was itself considered highly significant. It also demonstrates clearly that the production of sound-producing pipes was far from being a nascent technological innovation 40,000 years ago, but was a finely honed and sophisticated activity (see also Lawson and d'Errico, 2002, and d'Errico et al., 2003 for more discussion of the development of such technology).

It is worth bearing in mind, too, that suitable pipe-production raw materials are not limited to bone and ivory. Amongst many traditional and historical societies musical instruments, including pipes, have been produced predominantly from plant matter (see Chapter 2). Suitable materials include elder, willow, and fruit tree barks (see Lawson and d'Errico, 2002), which are relatively easily worked. These, unfortunately, very rarely preserve under normal archaeological conditions, but serve to remind us that the earliest bone and ivory examples that we find are unlikely to represent the *beginning* of a tradition of manufacture and, furthermore, that the choice of a complicated procedure such as that used to make the ivory examples is underlain by a very particular motivation and significance, far from being merely expedient.

Isturitz

Until the recent discovery of the well-preserved bone pipes from Geissenklösterle, Hohle Fels, and Vogelherd, the best-known source of Palaeolithic bone pipes was Isturitz, in the French Basque region of the Pyrenees (see Figure 3.6). This remains by far the richest source of intentionally produced sound-makers or, at least, fragments of them, with most being discovered during excavations by E. Passemard between 1912 and 1922 and R. and S. de Saint-Périer between 1928 and 1950. A selection of the finds was reported by Passemard in 1923 and 1944, and by the Saint-Périers in their monograph in 1952. The collections have also been subject to extensive examination, reanalysis, and reconstruction in more recent years (Buisson, 1990; Scothern, 1992; Le Gonidec et al., 1996; Lawson and d'Errico, 2002).

All of the pipe fragments were retrieved from layers associated with modern human technology, dating from the Aurignacian (the earliest Upper Palaeolithic occupation layers at the site) through to the Magdalenian (the latest Upper Palaeolithic layers). The nature of the early excavations and their reporting, and the cataloguing of the artefacts, has led to considerable confusion regarding the total number of pipe fragments, and the stratigraphic relationships of the layers within the site from which they came in the different

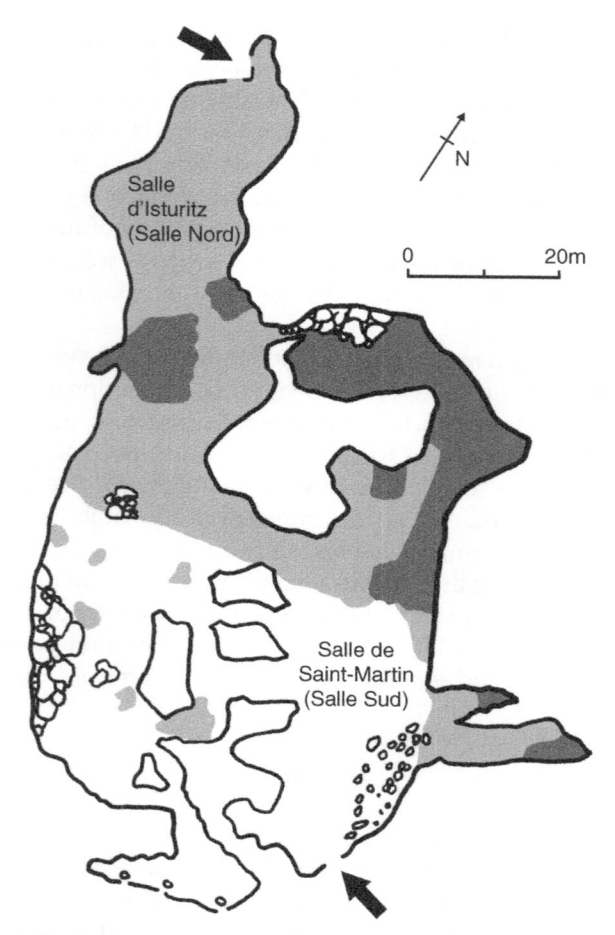

Fig. 3.6. Plan of Isturitz cave, France. Dark grey indicates areas excavated by Passemard (1912–22); light grey indicates areas excavated by Saint-Périer and Saint-Périer (1928–50); arrows indicate cave entrances (Redrawn from Saint-Périer and Saint-Périer 1952, Fig. 2, p. 5.)

excavations. More recent analyses of the records have helped to clarify these stratigraphical relationships (see especially Goutas, 2004; also Delporte, 1980–1; Esparza San-Juan, 1990; Foucher and Normand, 2004; Pétillon, 2004; Langlais, 2007; Normand et al., 2007) (see Table 3.2), and a careful examination of the (sometimes apparently contradictory) inventories of Saint-Périer and Saint-Périer (1952), Buisson (1990), Scothern (1992), and Lawson and d'Errico (2002) has made it possible to compile a database of pipe finds from Isturitz, which is presented in Table 3.3.

Isturitz cave is actually divided into two large chambers, in a line with each other, in total approximately 120 m long, 50 m wide, and 20 m tall (Buisson,

1990) (see Figure 3.6). The northernmost of these opens towards the town of Isturitz and is known as the Salle Nord, Salle d'Isturitz, or Grande Salle. The other, to the south and with its opening facing the town of Saint-Martin, is known as the Salle Sud, Petit Salle, or Salle de Saint-Martin. The Salle d'Isturitz and Salle de Saint-Martin were each excavated by both Passemard and the Saint-Périers. With the exception of the earliest pipe from the site (IA Sup 1921 77142) (Figure 3.7), all of the pipe fragments were retrieved from Salle d'Isturitz. This chamber also features a decorated stalagmite, incised with images of animals, and has striking acoustic properties (Reznikoff, 2008).

In addition to the pipe fragments the cave produced a rich record of other archaeological evidence for the use of the site, including hundreds of artefacts of shaped, incised, and pierced bone, ivory, antler, and teeth, worked shells, engraved stone plaques, cave art, and hearths, in addition to thousands of stone tools (Saint-Périer and Saint-Périer, 1952). These artefacts betray trading or transport relationships over considerable distances (Normand et al., 2007) and intensive use of the site, at least in some periods at certain times of year. The site was an important focal point for large groups of people throughout the Upper Palaeolithic (Bahn, 1983), and seems to have been a focus for major gatherings in spring and autumn in particular. Similarities in the stone and bone artefacts with those from the Dordogne area, and the transferral of flint over distances over 100 km (Gamble, 1983) suggest wide social contact (Scothern, 1992; Gamble, 1999). Neanderthals used the site before the arrival of modern humans, with Middle Palaeolithic (Mousterian) technologies being found in the lowest layers of Salle de Saint-Martin. The earliest Aurignacian layers at the site, associated with modern humans, are also in Salle de Saint-Martin, and have been radiocarbon-dated to 36,550 ± 610 BP (Turq et al., 1999). This date is from a layer corresponding with the bottom of Saint-Périer layer 'Base S. III' (Table 3.2); the Aurignacian pipe (Figure 3.7) came from the layer above this. It is worth bearing in mind that ultra-filtration C14 dating of the site, if carried out, may reveal this date to be an underestimate, as occurred at Geissenklösterle and at other sites (see Higham et al., 2012). Unfortunately any post-Palaeolithic remains at the site were removed through mining the cave's contents for phosphates prior to Passemard's excavations,[3] but the use of the site by modern humans during the Upper Palaeolithic spans more than 25,000 years. This is a staggering length of time, two-and-a-half times the period from the end of the last Ice Age to today, and this must be borne in mind when considering the record of sound-producers retrieved. The use of the site was certainly not continuous

---

[3] The Palaeolithic remains were only saved by the fact that the phosphate miners reached a stalagmitic layer formed during the early post-glacial period and, fortunately, thought they had reached the bedrock. The Palaeolithic remains were encased below this (Passemard, 1944; Buisson, 1990).

Table 3.2. Stratigraphy of the two chambers at Isturitz, France, showing the correspondence between the layers excavated by Passemard and the Saint-Périers

Material culture type/period	Salle d'Isturitz (Salle Nord, Grande Salle)		Salle de Saint-Martin (Salle Sud, Petit Salle)	
	Passemard	Saint-Périer	Passemard	Saint-Périer
Final Magdalenian-Azilian	B	Ist. Ia	xxxxxxxxxx	xxxxxxxxxx
Upper Magdalenian	F1	Ist. I	xxxxxxxxxx	xxxxxxxxxx
Middle Magdalenian	E α, β	Ist. II	Eω	S. I
Upper Solutrean	Base E	xxxxxxxxxx		Base S. I
Upper Solutrean		Ist. IIIb	xxxxxxxxxx	xxxxxxxxxx
Upper Solutrean	F2	**Ist. IIIa**	Z	
Gravettian	C	**Ist. III**	xxxxxxxxxx	xxxxxxxxxx
Gravettian	F3	**Ist. IV**	X	
Evolved Aurignacian	A α, β	Ist. V	xxxxxxxxxx	xxxxxxxxxx
Typical Aurignacian	xxxxxxxxxx	xxxxxxxxxx	Y	S. II
Typical Aurignacian	xxxxxxxxxx	xxxxxxxxxx	**Aω, σ**	S. III
Proto-Aurignacian	xxxxxxxxxx	xxxxxxxxxx	xxxxxxxxxx	Base S. III
Typical Mousterian	xxxxxxxxxx	xxxxxxxxxx	M	S. IV
Mousterian	xxxxxxxxxx	xxxxxxxxxx	P	S. V

Stratigraphy of the two chambers at Isturitz, showing the correspondence between the layers excavated by Passemard and the Saint-Périers. Bold indicates layers that produced pipe remains. (After Goutas, 2004; Esparza San-Juan, 1990; and incorporating information from Saint-Périer and Saint-Périer, 1952; Delporte, 1980–1; and Buisson, 1990.)

Note that with the exception of Aurignacian piece IA Sup 1921 77142, which came from Layer Aω in Salle de Saint-Martin (Passemard, 1944), all other pipe remains were retrieved from Salle d'Isturitz. [Buisson's (1990) inventory states that the two Magdalenian pieces found by Passemard in Layer E were from bed Eα (Salle d'Isturitz), not Eω (Salle de Saint-Martin).]

over that time, and it would be difficult to maintain that the finds represent a continuous tradition of musical practice. The fact that the site contained sound-producers from every period of its use does suggest, however, that it had particular properties, perhaps including the availability of certain raw materials (namely, raptor wing bones), its location, and its acoustics, which encouraged its use as a focus for such activity.

Having said that, the overwhelming majority of the pipe artefacts (seventeen of those described by Buisson, 1990, plus perhaps another five described by Scothern, 1992, and two by Lawson and d'Errico, 2002) came from the layers associated with Gravettian artefacts, and these are remarkably consistent in terms of raw material used and method of manufacture. These may indeed represent a consistent tradition over some time, although over quite how long is impossible to assert without absolute dating of the objects or the layers from which they came. The complete inventory of artefacts (inasmuch as can be established from the studies of Saint-Périer and Saint-Périer, 1952; Buisson, 1990; Scothern, 1992; and Lawson and d'Errico, 2002) is represented in Table 3.3.

Table 3.3. Inventory of Isturitz pipes discussed by Buisson (1990) in chronological order

Number	Period	Origin	Described as	Material (if known)	Details of Age	Description	Status & Comments	Location (if known)	Main References
Buisson's fig. 4.1 (Scothern 1)	Aurignacian	Isturitz (Salle de Saint-Martin)	Pipe fragment	Right ulna of diurnal raptor, probably vulture (distal end)	IA designates Passemard's Layer A. This refers to Layer Aω in Salle Aω in Salle Saint-Martin, typologically dated to Aurignacien typique	IA Sup 1921 77142 Length: 74.0 mm Diameter: 15.0–20.0 mm Bore: 14.0–15.3 mm 3 holes on anterior, diameters 5.0 × 8.0 mm and 4.0 × 8.5 mm, third hole broken Distal end heavily worked and may form embouchure, proximal end broken in region of third hole. Several traces of abrasive scraping, 9 short deep parallel incisions between second and third hole	This is the oldest of the pipes from Isturitz, the only one to come from Aurignacian contexts and the only one to be found in Salle Saint-Martin (Salle Sud). This was the first pipe to be published by Passemard (in 1923), although he had excavated the Magdalenian examples from Salle d'Isturitz earlier	Collection Passemard, Musée des Antiquités Nationales, Saint-Germain-en-Laye, Paris	Passemard, 1923, 1944; Buisson, 1990; Lawson and d'Errico, 2002

(Continued)

Table 3.3. Continued

Number	Period	Origin	Described as	Material (if known)	Details of Age	Description	Status & Comments	Location (if known)	Main References
						Periosteum (bone lining) still present, so not worked internally (Scothern, 1992, p. 94)			
Buisson's fig. 2(i) (Scothern 16)	Gravettian	Isturitz (Salle d'Isturitz)	Pipe fragment	Left ulna of diurnal raptor, probably *Gypaetus barbatus*—bearded lammergeier/vulture	IF3a designates Passemard's Layer F3 in Salle d'Isturitz, typologically dated to Gravettian. Corresponds with Saint-Périer layer IV	IF3a 1914 75252 A3	Virtually complete pipe reconstructed from two separate finds by Buisson (1990). Initial analysis and description carried out by Buisson (1990).	Collection Passemard/ Collection Saint-Périer, Musée des Antiquités Nationales, Saint-Germain-en-Laye, Paris	Buisson, 1990
Buisson's fig. 2(ii) (Scothern 2)	Gravettian (originally designated 'Final Aurignacian (Gravettian)' by Saint-Périers)	Isturitz (Salle d'Isturitz)	Pipe fragment		Ist. III 83888 designates items from Saint-Périer Layer III in Salle d'Isturitz (corresponding with Passemard Layer C, immediately above Layer F3); all recent chronologies agree Gravettian	Ist. III 1939 83888 (a) Combined dimensions: Length: 212.0 mm Diameter: 11.0–14.0 mm Bore: 10.8–11.8 mm	Detailed analysis of piercings and markings, and reconstruction, carried out by Lawson and d'Errico (2002) Highlights stratigraphical issues associated with the site, with one part coming from a layer		Saint-Périer and Saint-Périer, 1952; Buisson, 1990; Lawson and d'Errico, 2002

Buisson's fig. 3.3 (Scothern 10)	Gravettian	Isturitz (Salle d'Isturitz)	Pipe fragment	Left ulna of diurnal raptor, probably vulture (proximal end)	Ist. IV designates Saint-Périer's layer IV in Salle d'Isturitz, typologically dated to Gravettian cultural horizon (see Goutas, 2004). The original designation 'Final Aurignacian' derived from Breuil's typology (Lawson and d'Errico, 2002)	Ist. IV 1936 83889 Length: 56.5 mm Diameter: 13.6–15.0 mm Bore: 10.8–12.0 mm	4 holes on posterior, diameters 5.0 × 8.0 mm and 7.0 × 5.5 mm One end finely finished and complete, the other end slightly damaged. Longitudinal abrasive scraping and numerous fine incised parallel lines perpendicular to length Means of blowing possibly V-notched tongue duct (Scothern, 1992) equivalent to Ist. IV and the other from layer Ist. III (a) designation added by IM to differentiate from items illustrated in Buisson (1990), figs. 4.7, 4.8, and 4.9 (described later)	Collection Saint-Périer, Musée des Antiquités Nationales, Saint-Germain-	Saint-Périer and Saint-Périer, 1952; Buisson, 1990

(Continued)

Table 3.3. Continued

Number	Period	Origin	Described as	Material (if known)	Details of Age	Description	Status & Comments	Location (if known)	Main References
					Corresponds to Passemard's Layer F3	2 holes on anterior, diameters 5.0 mm. Breaks at the extremities. Traces of abrasive scraping		en-Laye, Paris	
Buisson's fig. 3.4 (Scothern 11)	Gravettian	Isturitz (Salle d'Isturitz)	Pipe fragment	Right ulna of diurnal raptor, probably vulture (proximal end)	Ist. IV designates Saint-Périer's layer IV in Salle d'Isturitz, typologically dated to Gravettian. Corresponds to Passemard's Layer F3	Ist. IV 1939 83889 (a) Length: 67.0 mm Diameter: 15.0–18.0 mm Bore: 12.3–13.0 mm 1 hole on posterior, 1 hole on anterior, diameters 4.0 × 6.0 mm. Breaks at the extremities, one in the region of the posterior hole. Traces of abrasive scraping; fine incisions perpendicular	(a) designation added by IM	Collection Saint-Périer, Musée des Antiquités Nationales, Saint-Germain-en-Laye, Paris	Saint-Périer and Saint-Périer, 1952; Buisson, 1990

Buisson's fig. 4.3 (Scothern 17?)	Gravettian	Isturitz (Salle d'Isturitz)	Pipe fragment	Ulna of small bird	Ist. IV designates Saint-Périer's layer IV in Salle d'Isturitz, typologically dated to Gravettian. Corresponds to Passemard's Layer F3	Ist. IV 1939 83889 (b) Length: 91.0 mm Diameter: 7.0 mm 2 holes on posterior, diameter 2.5 × 3.0 mm. Distal end heavily worked, proximal broken in the region of one of the holes. Worked end shows incisions that are traces of the removal of the distal epiphysis. Traces of abrasive scraping on surface and oblique to the longitudinal axis on the surfaces opposite the holes	(b) designation added by IM Unpublished before Buisson (1990)	Collection Saint-Périer, Musée des Antiquités Nationales, Saint-Germain-en-Laye, Paris	Buisson, 1990
Buisson's fig. 5.1(a)	Gravettian		Pipe fragment	Right ulna of a diurnal	Ist. IV designates Saint-Périer's	Ist. IV 1939 86757	Virtually complete pipe	Collection Favre,	Saint-Périer and Saint-Périer,

(Continued)

Table 3.3. Continued

Number	Period	Origin	Described as	Material (if known)	Details of Age	Description	Status & Comments	Location (if known)	Main References
		Isturitz (Salle d'Isturitz)		raptor, probably vulture (proximal end)	layer IV in Salle d'Isturitz, typologically dated to Gravettian. Corresponds to Passemard's Layer F3		reconstructed from three separate finds by Buisson (1990). Initial analysis and description carried out by Buisson (1990). Detailed analysis of piercings and markings, and reconstruction, carried out by Lawson and d'Errico (2002) (a) and (b) designations added by IM. On the basis of Buisson's (1990), fig. 5 it would appear that the dimensions of the three component parts of the pipe are as follows: Ist. IV 1939 86757:	earliest Saint-Périer collection, Musée des Antiquités Nationales, Saint-Germain-en-Laye, Paris	1952; Buisson, 1990

(Scothern 13)			L: 87 mm
D: 12–14 mm
2 complete holes (d: 7 × 6 mm and 7 × 7 mm), 2 partial holes
Buisson's fig. 5.1(b)	Pipe fragment	Means of blowing possibly V-notched tongue duct (Scothern, 1992)
Ist. IV 1939 (a)	Ist. IV 1939 (a): L: 44 mm
D: 4–10 mm
Bone fragment/splinter
Buisson's fig. 5.1(c)	Pipe fragment	Ist. IV 1939 (b)
Combined dimensions:
Length: 165.0 mm
Diameter: 12.0–24.0 mm
Bore: 11.0–17.0 mm	4 regular holes on anterior surface, diameters 6.0 mm × 7.0 mm and 7.0 mm, hole edges polished. Breaks at the extremities, one of which impinges on the

(Continued)

Table 3.3. Continued

Number	Period	Origin	Described as	Material (if known)	Details of Age	Description	Status & Comments	Location (if known)	Main References
(Scothern 4?)						body of the pipe to the region of one of the holes, the other of which affects the proximal epiphysis. The damage in the region of the hole closest to the proximal part of the bone is recent and probably caused by an excavation tool. Several traces of longitudinal scraping visible on diaphysis. Fine rectilinear and wavy short incised lines perpendicular to main axis on flattest surface; other clear incised lines on other surfaces	Ist. IV 1939 (b): L: 70 mm D: 18–25 mm No holes evident		

Buisson's fig. 6	Gravettian	Isturitz (Salle d'Isturitz)	Bone fragment	Detached epiphysis removed from bird ulna	Ist. IV designates Saint-Périer's layer IV in Salle d'Isturitz, typologically dated to Gravettian. Corresponds to Passemard's Layer F3	Ist. IV 1942 Detached epiphysis of bird bone ulna, removed by cutting a groove and then snapping off; suggests on-site manufacture of bird bone pipes (Lawson and d'Errico, 2002)	Collection Saint-Périer, Musée des Antiquités Nationales, Saint-Germain-en-Laye, Paris	Buisson, 1990; Lawson and d'Errico, 2002
Buisson's fig. 3.6 (Scothern 8)	Gravettian	Isturitz (Salle d'Isturitz)	Pipe fragment	Left ulna of diurnal raptor, probably vulture (proximal end)	Ist. IV designates Saint-Périer's layer IV in Salle d'Isturitz, typologically dated to Gravettian. Corresponds to Passemard's Layer F3	Ist. IV 1946 83889 Length: 93.0 mm Diameter: 15.0–27.0 mm Bore: 12.0–17.0 mm 1 hole on posterior and 2 on anterior, diameters 4.0 mm, one of which very close to another. Breaks at the extremities one of which close to the opposing holes. Traces of intense	Collection Saint-Périer, Musée des Antiquités Nationales, Saint-Germain-en-Laye, Paris	Saint-Périer and Saint-Périer, 1952; Buisson, 1990

(Continued)

Table 3.3. Continued

Number	Period	Origin	Described as	Material (if known)	Details of Age	Description	Status & Comments	Location (if known)	Main References
						longitudinal abrasive scraping, clear fine parallel incisions perpendicular to the longitudinal axis on posterior surface. Ist. IV 1946 83889 Length: 93.0 mm Diameter: 15.0–27.0 mm Bore: 12.0–17.0 mm 1 hole on posterior and 2 on anterior, diameters 4.0 mm, one of which very close to another. Breaks at the extremities one of which close to the opposing holes. Traces of intense longitudinal			

			abrasive scraping, clear fine parallel incisions perpendicular to the longitudinal axis on posterior surface. Possibly originally block & duct (Scothern, 1992)			
Buisson's fig. 5.2	Gravettian	Isturitz (Salle d'Isturitz)	Pipe fragment	Right ulna of large bird, possibly golden eagle (proximal end)	Ist. IV designates Saint-Périer's layer IV in Salle d'Isturitz, typologically dated to Gravettian. Corresponds to Passemard's Layer F3	Ist. IV 1946 86756 Length: 88.4 mm Diameter: 10.8–20.5 mm Bore: 8.6–13.0 mm 3 holes on posterior, diameters 4.0 × 6.0 mm and 5.0 × 6.5 mm. Both ends fractured, one break straight and in the region of a hole, the other irregular and in the region of the proximal	Collection Favre, earliest Saint-Périer collection, Musée des Antiquités Nationales, Saint-Germain-en-Laye, Paris	Saint-Périer and Saint-Périer, 1952; Buisson, 1990

(Continued)

Table 3.3. Continued

Number	Period	Origin	Described as	Material (if known)	Details of Age	Description	Status & Comments	Location (if known)	Main References
						epiphysis, due to crushing. Traces of scraping or fine abrasion			
Buisson's fig. 3.7	Gravettian	Isturitz (Salle d'Isturitz)	Pipe fragment	Left ulna of diurnal raptor, probably vulture (proximal end)	Ist. IV designates Saint-Périer's layer IV in Salle d'Isturitz, typologically dated to Gravettian. Corresponds to Passemard's Layer F3	Ist. IV 83889 (a) Length: 108.0 mm Diameter: 13.0–17.0 mm Bore: 11.0 mm 1 hole on posterior, 1 hole on anterior, diameters 4.0 mm Breaks at the extremities, one in the region of the posterior hole. Several traces of abrasive scraping, fine parallel incisions perpendicular to the longitudinal	(a) designation added by IM. Unpublished before Buisson (1990)	Collection Saint-Périer, Musée des Antiquités Nationales, Saint-Germain-en-Laye, Paris	Buisson, 1990

Buisson's fig. 4.4	Gravettian	Isturitz (Salle d'Isturitz)	Pipe fragment	Bird bone, otherwise unidentifiable	Ist. IV designates Saint-Périer's layer IV in Salle d'Isturitz, typologically dated to Gravettian. Corresponds to Passemard's Layer F3	Ist. IV 83889 (b) Length: 52.0 mm Diameter: 7.5 mm Bore: 6.0 mm 3 holes, one of which is quadrangular, diameter 2.0 mm, other two on opposite surface at broken ends. Broken at both ends in the region of the holes that are on the same face; traces of abrasive scraping on surface axis, on posterior and lateral surfaces	(b) designation added by IM. Unpublished before Buisson (1990)	Collection Saint-Périer, Musée des Antiquités Nationales, Saint-Germain-en-Laye, Paris	Buisson, 1990
Buisson's fig. 4.2 (Scothern 6)	Gravettian	Isturitz (Salle d'Isturitz)	Pipe fragment	Ulna of diurnal raptor, probably vulture (distal end)	IF3α designates Passemard's Layer F3 in Salle d'Isturitz, typologically dated to Gravettian. Corresponds to	IF3α 1914 75253 A Length: 74.5 mm Diameter: 14.0–15.0 mm Bore: 12.5–13.2 mm		Collection Passemard, Musée des Antiquités Nationales, Saint-Germain-en-Laye, Paris	Passemard, 1944; Buisson, 1990

(Continued)

Table 3.3. Continued

Number	Period	Origin	Described as	Material (if known)	Details of Age	Description	Status & Comments	Location (if known)	Main References
					Saint-Périer's layer IV	1 hole on posterior, diameter 3.5 mm. One end is heavily worked and may form embouchure, the other is broken in the region of the hole. Numerous traces of abrasive scraping; fine incised wavy lines along length on lateral and anterior surfaces			
Buisson's fig. 4.6 (Scothern 7)	Gravettian	Isturitz (Salle d'Isturitz)	Pipe fragment	Ulna of diurnal raptor, probably vulture (diaphysis only)	IF3β designates Passemard's Layer F3 in Salle d'Isturitz, typologically dated to Gravettian. Corresponds to Saint-Périer's layer IV	IF3β 21 75253 B Length: 51.0 mm Diameter: 13.0 mm Bore: 12.0 mm 2 holes, one in posterior surface, one in anterior. Broken at both ends in region of holes; several traces of abrasive scraping		Collection Passemard, Musée des Antiquités Nationales, Saint-Germain-en-Laye, Paris	Passemard, 1944; Buisson, 1990

Buisson's fig. 4.7	Gravettian (originally designated 'Final Aurignacian (Gravettian)' by Saint-Périers)	Isturitz (Salle d'Isturitz)	Pipe fragment	Right ulna of diurnal raptor, probably vulture (diaphysis only)	Ist. III 83888 designates items from Saint-Périer Layer III in Salle d'Isturitz (corresponding with Passemard Layer C, immediately above Layer F3); all recent chronologies agree Gravettian cultural horizon (see Goutas, 2004). The original designation 'Final Aurignacian' derived from Breuil's typology (Lawson and d'Errico, 2002)	Ist. III 1939 83888 (b) Length: 74.0 mm Diameter: 15.0 mm. Bore: 14.0 mm 3 holes, two on anterior surface, one of which complete and has diameter 4.5 × 5.2 mm, one hole on posterior. Broken at ends in region of anterior and posterior holes. Several traces of abrasive scraping; fine short parallel incisions, perpendicular or slightly oblique to longitudinal axis appear on all surfaces	(b) designation added by IM to differentiate from item illustrated in Buisson (1990), fig. 2 (described earlier), and figs. 4.8 and 4.9 (described later)	Collection Saint-Périer, Musée des Antiquités Nationales, Saint-Germain-en-Laye, Paris	Saint-Périer and Saint-Périer, 1952; Buisson, 1990
Buisson's fig. 4.8	Gravettian (originally designated 'Final Aurignacian	Isturitz (Salle d'Isturitz)	Pipe fragment	Bird bone, otherwise unidentifiable	Ist. III 83888 designates items from Saint-Périer Layer III in Salle d'Isturitz	Ist. III 1939 83888 (c) Length: 68.4 mm Diameter: 12.5–13.0 mm	(c) designation added by IM to differentiate from items illustrated in Buisson	Collection Saint-Périer, Musée des Antiquités	Saint-Périer and Saint-Périer, 1952; Buisson, 1990

(Continued)

Table 3.3. Continued

Number	Period	Origin	Described as	Material (if known)	Details of Age	Description	Status & Comments	Location (if known)	Main References
	(Gravettian)' by Saint-Périers)				(corresponding with Passemard Layer C, immediately above Layer F3); all recent chronologies agree Gravettian cultural horizon (see Gouttas, 2004). The original designation 'Final Aurignacian' derived from Breuil's typology (Lawson and d'Errico, 2002)	Bore: 10.5 mm 1 hole on flattest face, diameter 5.5 mm Broken at the extremities, of which one break in the region of hole	(1990), figs. 2 and 4.7 (described earlier), and 4.9 (described later)	Nationales, Saint-Germain-en-Laye, Paris	
Buisson's fig. 4.9	Gravettian (originally designated 'Final Aurignacian (Gravettian)' by Saint-Périers)	Isturitz (Salle d'Isturitz)	Pipe fragment	Bird bone, otherwise unidentifiable	Ist. III 83888 designates items from Saint-Périer Layer III in Salle d'Isturitz (corresponding with Passemard Layer C, immediately above Layer F3);	Ist. III 1939 83888 (d) Length: 104.0 mm Diameter: 8.5 mm Bore: 10.0 mm 2 holes, one on one surface, diameter 2.0 ×	(d) designation added by IM to differentiate from items illustrated in Buisson (1990), figs. 2, 4.7, and 4.8 (described earlier)	Collection Saint-Périer, Musée des Antiquités Nationales, Saint-Germain-en-Laye, Paris	Saint-Périer and Saint-Périer, 1952; Buisson, 1990

Buisson's fig. 4.5 (Scothern 5)	Gravettian (originally designated 'Final Aurignacian (Gravettian)' by Saint-Périers)	Isturitz (Salle d'Isturitz)	Pipe fragment		

3.0 mm, the other on opposite surface, diameter 2.0 × 3.0 mm. Broken at the extremities, of which one break is close to a hole. Several traces of intense abrasive scraping along length

all recent chronologies agree Gravettian cultural horizon (see Goutas, 2004). The original designation 'Final Aurignacian' derived from Breuil's typology (Lawson and d'Errico, 2002)

Bird bone, otherwise unidentifiable (ulna?)

Ist. III 83888 designates items from Saint-Périer Layer III in Salle d'Isturitz (corresponding with Passemard Layer C, immediately above Layer F3); all recent chronologies agree Gravettian cultural horizon (see Goutas, 2004). The original designation 'Final

Ist. III 83888
Length: 34.5 mm
Diameter: 7.8–8.3 mm
Bore: 6.6 mm

3 holes, one on one surface, diameter 3.0 mm, two on opposite surface at broken ends. Broken at both ends in the region of the holes that are on the same

Originally published (Saint-Périer and Saint-Périer, 1952) in more complete form than now exists. Complete version pictured in Saint-Périer and Saint-Périer (1952), plate IV, top left. Scothern gives original length as 125 mm, and notes only two holes, presumably on

Collection Saint-Périer, Musée des Antiquités Nationales, Saint-Germain-en-Laye, Paris

Saint-Périer and Saint-Périer, 1952; Buisson, 1990

(Continued)

Table 3.3. Continued

Number	Period	Origin	Described as	Material (if known)	Details of Age	Description	Status & Comments	Location (if known)	Main References
					'Aurignacian' derived from Breuil's typology (Lawson and d'Errico, 2002)	face; several traces of abrasive scraping on surface	basis of Saint-Périer, plate IV		
Buisson's fig. 3.2	Solutrean	Isturitz (Salle d'Isturitz)	Pipe fragment	Ulna of diurnal raptor, probably vulture (diaphysis only)	Ist. IIIa designates Saint-Périer's layer IIIa in Salle d'Isturitz, typologically dated to *Solutréen supérieur*	Ist. IIIa 1939 83887 (a) Length: 61.5 mm Diameter: 13.0 mm Bore: 12.3 mm 2 holes, one on posterior, diameter 4.2 mm, one on anterior at point of break. Broken at both ends. Longitudinal abrasions	(a) designation added by IM	Collection Saint-Périer, Musée des Antiquités Nationales, Saint-Germain-en-Laye, Paris	Saint-Périer and Saint-Périer, 1952; Buisson, 1990
Buisson's fig. 3.8	Solutrean	Isturitz (Salle d'Isturitz)	Pipe fragment	Right ulna of diurnal raptor, probably vulture	Ist. IIIa designates Saint-Périer's layer IIIa in Salle d'Isturitz, typologically	Ist. IIIa 1939 83887 (b) Length: 74.0 mm Diameter: 15.0–20.0 mm	(b) designation added by IM	Collection Saint-Périer, Musée des Antiquités Nationales,	Saint-Périer and Saint-Périer, 1952; Buisson, 1990

Buisson's fig. 3.1 (Scothern 15?)	Magdalenian	Isturitz (Salle d'Isturitz)	Pipe fragment	Ulna of diurnal raptor, probably vulture (diaphysis only)	IEα designates Passemard's layer Eα in Salle d'Isturitz, typologically dated to *Magdalénien moyen*	IEα 1914 P2 77153 Length: 68.0 mm Diameter: 14.4 mm Bore: 13.0 mm 2 holes, one on posterior, diameter 3.0 mm, one on anterior at point of break. Broken at both ends	Collection Passemard, Musée des Antiquités Nationales, Saint-Germain-en-Laye, Paris	Passemard, 1944; Buisson, 1990; Lawson and d'Errico, 2002

Additional details (first row, continued): (proximal end) | dated to *Solutréen supérieur* | Bore: 14.0–15.3 mm. 1 hole on anterior surface, diameter 3.5 mm and possible additional hole on same surface evidenced by trace of cupule at one end. Breaks at the extremities, one in the region of supposed second hole. Traces of abrasive scraping | Saint-Germain-en-Laye, Paris

(Continued)

Table 3.3. Continued

Number	Period	Origin	Described as	Material (if known)	Details of Age	Description	Status & Comments	Location (if known)	Main References
						Longitudinal abrasive scraping and numerous fine incised parallel lines perpendicular to length			
Buisson's fig. 3.5 (Scothern 9)	Magdalenian	Isturitz (Salle d'Isturitz)	Pipe fragment	Left ulna of diurnal raptor, probably vulture (proximal end)	IE*a* designates Passemard's layer E*a* in Salle d'Isturitz, typologically dated to *Magdalénien moyen*	IE*a* 1914 P1 77153 Length: 111.0 mm Diameter: 15.0–23.0 mm Bore: 13.0–15.0 mm. 1 hole on posterior, 1 hole on anterior, diameters 3.5 mm. Breaks at the extremities, one in the region of the posterior hole. Several traces of intense abrasive scraping,		Collection Passemard, Musée des Antiquités Nationales, Saint-Germain-en-Laye, Paris	Passemard, 1944; Buisson, 1990; Lawson and d'Errico, 2002

and fine parallel incisions perpendicular and oblique to the longitudinal axis on anterior, lateral, and posterior surfaces

Buisson's (1990) inventory includes a total of twenty-four artefact fragments (counting individually the five component parts refitted by him into two virtually complete objects), or twenty-one (counting the reconstructed artefacts singly).

Scothern's (1992) inventory and table 7 (p. 92) includes a total of eighteen artefact fragments. Twelve of these are directly matchable to items in Buisson's (1990) inventory; the remaining six of them appear to have featured in Saint-Périer publications but to have been missing from the collections at the time she examined them, and do not appear to correlate with any of the remaining thirteen in the Buisson (1990) inventory. There is some confusion, however, as there are several contradictions between Scothern's inventory, text, and table 7 (p. 92).

Scothern's analysis of the collections appears to have been carried out before Buisson's reconstruction of the artefacts he illustrates in his fig. 2 and fig. 5.1. The component parts of the pipe he illustrates in fig. 2 are listed separately in Scothern's inventory; her 'lost' item Ist. SP52D corresponds with artefact Ist. IV 1939 86757 (Buisson's fig. 5.1 (a)). Her 'lost' item Ist. SP52B would appear to be the complete version of Ist. III 83888 (Buisson, 1990, fig. 4.5); the complete version is pictured in Saint-Périer and Saint-Périer (1952), plate IV, top left, although this was incomplete by the time Buisson and Scothern examined the artefacts.

The six remaining 'lost' examples are presented in Table 3.4 as described by Scothern (1992).

Buisson's (1990) inventory thus includes a total of twenty-four artefact fragments (counting individually the five component parts refitted by him into two virtually complete objects), or twenty-one (counting the reconstructed artefacts singly).

Scothern's (1992) inventory and her table 7 (1992, p. 92) includes a total of eighteen artefact fragments. Twelve of these are directly matchable to items in Buisson's (1990) inventory; the remaining six of them appear to have featured in Saint-Périer publications but to have been missing from the collections at the time she examined them and, on the basis of their given dimensions and codes, do not straightforwardly correlate with any of the remaining twelve in the Buisson (1990) inventory. There is some confusion, however, as there are several contradictions between Scothern's inventory, text, and her table 7 (p. 92).

Scothern's analysis of the collections appears to have been carried out before Buisson's reconstruction of the artefacts he illustrates in his fig. 2 (Figure 3.8, this volume) and fig. 5.1 (see Lawson and d'Errico, 2002, plates II, III, and V for photographs of these artefacts). The component parts of the pipe Buisson illustrates in his fig. 2 are listed separately in Scothern's inventory; her 'lost' item Ist. SP52D corresponds with artefact Ist. IV 1939 86757 (Buisson's fig. 5.1 (a)). Her 'lost' item Ist. SP52B would appear to be the complete version of Ist. III 83888 (Buisson, 1990, fig. 4.5); the complete version is pictured in Saint-Périer and Saint-Périer (1952) plate IV, top left, although this was incomplete by the time Buisson and Scothern examined the artefacts.

The six remaining 'lost' examples are presented in Table 3.4 as described by Scothern (1992). Of these, Ist SP52A *may* represent Ist. IV 1939 (b) (Buisson, 1990, fig. 5.1 (c)), and Ist 1939 *may* represent Ist. IV 1939 83889 (b) (Buisson, 1990, fig. 4.3). Ist SP52 *may* correlate with the item illustrated in Lawson and d'Errico (2002) plate I.6 (not analysed by Buisson, 1990), and Ist SP52E certainly correlates with the item illustrated by Saint-Périer and Saint-Périer (1952) at the bottom left of plate VII (also missing at the time of Buisson's analysis). This being the case, only items Ist SP52C and Ist SP52F from Scothern's inventory would remain unaccounted for, the latter perhaps because Buisson (1990) did not analyse bones without piercings unless they were part of items he reconstructed.

A further two artefacts are illustrated by Lawson and d'Errico (2002) which do not feature in Buisson's (1990) publication. As noted, the former *may* correlate with Scothern's item SP52 (Scothern 3), although this is unlikely given the catalogue number, and the latter does not appear to correlate with any of the 'lost' examples from Scothern's (1992) inventory (see Table 3.5).

If it is genuinely the case that none of the artefacts presented in Tables 3.3–3.5 correlate with each other, as the measurements given appear to suggest, then this brings the total number of Isturitz fragments reported by Buisson (1990), Scothern (1992), and Lawson and d'Errico (2002) to thirty-four (or thirty-one counting Buisson's reconstructed artefacts singly). If the suggested

Table 3.4. Isturitz pipes listed as 'lost' by Scothern (1992)

Number	Period	Origin	Described as	Material (if known)	Details of Age	Description	Status & Comments	Location (if known)	Reference
Scothern 3	Gravettian	Isturitz (Salle d'Isturitz)	Flute fragment	Bird (p. 89)	'Final Aurignacian' (Gravettian)	Ist SP52 [may be Lawson and d'Errico, 2002, plate I.6] Length: 52 mm Width: 12 mm Blow hole diameter: 12 mm	SP52 numbers are Scothern's (1992) own, designating artefacts featured in the Saint-Périer and Saint-Périer (1952) publication but now apparently missing from the collections. May be Lawson and d'Errico (2002), plate I.6 (dimensions match closely), though this is a Passemard find so unlikely to have been published in Saint-Périer and Saint-Périer (1952) as	Lost	Saint-Périer and Saint-Périer, 1952; Scothern, 1992; Turk and Kavur, 1997

(Continued)

Table 3.4. Continued

Number	Period	Origin	Described as	Material (if known)	Details of Age	Description	Status & Comments	Location (if known)	Reference
Scothern 4	Gravettian	Isturitz (Salle d'Isturitz)	Flute fragment	Bird (p. 89)	'Final Aurignacian' (Gravettian)	Ist SP52A [may be Ist. IV 1939 (b), Buisson, 1990, fig. 5.1 (c)] Length: 76 mm Width 15–16 mm No finger holes Blow hole diameter: 8 mm	Scothern's label suggests May correlate with Ist. IV 1939 (b), illustrated by Buisson as fig 5.1 (c)—part of a virtually complete pipe reconstructed by Buisson (fig. 5.1)	Lost	Saint-Périer and Saint-Périer, 1952; Buisson, 1990; Scothern, 1992; Turk and Kavur, 1997
Scothern 12	Gravettian	Isturitz (Salle d'Isturitz)	Flute fragment	Bird (p. 89)	Gravettian. 28–22 kya	Ist SP52C Length: 132 mm Width: 17–27 mm Bore: 15 mm		Lost (Scothern, 1992, p. 92), although in inventory Scothern lists location as	Saint-Périer and Saint-Périer, 1952; Scothern, 1992

Scothern 14	Gravettian	Isturitz (Salle d'Isturitz)	Flute fragment	Bird (p. 89)	Gravettian. 28–22 kya	Blow hole diameter: 1.6 mm Finger hole diameter: 8 mm Possibly originally block & duct (Scothern, 1992)	Saint-Germain, Paris		
						1st SP52E [Saint-Périer and Saint-Périer, 1952, plate VII, bottom left] Length: 94 mm Width: 12–21 mm Finger hole diameter: 6 mm		Listed in Scothern, 1992, table 7, p. 92, but not listed in Scothern's inventory. Scothern's text describes this as having 2 complete and 1 truncated finger hole (p. 94), whereas table 7 describes as left	Lost (Scothern, 1992, p. 92)	Saint-Périer and Saint-Périer, 1952; Scothern, 1992

(Continued)

Table 3.4. Continued

Number	Period	Origin	Described as	Material (if known)	Details of Age	Description	Status & Comments	Location (if known)	Reference
						Bore ? (p. 92) Retains distal epiphysis (p. 94) Means of blowing possibly V-notched tongue duct (Scothern, 1992)	This correlates with artefact pictured in Saint-Périer and Saint-Périer, 1952, plate VII, bottom left. This was missing also when Buisson examined the collection		
Scothern 17	Gravettian	Isturitz (Salle d'Isturitz)	Flute fragment	Bird (p. 89)	Gravettian. 28–22 kya. [83889 is Saint-Périer layer IV/ Passemard Layer F3]	Ist 1939 [MAY be Ist. IV 1939 83889 (b), Buisson, 1990, fig. 4.3] Length: 89 mm Width: 9–10 mm	Two items with the code Ist 1939 are listed by Scothern (1992) in inventory and table 7 (p. 92). One is Ist. III 1939 83888 (see earlier), and this may be Ist. IV 1939 83889 (b). (Buisson, fig. 4.3)	Saint-Germain, Paris (Scothern, 1992, p. 92), although in inventory Scothern lists location as lost	Buisson, 1990; Scothern, 1992

| Scothern 18 | Magdalenian | Isturitz (Salle d'Isturitz) | Flute fragment | Bird Ulna (Scothern, 1992, p. 124) | Magdalenian. 18–12 kya | Ist SP52F Length: 175 mm Width: 10–15 mm Notch diameter: 9 × 2 mm. (although inventory says 'blow hole, 0.2 cm diameter'). Bore ? | 3 finger holes, diameters: 6 mm, 7 mm, 8 mm | Different width and finger hole diameters given in Scothern's table 7, p. 92: 0.6 cm, 0.8 cm, and 0.8 cm | Capable of being a sound-producer | Lost | Saint-Périer, 1947; Scothern, 1992 |
| | | | | | | | | | Appears not to feature in Buisson's analysis, perhaps because no finger holes | | |

The six remaining 'lost' examples described by Scothern (1992). Of these, Ist SP52A *may* represent Ist. IV 1939 83889 (b) (Buisson, 1990, fig. 4.3). Ist. SP52 *may* correlate with the item illustrated in Lawson and d'Errico (2002), plate I.6 (not analysed by Buisson, 1990), and Ist. SP52E certainly correlates with the item illustrated by Saint-Périer and Saint-Périer (1952) at the bottom left of plate VII (missing at the time of Buisson's analysis). This being the case, only items Ist. SP52C and Ist SP52F from Scothern's inventory would remain unaccounted for, the latter perhaps because Buisson (1990) did not analyse bones without piercings unless they were part of items he reconstructed.

Table 3.5. Two more objects from Isturitz described by Lawson and d'Errico (2002)

Number	Period	Origin	Described as	Material (if known)	Details of Age	Description	Status & Comments	Location (if known)	Main References
+ (Lawson and d'Errico, 2002, plate I.6)	Gravettian?	Isturitz (Salle d'Isturitz)	Pipe fragment	Ulna of diurnal raptor, probably vulture (distal end)	From Passemard 75252 collection (catalogued in 1929) so probably from the same excavation and context as IF3α 1914 75252 A3, i.e. Passemard's Layer F3 in Salle d'Isturitz, typologically dated to Gravettian	75252 A1 Worked bird-bone tube fragment not published by Buisson (1990) (Lawson and d'Errico, 2002). Length: c.50 mm Diameter: c.8.5–11.5 mm (On basis of Lawson and d'Errico, 2002, plate I.6)	May be Scothern's Ist. SP52 (Scothern 3) (dimensions match closely), though this is a Passemard find not a Saint-Périer and Saint-Périer (1952) find as Scothern's label implies	Collection Passemard, Musée des Antiquités Nationales, Saint-Germain-en-Laye, Paris	Lawson and d'Errico, 2002
+ (Lawson and d'Errico,	Gravettian	Isturitz (Salle d'Isturitz)	Pipe fragment	Ulna of diurnal raptor, probably vulture	Ist. IV designates Saint-Périer's layer IV in Salle d'Isturitz, typologically dated	Ist. IV 1935 Worked bird-bone tube fragment		Collection Saint-Périer, Musée des Antiquités	Lawson and d'Errico, 2002

A further two artefacts are illustrated by Lawson and d'Errico (2002) which do not feature in Buisson's (1990) publication. As noted, the former *may* correlate with Scothern's item SP52 (Scothern 3), although this is unlikely given the catalogue number, and the latter does not appear to correlate with any of the 'lost' examples from Scothern's (1992) inventory.

correlations do apply, then there is a total of twenty-seven fragments (or twenty-four counting Buisson's reconstructed artefacts singly).

Of those analysed by Buisson (1990), their form and dimensions indicated that they were from adult diurnal raptors of considerable size, in most cases almost certainly vultures; in the six specimens where it was possible to match bones directly, one is possibly from an eagle (Buisson's fig. 5.2), and the remainder appear all to be from vulture (table, p. 422). Buisson notes that the dimensions of the pieces are certainly smaller than the bones from which they came because they had been subjected to scraping and abrasion to regularize their forms; the abrading of the body of the flute (diaphysis) results in fine striations along its length. The distal epiphysis was then removed by sawing multiple times perpendicular to the length, and then probably snapped off; these were then also subject to fine abrasion to smooth them. The distal part of the bone appears to correspond with the proximal part of the flute, and Buisson says the proximal end seems not to have been deliberately removed, but is more fragile and so was broken in deposition. Only the complete flute features a sawn proximal end. It seems that the holes were made with a similar but not entirely consistent technique. The location seems first to have been marked with incisions; in the case of the Aurignacian example, these positions seem subsequently to have been adjusted before the holes were produced, though Buisson points out that the remaining incisions may instead be a form of decoration (see Figure 3.7). The hole was created by effecting a thinning of the outer surface of the bone by notching, scraping, or grooving, creating a 'cupule' or hollow in the surface, up to the point where the bone was pierced, or partially pierced, and then punctured through the thinned surface. This leaves a slightly irregular hole, which tends to be oval; in some cases the hole was then finished off by rotational drilling. The size of the holes appears not to be proportional to the dimensions of the bone, with the exception that the smallest holes are found in the weakest diaphyses. The locations of the holes also vary, in terms of being on the anterior or posterior face of the bone—in one case it is on the flatter (posterior) side, in four cases they are on the anterior side, which is of variable curvature, in two cases on the surface marked with the tubercules, but in eleven cases both the anterior and posterior surfaces feature holes. Most often two holes are preserved, but the number varies between one and four. Breakage tends to be at a point with a hole, which constitutes a weak point in the length of the diaphysis (of thirty-two broken extremities, twenty-six breaks are at a hole), but it is not possible to say whether such breakage was deliberate prior to deposition or post-depositional.

Of course, whilst Buisson succeeded in reconstructing two artefacts from fragmentary components, where the joins between the broken ends were evident, it is quite possible that some of the other fragments reported also constitute parts of single artefacts, albeit having suffered breaks that do not directly refit. Lawson and d'Errico (2002) carried out microscopic analysis of

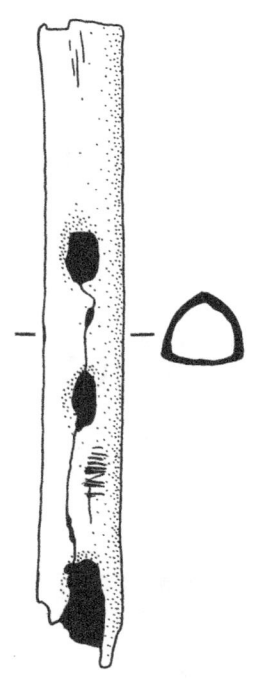

Fig. 3.7. Pipe from Salle Saint-Martin, Isturitz cave, France; right ulna of diurnal raptor, probably vulture (distal end). Aurignacian. Length: 74 mm. (Redrawn from Buisson 1990, Fig. 4.1, p. 426.)

the incised 'decorative' lines on the two artefacts reconstructed by Buisson, and were able to draw detailed conclusions regarding the order in which the lines were made and the tools used for their creation (discussed later); similar analysis of the other, fragmentary, pieces could potentially allow the identification of matching pieces, whose incised lines were made on the same occasion with the same tool, thus allowing their recognition as component parts of the same artefact.

The Aurignacian example, the only one to come from Salle de Saint-Martin (Figure 3.7), was discovered by Passemard in 1921 and reported in 1923 (Passemard, 1923). This has much in common with the bird bone pipes from Geissenklösterle and Hohle Fels, as well as a Gravettian example from Pair-non-Pair, Gironde (France) (Roussot, 1970). The remains of three holes are preserved, also produced by scraping a depression and piercing it, after the exterior of the bone had been smoothed by scraping. Like the Hohle Fels, Geissenklösterle and Pair-non-Pair pipes, it also features a series of incised lines between the second and third holes, possibly related to the placing of the hole. This pipe, plus the two most complete Gravettian pipes from Isturitz (see Figure 3.8, this volume; Lawson and d'Errico, 2002, plates II, III and V; and

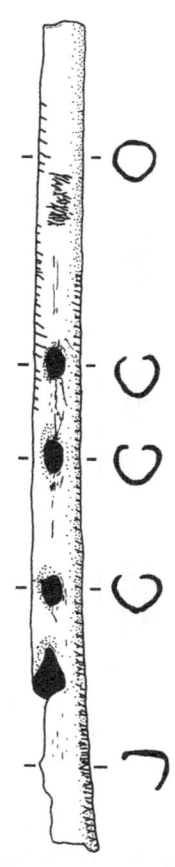

Fig. 3.8. Pipe from Salle d'Isturitz, Isturitz cave, France; left ulna of diurnal raptor, probably *Gypaetus barbatus* (bearded lammergeier/vulture). Reconstructed by Buisson from two fragments. Gravettian. Length: 212 mm. (Redrawn from Buisson 1990, Fig. 2, p. 424, and d'Errico et al. 2003, Fig. 10, p. 40.)

Buisson, 1990, fig. 2 and fig. 5.1), 'Flute 1' from Hohle Fels (Figure 3.2, this volume) and 'Flute 1' from Geissenklösterle (Figure 3.4, this volume) are also characterized by two of the holes forming a pair a greater distance from the other hole(s) than they are from each other.

Parallel notches and lines feature widely on bone and stone artefacts from contexts associated with Gravettian tools, and are a significant feature too of the bone pipes from Isturitz. Lawson and d'Errico's (2002) analysis showed that the incised lines on the reconstructed artefacts could not be considered merely to be decorative, but are actually the result of a series of different sessions of incising activities, using different tools, and creating different types of lines. It seems in some cases that a particular number of lines were intended, with decreased spacing between lines towards one end of the bone reflecting a need to accommodate more lines in the diminishing space. Some, whilst being precisely and carefully made, are virtually invisible to the naked eye. They also appear not to have had a purely functional purpose of improving grip, as some of the densest series of lines are in locations where there is no need to hold the object. Lawson and d'Errico compare the marking sequences on the artefacts to those used to convey information in Artificial Memory Systems, in which they carry meaning through their relationships to each other, or through the process of their creation itself. The idea that they might represent a notational system relating to sound production specifically may be belied by the fact that similar incised-line sequences are characteristic features on many other types of bone and stone artefacts associated with Gravettian technological assemblages. But it does seem to be possible to rule out purely decorative or functional roles for these carefully executed markings.

Graeme Lawson and Francesco d'Errico's extensive analyses of the most complete examples (Lawson and d'Errico, 2002; d'Errico et al., 2003) suggest that at least two of the examples seem to have been designed to be played as end-blown trumpet- or reed-voiced pipes, rather than as flutes. The presence of a sawn-off ulna epiphysis at the site (Buisson, 1990, fig. 6; Lawson and d'Errico, 2002, plate I.5) suggests that manufacture of these artefacts occurred on site, whilst polish on the surface of the pipes, created by handling around the holes and on the opposite side, suggests that the pipes were not only manufactured there, but also used there.

It can be seen from the inventories (Tables 3.3–3.5) that the material used for the pipes is remarkably consistent. Radius and ulna bones of large birds are particularly suitable for pipe manufacture, being of suitable size, naturally already hollow, light, and relatively easily worked, so it is perhaps not surprising that there should be recurrence in their use—but the consistency at Isturitz goes beyond that. The overwhelming majority of the pipe fragments are made from ulna bones coming from diurnal raptors, almost certainly vultures given their dimensions (*Gyps fulvus* and/or *Aegypius monachus* and/or *Gypaetus barbatus*; griffon vulture, black vulture, and bearded vulture/lammergeier respectively) (Buisson, 1990; Lawson and d'Errico, 2002). The choice of diurnal raptors and, in particular, carrion birds, could be significant. Carrion birds circling can provide a very effective signal of the presence of injured or dead animals in the landscape that would otherwise be invisible to a terrestrial observer. Furthermore, upon encountering a wounded or dead animal, they would constitute potentially formidable competitors for the resource. It seems possible that the overwhelming predominance of these birds as the providers of raw material for the pipes made by the Gravettian occupants of Isturitz betrays a 'special relationship' between the humans, the carrion raptors, and the provision of food at this time. At the least, it might be a consequence of vultures being the most frequently encountered raptors in the course of overlapping subsistence activities.

Whilst many of the pieces from Isturitz are very fragmentary and impossible to analyse musicologically, many of them provide a real indication of the level of sophistication of which their manufacturers were capable, and are indicative of a level of importance attributed to musical activity that ranks at least equivalent to other, more familiar, symbolic activities. The Isturitz examples are particularly significant in indicating the importance of musical activity at a site that was clearly the focus for large-scale communal gatherings and extended social and trade networks, over considerable tracts of time.

Other sites

According to Scothern's analysis, the earliest definite example of a 'block and duct' flute comes from a Gravettian context at Goyet, in Belgium. This 105-mm-long bone has the first example of an intentionally produced 'sound window' (Scothern, 1992), a feature designed to direct the breath against a block in the inside of the bone, creating the vibrating air column necessary to produce sound. The block in the middle of the bone might have been made of wood, or resin, as is the case in one of the examples from Dolní-Věstonice, Moravia, where the resin plug is still, remarkably, preserved. Modern recorders (or vertical flutes) still use the same block and duct principle, essentially unchanged for 25,000 years. Three whistles from Le Roc de Marcamps, France, all made from bird bone and dated by association with Magdalenian V–VI tools (Roussot, 1970), also appear to make use of this principle (Scothern, 2002). In this case all three seem to have been designed as small whistles (one of the three is complete) rather than representing recorder-like pipes.

As for other less certain artefacts, Scothern (1992) concludes, from the literature and personal observation, that the early Aurignacian pieces from Saint-Avil, Istállóskö, Bukovač, and Potočka Zijalka are more likely to be the products of natural processes than the result of human agency. Although all can produce sound, it is impossible to demonstrate that the modifications are intentional or man-made. Later Aurignacian and Gravettian artefacts from Spy and Maisières Canal are more likely to be intentional sound-producers according to Scothern (1992). Where commentary and assessment is available, the reputed artefacts in Appendix Tables 1 and 2 include notes regarding their likely status and capabilities.

Almost all of the pieces discussed of which we can be confident of their status as intentional sound-producers were made using bird bone. With the exception of the mammoth-ivory artefacts from the Swabian Alb, objects reputed to be sound-producers and made on materials other than bird bone have tended to be of far more dubitable status. A final piece, particularly worthy of note here as it is made from a reindeer tibia and can be confidently identified as an intentional sound-producer, comes from Grubgraben bei Kammern, in Austria (Einwögerer and Käfer, 1998; Käfer and Einwögerer, 2002). This 165-mm-long artefact, C14 dated to 18,920 ± 90 BP, has three very neat, evenly-spaced, circular holes drilled in the diaphysis of the bone, $c.5$ mm in diameter. The epiphyses had been carefully removed with a silex tool, although the ends of the artefact are now damaged making it impossible to say whether they were further modified in any way for the production of sound, for example, by the addition of block or reeds. Nevertheless, an exact replica of the artefact proved capable of producing eight pitches when reconstructed and played as a bevelled flute (Käfer and Einwögerer, 2002). It is clear that whilst bird bone was clearly a very important medium for the production of sound-producing pipes, we should certainly not be closed to the possibility of the use of other materials. In fact, given the abundance of reindeer bone at many Palaeolithic sites, and its use for other bone artefacts (including phalangeal whistles, Chapter 4), it is surprising that it is apparently not used more frequently for the production of sound-producing pipes in the Palaeolithic. The Grubgraben example currently constitutes a noteworthy exception.

REPRESENTATIONS OF INSTRUMENTS

Representations of musical instruments in Palaeolithic art are sparse, as are direct representations of people in parietal art before the Holocene. Apart from the possible representation of a rasp in the hand of the Venus of Laussel (Chapter 4), there is one example of an illustration of an anthropomorphic figure with an object protruding from the face area which may represent a nose flute. This engraved image is at Grotte des Trois Frères, Ariège, and depicts a figure with the head of a bison, who appears to be dancing, and from whose nose appears to protrude a long linear object. It has been suggested that this may represent a nose flute (Dauvois, 1989; Bahn, 1997); nose flutes are a well-documented instrument in many ethnographic groups (see, for an overview, Sadie, 2001; Moyle, 1990). Whilst this is a possible interpretation of this picture and, potentially, of some artefacts, in this instance the picture is so ambiguous and stylized that it is difficult to be certain as to what is intended to be represented. If the image is to be interpreted absolutely literally, then it is possible that it depicts a person wearing a bison head, with a pipe to the nose of the bison; in this case it is difficult to imagine that it would be possible to play the pipe. Alternatively, it may be that it is a pipe and the image is supposed to depict it merely being held to the face of the bison mask, or it may be something else entirely. On the other hand, if it is an abstract and symbolic image representing, for example, a 'bison spirit', then it is difficult to make any confident assertions about the meaning of any of the extraneous lines and features of the engraving.

THE USE OF BONE FOR INSTRUMENT MANUFACTURE IN THE MIDDLE AND UPPER PALAEOLITHIC

As will be seen in Chapter 4, the material most commonly used for the production of the reputed bullroarers, rasps, and phalangeal whistles was reindeer, either antler or bone. As a major food subsistence resource, reindeer bones and antler were abundant and their use as a material was widespread amongst modern humans. Sound-producing pipes are an entirely different matter, however. The bone of which the Palaeolithic pipes consist varies. Of those instances in which the faunal origin of the bone is specified (and for more than a third of the reputed pipe fragments it is not), the overwhelming majority are from birds, although there are a few examples of other animals such as bear, hare, red deer, one reindeer, and a few cases of unspecified mammal species. Not only do bird-bone objects have a considerably greater numerical representation, but it is also conspicuous the proportion of these that are considered likely to be genuine sound-producers, as opposed to others made on bear or ungulates, for example, of which many more are of doubtful anthropogenic origin or musical function. There is also a geographical split in the use of raw material for the manufacture of the flutes/pipes. The majority of the bird-bone artefacts come from western Europe, whereas those made from other materials such as ungulates, or more commonly bear bone, come from central and eastern Europe. This is clear from Appendix Tables 1 and 2, and is also observed by Leocata (2001).

In many instances the birds' species is not specified, but of those that are, there are vulture (Passemard, 1923; Saint-Périer and Saint-Périer, 1952; Buisson, 1990; Lawson and d'Errico, 2002), eagle (Piette, 1874, 1875, 1907; Seewald, 1934), swan (Seewald, 1934; Bergounioux and Glory, 1952; de Bayle des Hermens, 1974; Hahn and Münzel 1995; Conard et al., 2009), and greylag goose (Absolon, 1936). In any case, the bird species used has to be very large in order to possess wing bones of suitable dimensions for sound production, meaning that there are few species beyond those mentioned that would indeed be suitable. Interestingly, in the case of the mammoth ivory examples from the Swabian Alb they appear to have been manufactured to replicate the form of bird bone examples though larger.

This situation raises several important and intriguing questions. Why do we see musical instruments associated with anatomically modern humans for the first time in Europe, rather than elsewhere? Is the availability of suitable faunal resources the determining factor in the incidence of instruments in the Upper Palaeolithic—is it that avian faunal resources were available in Europe that were not available elsewhere? Did they become available in Europe at the same time that modern humans arrived? If such resources were available in Europe before the arrival of modern humans, did Neanderthals make use of them? If not, why not? Is this a behavioural change representative of a cultural or creative revolution when modern humans reached Europe? Or is it the result of a lack of preservation or excavation elsewhere?

Raw material availability?

The first question to tackle concerns the availability of these avian resources to modern humans and Neanderthals. Was there a change in their availability in continental Europe at the time modern humans arrived, around the transition between Oxygen Isotope Stage 4 (OIS4) and OIS 3? Were they available to hominins outside Europe in OIS4 and OIS3? Or does their appearance as a raw material in the archaeological record mark a change in *use* of an existing resource?

In addition to the various archaeological excavation reports of sites from these areas, an extraordinarily useful resource is provided by Tyrberg's (1998 and 2008) collation of all of the avian fauna data from sites in greater Europe, the Near East, and North Africa. This extraordinary resource draws upon all the available published data for the presence of bird remains at sites of Palaeolithic age. This is particularly useful as it deals not with the *use* of birds by past populations, but the *presence* of birds in their environments, drawing as it does not just on archaeological sites but all sites of that age.

Tyrberg's (1998, 2008) survey shows that during the period prior to the arrival of modern humans in Europe, whilst Europe was occupied by Neanderthals and modern humans occupied Africa and, periodically, the Near East (OIS5, OIS4, and OIS3 in Mousterian contexts—i.e. the Middle Palaeolithic in Europe and MSA in Africa), swan, vulture, goose, and eagle species were present in the following European, Near Eastern, and North African areas:

Swan (*Cygnus* sp.): Azerbaijan, France, Germany, Netherlands, and Russia

Vulture (*Aegypius* sp., *Gypaetus* sp., *Gyps* sp.): Azerbaijan, France, Georgia, Germany, Greece, Portugal, Romania, Spain

Goose (*Anser* sp., *Branta* sp.): Azerbaijan, Bulgaria, France, Georgia, Germany, Italy, Netherlands, Poland, Russia, Spain, UK

Eagle (*Aquila* sp.): Azerbaijan, Bulgaria, France, Germany, Greece, Italy, Portugal, Romania, Russia, Spain, Libya (one example of *Circaetus* sp., OIS4).

During the early Upper Palaeolithic (OIS3 in Aurignacian and Gravettian contexts), these species show up in the following areas:

Swan (*Cygnus* sp.): Germany, Poland

Vulture (*Aegypius* sp., *Gypaetus* sp., *Gyps* sp.): Bulgaria, Portugal, France, Romania, Italy, Spain

Goose (*Anser* sp., *Branta* sp.): France, Greece, Italy, Moldova, Norway, Spain, Austria

Eagle (*Aquila* sp.): Bulgaria, France, Germany, Italy, Spain

Note that with the exception of one example of *Circaetus* (Eagle) from Libya from OIS4 Middle Stone Age contexts, there appears to be a dearth of avian species suitable for bone pipe manufacture in North Africa and the Near East until OIS2. At this time, the site of Ohalo II in Jordan produces some examples of various *Anser* (geese) and *Cygnus* (swans), dated to the Early Epi-Palaeolithic (*c.*19,400 BP). Apart from these two exceptions, Libya, Jordan, Lebanon, Israel, Syria, Morocco, Tunisia, Iran, Iraq, and Saudi Arabia appear to exhibit none of these species throughout the Middle Palaeolithic/MSA and Upper Palaeolithic/LSA before the late glacial period, just before the end of the last Ice Age. So it seems that during OIS5, OIS 4, and OIS 3 there was no practical presence of these species suitable for pipe-making in the area immediately outside of greater Europe. A change with the onset of OIS 2 brought some species to the region of Ohalo, but it is only at the end of OIS 2 and the beginning of OIS 1 (the postglacial period) that the species under investigation here appear in the records of the Near East and North Africa. The record of continental Europe is a marked contrast. These species seem to have been present throughout the Middle and Upper Palaeolithic in various locations throughout the continent. In other words, it seems that Neanderthals in Europe had potential access to all of these species, as did modern humans upon their arrival, but Mousterian/MSA populations outside of Europe (i.e. Near East and North Africa) did not.

This means that, for modern humans, we cannot categorically attribute the adoption of bird bone for the production of sound-producers upon their arrival in Europe to a change in behavioural *capability*, but to a change in behavioural *opportunity*.

Neanderthal use of avian fauna for subsistence and as a raw material

Whilst there are numerous sites in which Mousterian-type technologies are found in association with avian faunal remains, the prevailing view for many years has been that Neanderthals made little or no use of avian resources for subsistence purposes. This view has increasingly been challenged, however

(Hardy and Moncel 2011), and it is now clear that in at least some locations Neanderthals exploited avian resources for subsistence from an early time (125,000–250,000 years ago). However, the use of any bone as a raw material by Neanderthals is comparatively rare and it appears at this stage that they did not exploit avian bone as a raw material resource for technological purposes.

It is worth noting, however, that some site reports make no mention of avian fauna, focusing instead upon mammalian fauna (e.g. Haua Fteah, Libya, McBurney, 1967; Divje babe I, Turk et al., 1997). This does not necessarily mean that no avian fauna were present at these sites; indeed, MacDonald's (1998) study of the avian fauna of the Haua Fteah illustrates as much. This does beg the question of how many Mousterian sites might contain evidence for the deliberate use of birds as a resource (subsistence or material) which has never been published or even analysed, and it is increasingly evident that Neanderthal were indeed acquiring birds for other purposes (Hardy and Moncel, 2011; Finlayson et al., 2012).

Nevertheless it does appear at this stage that there is no strong evidence that the avian bone accumulations at Mousterian sites feature modification indicating that they made use of avian bone for technological purposes.

Use of avian fauna and technological limitations

Was this through a lack of appropriate technology for capturing avian resources? It is often asserted that a principal limiting factor in the use of avian resources is the presence or absence of the technologies required to capture them, as birds are generally fast-moving and relatively small targets. Projectile weapons or sophisticated net-traps are often deemed to be necessary prerequisites for the exploitation of this source of food and raw material. However, Cassoli and Tagliacozzo (1997) point out that during moulting, ducks, geese, and swans cannot fly, so are relatively easy to approach and kill 'using simple sticks or clubs' (p. 308). These birds moult and become flightless during the summer months, from May in the case of geese, through July and August in the case of swans and ducks, and could at this time provide a relatively accessible and substantial resource for food and bone. It seems also that the procurement of substantial food (and tool) resources from large birds need not be dependent upon sophisticated tool use (such as projectiles for example), especially during the summer months when geese, swans, and ducks are moulting; clubs and nets would be quite adequate. It seems that technological limitations alone are not enough to explain the lack of use of these resources by Neanderthals, for these species at least. The situation with regard to vultures, used by modern humans to provide the bone for the vast majority of the Gravettian finds from Isturitz, for example, may have been more challenging.

Use of avian fauna due to environmental stress

It seems that the technologies and behaviours, motivations, or environmental demands, necessary to trap or hunt birds were sometimes present in Europe with Neanderthals, but that between c.80,000 and 20,000 years ago, large avian species suitable for food and raw material subsistence were largely absent from the areas occupied by modern humans in North Africa and the Near East.

Stiner et al. (2000) suggest that the use of avian fauna (and small game) was an innovation especially undertaken by modern humans in times of limited large game resources and population/resource pressure. This is echoed by observations, both archaeological and ethnographic, by Gotfredsen (1997), Lefèvre (1997), and Serjeantson (1997) that the importance of birds as a subsistence resource depends upon the availability of other resources, and that this is dependent largely upon season and latitude. Consequently, avian fauna seem to be most important to subsistence at high latitudes, under arctic and subarctic conditions, although they can also form a major part of the diet in coastal and lakeside locations where large numbers of seabirds accumulate. Stiner et al. (2000) hypothesize that factors of environmental stress and human demand for resources resulted in increased use of relatively small avian food resources, such as partridge and quail, at the start of the Upper Palaeolithic; they cite major use of these avifauna as a subsistence resource in early Upper Palaeolithic contexts in Italy and Turkey. But in the case of the use of large avian fauna at early Upper Paleolithic sites such as those of the Ach Valley in Germany, and Isturitz in the Pyrenees, this does not appear to be marginal subsistence; whilst these are certainly relatively high latitude subarctic conditions at this time, the birds seem to have formed a part of a wider package of subsistence typical of humans at that time. They seem to have been incorporated into the resource acquisition repertoire by choice rather than necessity. It is, however, the case that the greater population density in these locations, or periodic large aggregations of people, may have stimulated the exploitation of this newly discovered resource.

There is certainly evidence of large avian fauna being used as a main subsistence resource themselves. Grotta Romanelli, in Italy, shows that towards the end of the Upper Palaeolithic (Epigravettian), forty species of birds were being used to supplement the main diet of large mammals, but five large species were represented in numbers great enough to indicate their forming a subsistence diet in their own right. These were white-fronted goose (*Anser albifrons*), bean goose (*Anser fabalis*), brent goose (*Branta bernicla*), great bustard (*Otis tarda*), and little bustard (*Otis tetrax*). In addition, swans, ducks, eagles, hawks, owls, and ravens also all displayed cut-marks on the bones, and in some cases, decoration (Cassoli and Tagliacozzo, 1997).

As noted, whooper swan, greylag goose, and white-tailed eagle (all of which have been used for creating sound-producers during the Upper Palaeolithic) have been found in Ohalo II in Jordan at the last glacial maximum (LGM) (the site is dated 19,000–23,000 BP) (Simmons and Nadel, 1998), in an apparently rare period of availability of these resources in this part of the world. It is possible that the extreme conditions of the LGM at latitudes further north resulted in a shift south of these avian species. Indeed, no bird bone pipes at all have been retrieved from contexts associated with Solutrean technology in western Europe (corresponding with the time of the LGM), suggesting a reduction in the use or availability of that resource at those latitudes at that time.

The use of large avian species for subsistence results, as a by-product, in the ready availability of large, relatively easily worked, hollow bones. In addition, species may be selected specifically for their potential as a source of raw material for tools, and such activity is documented in later archaeological and ethnographic contexts. For example, Van Wijngaarden-Bakker (1997) concludes from the study of Dutch Neolithic sites that large avian species were actively selectively hunted primarily as a raw material, rather than subsistence, resource; at Aartswoud for example, the five bone tools retrieved were made from white-tailed sea eagle and crane, neither of which species are found in the food remains of that site. So it would seem that greater subsistence pressures at more northern latitudes may increase the demand to make use of such resources, although would appear not to be the sole motivation for acquisition of these resources. Equivalent analyses of more Palaeolithic assemblages would be invaluable in refining our understanding of the motivations of the people responsible for them.

It is worth noting here, perhaps, that at none of the sites described in the preceding paragraphs have bird bone flutes or pipes been identified, despite the presence of suitable raw material, suggesting that even when such materials were readily available, their use for the manufacture of sound-producers was rare (as is indeed the case amongst many hunting and gathering societies of recent history, as discussed in Chapter 2).

So the apparently sudden start of the use of avian bone as a raw material (and large avian fauna as a resource more generally) at the start of the Upper Palaeolithic in Europe may be due in part to a relative lack of availability of such resources outside of Europe, in the Near Eastern and North African areas occupied by modern humans. There is also the possibility that the apparent increase in the use of bone in the archaeological record of modern humans between the Middle Palaeolithic/MSA and the Upper Palaeolithic is a consequence of differential recovery of such materials. Upper Palaeolithic Europe has been subject to such a level of intensity of excavation in comparison to areas outside of Europe, and generally comparatively favourable conditions of preservation, that it seems distinctly possible that future excavations from

Africa pre-dating the Upper Palaeolithic will result in the discovery of avian faunal use for subsistence and tools.

CULTURAL REVOLUTION?

One possibility that has long been entertained is that the appearance of such behaviours in the European record represents a genuine 'revolution' in capability amongst modern humans upon their arrival in Europe, part of a package of behaviours that has traditionally been taken to include symbolism, in the form of representation ('art') and ornamentation, as well as technologies such as bone points, harpoons, and flint blades. Increasingly, however, it seems that the capacities for these behaviours did not emerge in Europe, but were exhibited under certain circumstances throughout at least the preceding 50,000 years in Africa (e.g. McBrearty and Brooks, 2000; Henshilwood et al., 2001; d'Errico et al., 2001; d'Errico et al., 2003; Mellars et al., 2007). The record in Europe is unique in the density and proliferation of the evidence for such behaviours, and more needs to be understood about the particular social and ecological conditions that led to this situation, but it is now clear that whilst the conditions for the perpetuation and proliferation of these behaviours might have been ideal (perhaps even uniquely so for some time) in Europe, it cannot be considered the birthplace of the capabilities for them.

From the earliest occupation of Europe by *Homo sapiens*, there is evidence of the use of bone for the creation of tools and other objects (amongst the earliest being the Geissenklösterle flutes; Richter et al., 2000; Higham et al., 2012), suggesting that the use of this raw material was a technology brought with these earliest modern humans to Europe, and not something 'discovered' upon arrival. Finds from Blombos cave in South Africa confirm that the use of bone as a raw material has a long time-depth in modern humans; these examples of spear points and awls date to around 70,000 years ago, and are made on fragments of ungulate long bones and mandible (Henshilwood et al., 2001; d'Errico et al., 2001). Whilst this technology may have been lost and regained many times during the time span of *Homo sapiens*, its early incidence indicates that the use of bone to make other items is not dependent upon a single cognitive revolution amongst modern humans (see also McBrearty and Brooks, 2000). The incidence of such innovations, and their perpetuation, are two separate concerns. This evidence does not undermine the possibility that the choice to use bone, and in particular, avian bone, to produce *musical instruments* was something to arise in Europe, but if so, this was an innovation virtually contemporaneous with human arrival in Europe.

Traditionally, if Palaeolithic music archaeology is mentioned at all it tends to be in the context of the emergence of other art and culture. Certainly there is a revolution in evidence for creative activity in the *archaeological record* of Europe after the arrival of *Homo sapiens*, which contrasts in many respects with the record of *Homo neanderthalensis*, but this does not necessarily mean that it was at this time that there was a revolution or explosion of creativity in the modern human *brain*; it may merely indicate the arrival in Europe of such behaviours with modern humans, not the time of their arising in modern humans.

Alternatively, it could indicate the spread and popularization of such behaviours, and this need not be indicative of a change in the cognitive capacity for the behaviours; there are many people in the world today with the cognitive *capacity* to use or programme a computer, for example, but who will never do so, along with the rest of their contemporaries in their culture. Likewise, few would argue that a citizen of the Roman Empire lacked the cognitive capacity to drive a motor car. As amply discussed by McBrearty and Brooks (2000), an increasing amount of evidence such as that from Blombos and other sites in southern and northern Africa suggests that the idea of the sudden emergence in *Homo sapiens* of an *entire suite* of artistic and symbolic capabilities upon their arrival in Europe is unlikely to be accurate, and that aspects of those behaviours were being exhibited at various times and places by *Homo sapiens* considerably prior to their arrival in Europe. The capacity can be there without it being used to fulfil that purpose. So if there is a revolution, it can be a cultural one without being a cognitive one.

Such models of a creative explosion had hitherto frequently led to the equation of musical behaviours with other artistic behaviours, supposedly being part of a package of creative and artistic capabilities and sensibilities that emerged simultaneously in modern humans. The evidence considered in the following chapters illustrates that important elements of the physiological and neurological foundations of musical behaviours were in place considerably before the advent of anatomically modern *Homo sapiens*, and it is very likely that all such foundations were in place long before the European Upper Palaeolithic.

A revolution we may be seeing, however, is in the bone resources used. The Upper Palaeolithic of Europe appears to show a great increase in the working of bone as a raw material for the production of other items, such as containers, needle-holders and needles, pipes for blowing paint, as well as sound-producers; as has been discussed, it appears that a greater proliferation of these resources was available to modern humans in Europe than had been available in the areas they had inhabited previously.

Whilst indicating an apparent increase in the *use* of avian faunal resources, and whilst such resources provide an easily worked material for the manufacture of sound-producers, caution should be applied before concluding that these artefacts necessarily represent the genesis of anthropogenic sound-producers. As has been noted in Chapter 2 the incidence of the use of bone in the manufacture of sound-producers in contemporary hunter-gatherer societies is rare, and the vast majority of such objects are made using other materials. Furthermore, pipe-type instruments form a small proportion of those used. There is no reason, on this basis, to conclude that the pipe-type sound-producers made of bone that we see in the Upper Palaeolithic of Europe represent the first musical instruments, and even less so, the first incidence of musical behaviour. What seems more likely is that the record we have is representative of a change in the production in instruments, either in the form of the raw materials chosen for—or available for—the purpose, the capability to work bone materials or to catch certain fauna whose bone could then be used, or by a change in musical behaviour that resulted in an increased dominance of melodic instruments (relatively rare in the modern hunter-gatherer groups discussed in Chapter 2).

What these artefacts do frequently show, however, even from the earliest examples, is an already well-developed and sophisticated mode of production, superior to many mediaeval examples. What they are most valuable for, in contrast to identifying the earliest incidence of musical behaviour, is to confirm to us that recognizable musical behaviours were well developed and played an important role in the lives of anatomically modern humans by the time of the early Upper Palaeolithic.

4

Palaeolithic Music Archaeology 2: Other Sound-Producers

INTRODUCTION

Our background in the Western musical tradition can easily skew our expectations as to what should constitute the principal material evidence for musical activity—namely, the instrumentation—and we need to be prepared to identify other forms of sound-producing objects that may be less easily recognized by us than pipes. As the discussion of recent hunter-gathering societies in Chapter 2 illustrates, in many traditional cultures melodic content in musical activities is provided not by artificial instrumentation, but by the voice, and instruments, if used, are principally rhythmic and/or percussive (e.g. Johnston, 1989; Nettl, 1992; Breen, 1994; McAllester, 1996; Locke, 1996). These can include drums (struck idiophones), bullroarers (free aerophones), rasps (scraped idiophones), rattles, and features of the natural environment. The use of bone, ivory, and antler is relatively rare in these societies, and instruments are made predominantly from biodegradable matter such as gourd, wood, hollow logs, and dried spiders' nests (e.g. Johnston, 1989; Nettl, 1992; Breen, 1994; McAllester, 1996; Locke, 1996) which would be very unlikely to preserve had they been used in archaeological contexts.

In short, on the basis of a more 'worldly' conception of musical activity it is clear that we should not expect human music to be predominantly instrumental, nor that when it is instrumental, that it be predominantly flute-orientated or even widely preserved. So the record that we do find is likely to represent the merest tip of the iceberg of the diversity of sound-producing activities to have been carried out in the Upper Palaeolithic—and this means we need to look very carefully at the full potential diversity of objects that may have been used and activities that may have been carried out.

Whilst a high proportion of the sound-producers used by traditional societies (and indeed, Western societies) are made from biodegradable materials, we are fortunate that certain materials other than wood (and other plant matter) also have the potential to make excellent raw materials for many types of sound-producers, and preserve better. As a consequence, in addition to the

100 *The Prehistory of Music*

relatively prolific pipes, other types of potential sound-producing objects *have* been identified from the Upper Palaeolithic. There are pierced phalanges (toe bones) interpreted as 'phalangeal whistles', objects which have been interpreted as bullroarers (Harding, 1973; Dams, 1985; Dauvois, 1989, 1999; Scothern, 1992), sections of bone which have been notched with parallel grooves, which may be rasps (Huyge, 1990, 1991; Kuhn and Stiner, 1998; Dauvois, 1989, 1999), and mammoth bones which may have been used as percussion instruments (Bibikov, 1978; Hadingham, 1980; Soffer, 1985). Finally, some authors have suggested that caves themselves, and features of caves, were used as sounding devices (lithophones) (Glory, 1964, 1965; Dams, 1984, 1985; Reznikoff and Dauvois, 1988; Dauvois, 1989, 1999; Reznikoff, 2008).

With the exception of lithophones, all of the known or reputed examples of sound-producers from the Palaeolithic are made from bone, ivory, or antler. It is worth noting at this early stage of the discussion that the apparent exclusivity of these raw materials for the manufacture of musical instruments in the Palaeolithic is very likely to be the result of differential preservation, rather than the exclusive choice of those materials on the part of the peoples carrying out the manufacture.

Having given an overview of these possible Palaeolithic sound-producers this chapter goes on to highlight some key questions we must ask if we are to assess the significance of these objects for our understanding of the behaviour of modern humans, and to discuss the Upper Palaeolithic record of sound-producers as a whole.

OTHER AEROPHONES

In addition to pipes there are two other types of aerophone which may be represented in Palaeolithic contexts: phalangeal whistles and 'bullroarers' (known as free aerophones).

Phalangeal whistles

A form of aerophone sound-producer that is almost as prolific as the reputed pipes and flutes is the 'phalangeal whistle'. These objects consist of a phalanx bone pierced, in most cases, with a single hole at the proximal end of the posterior surface (Harrison, 1978) (Figure 4.1). The product of this is a small hollow vessel-flute; by placing the proximal end of the phalanx against the lower lip, one can blow over the top of the hole to produce a clear tone. Some examples pierced nearer the centre of the diaphysis may have been played by placing the bone lengthwise against the lips (see Figure 4.2 for examples). The

Palaeolithic Music Archaeology 2: Other Sound-Producers   101

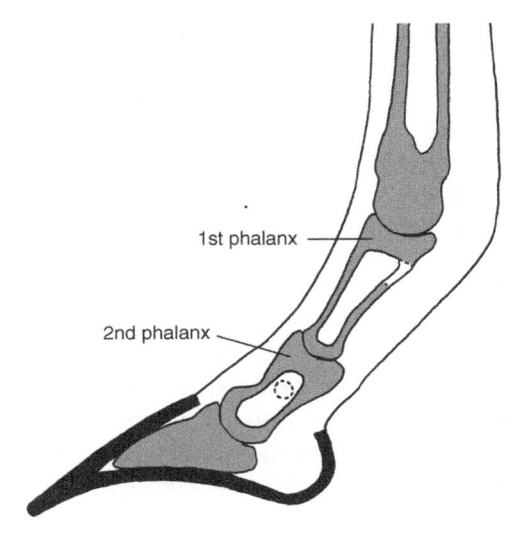

Fig. 4.1. Right foreleg of reindeer, showing first, second and third phalanges. The first and second phalanges are naturally hollow and when pierced with a single hole can function as whistles. Many are known from throughout the Palaeolithic. (Redrawn from Dauvois 1989, p. 4.)

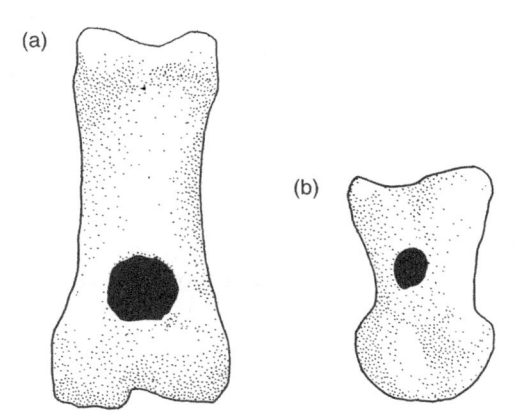

Fig. 4.2. Phalangeal whistles.
(a) Saint-Marcel, Indre, France; 1st phalanx of reindeer. Magdalenian. Length not given, but approx. 50mm. (Redrawn from Dauvois 1989, p. 9, image 2.)
(b) Tuto de Camalhot, Saint-Jean-de-Verges, France; 2nd phalanx of reindeer. Early Aurignacian. Length: 32.8mm. (Redrawn from Dauvois 1989, p. 6, image 1.)

tone produced depends upon the size of the phalanx and the dimensions of the hole (Harrison, 1978; Dauvois, 1989). Known details of these reputed artefacts are outlined in Appendix Table 3.

Many such objects have been found in Palaeolithic contexts dating from the Mousterian through to the Magdalenian, the Mesolithic, and beyond to recent history. There are records of their manufacture and use in various ethnographic contexts (e.g. amongst North American Indians; Sollas, 1924) as an aid to signalling in hunting, or in shamanistic rituals (Harrison, 1978). Dauvois (1989) has compared their use to the whistling language of the shepherds of the Vallée d'Aas in the Pyrenées-Atlantiques. It should also be noted, however, that similar objects have, in some cultures, also been used as anthropomorphs, to represent stylized human figures (Caldwell, 2009), and this possibility for their interpretation should also be borne in mind (although the two possibilities are not necessarily mutually exclusive). If long-term use as whistles creates characteristic use-wear polish on the surface near the hole, where they are held to the mouth (as is the case on some flutes), then identification of such wear could help positively diagnose use as whistles, but work to identify such wear has not yet been carried out.

Of those archaeological phalanges of which the faunal origin is stated, the overwhelming majority are from reindeer (e.g. Megaw, 1960; Harrison, 1978; Dauvois, 1989), and consist generally of the first phalanx, although the smaller second phalanx is also sometimes used (Dauvois, 1989). Harrison (1978), who carried out a series of experiments regarding the process of creating punctured phalanges (discussed later in this chapter), found that the creation of a phalangeal whistle using a stone tool took on average between three and four minutes.

That such whistles have had an anthropogenic origin and important roles in human society at various points is not in doubt, and many of those phalangeal whistles found are clearly the result of human agency. However, the interpretation of many of these objects as intentionally created sound-producers is complicated by the fact that damage resembling deliberate piercing can occur naturally very easily under the process of deposition (Harrison, 1978), or as a consequence of carnivore action (Chase, 1990). Although a phalanx itself will not provide much sustenance, they can be swallowed by a carnivore in the course of consuming neighbouring areas, and suffer damage through chewing and digestion. Chase (1990) illustrates this with an example of a phalanx found in the coprolite of a coyote, which closely resembles many punctured phalanges found in archaeological contexts (Enloe et al., 2000). The area at the posterior surface of the proximal end of a reindeer phalanx is very fragile relative to the rest of the phalanx, being about half the thickness of the bone elsewhere, and when pressure is exerted upon a phalanx by a surrounding gravel matrix it is invariably this area that punctures, whether the bone is fresh or old (Harrison, 1978). Furthermore, the hole that results is not dependent upon the shape of the stone that pierced it, but upon the properties of the bone itself, meaning that very consistent-shaped holes are produced under different conditions, dependent only upon the age of the bone.

Harrison (1978) found that fresh bone punctures consistently, resulting in a hole with a smooth curved outline, the only irregular section being where the puncture abuts the superior epiphysis. As much as seven-eighths of the circumference of the hole may be smooth and round (p. 10). Such damage has the appearance of a deliberately created hole, and punctured phalanges created naturally in this way also tend to function well as whistles.

An old bone under these conditions tends to puncture in a more fragmentary way, producing a rough-edged hole more easily recognizable as natural damage. However, there is conversely the possibility that genuine anthropogenic phalangeal whistles would suffer the same type of damage upon deposition if they were old when deposited.

The implications of Harrison's (1978) work are several. In terms of determining the authenticity of punctured phalanges as artefacts, there are three main problems:

1) Reindeer phalanges puncture easily and consistently at the proximal end of the posterior surface, and naturally produced punctures in fresh bone can look intentionally produced;

2) Intentionally produced punctures may be damaged in deposition so as to be unrecognizable as such;

3) Both naturally-produced and intentionally-produced punctured phalanges tend to be capable of functioning well as whistles. Thus, functionality cannot be taken as an indicator of anthropogenesis.

There are some important indicators as to authenticity that can be applied, however:

1) Phalanges punctured in deposition will tend to include pieces of the bone wall inside the cavity; an intentionally produced whistle would have had these removed as they would impede its function;

2) Intentionally produced punctures often have smooth edges where they have been smoothed by the rotation of the punching tool or burin, and are often slightly bevelled as a consequence;

3) Any smooth hole that is *not* located at the proximal end of the distal surface is likely to be intentionally produced, as the bone diaphysis does not puncture easily anywhere else.

It is on the basis of these criteria that those examples subjected to close scrutiny, detailed in Appendix Table 3, have been judged, in particular by Harrison (1978), Scothern (1992), and to a lesser extent by Dauvois (1989). Note that not all of the items included in Table 3 have been subject to this level of scrutiny and, as with the pipes/flutes, several have not been re-examined closely since their original discovery and/or publication.

Whilst the status of many individual reputed phalangeal whistles must be subject to question, their existence and use by Palaeolithic humans is not at issue; there are numerous examples of which their anthropic origin is well established. Reindeer phalanges were a very readily available raw material throughout much of the Pleistocene, easy to work, and would have been plentiful in living environments as reindeer formed a major subsistence staple in both the Middle and Upper Palaeolithic. For what they were used is open to greater question.

Whilst many of these objects can be confidently classified as intentionally produced sound-producers, they should not necessarily be classed as musical instruments as there are many potential uses for a small whistle. In his analysis Dauvois (1989) focused more on the acoustic properties of various French examples of phalangeal whistles, than on their provenance. He noted that the whistle-language of the Pyrenean Vallée d'Aas shepherds uses frequencies between 2,000 and 2,650 Hz—sounds within these frequencies travel well over long distances, and the human ear is most sensitive to sounds between 1,500 and 4,000 Hz. It is especially sensitive to sounds around 3,800Hz as this is the frequency range at which the *auditory meatus* (ear canal) resonates (Campbell and Greated, 1987), and at which the ossicles (middle-ear bones, *malleus*, *incus*, and *stapes*) operate most effectively. Dauvois analysed the frequencies produced by various examples of phalangeal whistles to see how they compared with this range. Of the thirty-five examples he tested, twenty-one produced tones within the range of the shepherds' whistle-language, and all thirty-five fell within the 1,500–4,000-Hz range of sensitive human hearing. Harrison (1978) also carried out tests of the audibility of the phalanges' sound over long distances. Some of the examples he tested (e.g. one from La Madaleine, Appendix Table 3) were audible with 100 per-cent reliability over distances up to 1.25 km, and clearly have the potential to be used as communicating devices between individuals separated in the environment. Those with small holes proved poor at making sounds that carried over long distances (Harrison, 1978), although small holes have often been intentionally produced. An explanation other than long-distance communication or hunting coordination must be sought for such artefacts.

An intriguing finding from a later study is reported by Dauvois (1999), who used reproduction whistles made on modern reindeer phalanxes (identical to those of the past) to test animal reactions to the sounds produced by them. These experiments were carried out on the Russian-Finnish border in the eastern-central part of Finland. Dauvois reports that 'Herds of reindeer at first intrigued by the whistling came near the sound source, but after repeating the whistling (the acoustic forms of which are given on [Dauvois's 1999] fig. 3), laid down on the snow or ice. This peculiarity occurred on various occasions, at several days interval and with various herds' (1999, p. 165). This effect was observed identically with both a replica Palaeolithic reindeer phalangeal whistle and with a genuine reindeer phalangeal whistle dating to the early twentieth century made by natives of Arctic Canada.[1] It seems from these findings and those of Dauvois (1989) that the whistles would thus allow humans to communicate with each other using a frequency register that carries over large distances yet does not actually alarm the animals that formed their principal hunting prey—an extremely useful combination. Indeed, if the sound of the whistle could actually induce reindeer to lie down and be approached, then its usefulness in hunting would extend far beyond communication with other hunters.

Bullroarers (free aerophones)

Another sound-producer of which there may be Palaeolithic evidence is the bullroarer. This instrument consists of a flat perforated piece of wood, stone, bone, antler, or ivory on the end of a cord, which creates a whirring sound when spun in a circular motion. The sound can vary according to the size of the bullroarer, the length of the string and its rate of rotation, from a low rumble like thunder or a bull when spun slowly, to a high-pitched scream when spun very fast. Bullroarers are, or have been, used in a great diversity of cultures throughout the world, by populations as far apart as Australia, Africa, and North America, amongst the Maori of New Zealand, in New Guinea, and amongst the Sami of Scandinavia, and their powerful religious roles in ancient Greece are also well documented (Harding, 1973; Lang, 1884; Haddon, 1898; Maioli, 1991; Victor and Robert-Lamblin, 1989; Montagu, 2007). Very often their use has sacred and religious associations—amongst some Australian Aboriginal groups, for example, bullroarers are used to imitate the spirits occurring in the natural sounds of nature—but they can also have non-symbolic applications, as is the case in Malaysia where they are used to scare animals away from plantations (Scothern, 1992).

In the Palaeolithic record objects that may be bullroarers are less prolific than those reputed to be pipes or flutes. This may be partly as a consequence of their being less likely to be identified as such, but may also be due to the fact that effective bullroarers are very readily made from materials that are biodegradable; there is no particular functional reason to manufacture a bullroarer out of bone, stone, or antler as opposed to wood. We do, however, have

---

[1] These experiments are described in French in Dauvois (1999), p. 173; this description and Dauvois's footnote 5 imply an indebtedness to a research project carried out by the University of Manitoba, Canada, under the direction of Prof. M. William Pruitt in the department of zoology.

several examples of what could be bullroarers, fortunately made on robust materials that do preserve archaeologically.

We do need to be careful about the interpretation of pierced bone, stone, and antler objects, however, as there are numerous purposes to which a pierced flat object can be put, and numerous ways in which a piercing can occur, not all of which are the consequence of human intention. Various possible functions for pieces of pierced bone from Palaeolithic contexts have been proposed on the assumption that the perforations were man-made, including use as fishing weights (Zervos, 1959), pendants, and as possible bullroarers (Scothern, 1992). As the preceding discussions of phalangeal whistles and pipes have highlighted, however, holes in bone can be created by processes other than human action. Some of the most important work regarding the anthropic origin of pierced Palaeolithic bones, and thus by implication the 'authenticity' of potential Palaeolithic musical instruments, has been carried out by Francesco d'Errico and colleagues (e.g. d'Errico and Villa, 1997; d'Errico et al., 1998). They compared natural assemblages of faunal bones found at hyena dens and caves with those found in a variety of Palaeolithic contexts which have produced items variously claimed as worked bone, pendants, or pipes. They also examined the relative quantities of different bones and parts of bone which survived in each case, and they looked closely at the effects on bones of carnivore activity, in terms of both gnawing action and digestive action. This included the production of lines and 'incisions', smoothed apparently 'polished' surfaces and holes in bones. As a result of these comparisons they convincingly argue that a large number of Palaeolithic 'pendants' are in fact the product of hyena activity, not human behaviour; this includes objects such as those from Kulna, Bois Roche, Pech de l'Azé, and Bocksteinschmeide. None of these pendant-like pieces had been claimed to be musical instruments, but we must clearly apply these analytical criteria carefully to pieces in which the human agency in their creation is ambiguous. As many small pendant-like pieces of bone with a single perforation seem to be the product of carnivore gnawing and digestion, interpretation of these would have to be made in light of d'Errico and Villa's criteria for the identification of human agency in their construction, even before any interpretation regarding their function could be made.

Whilst almost any object attached to the end of a cord will make a sound when it is spun fast through the air, the frequency and volume of the sound will be related to the size and shape of the object (and, to a lesser degree, the cord). Objects of the small size of the pierced pieces of bone that formed the focus for d'Errico and Villa's study discussed here would not make the most effective bullroarers, and even if such objects could be effectively identified as being anthropogenic, their use as bullroarers might not be the most parsimonious interpretation in any case.

There *are*, however, examples of reputed bullroarers of which there can be no doubt of their human origin, and which furthermore are of much more suitable dimensions to function as bullroarers. A particularly spectacular example is an artefact from Magdalenian layers at La Roche de Birol, Lalinde, in the Dordogne (excavated by Peyrony in 1930 and analysed by Schaeffner, 1936; see Figure 4.3). This oval carved piece of reindeer antler is approximately 180 mm long and 40 mm wide at its broadest point. It is incised all over with long lines running along its length, and shorter lines perpendicular to these. It is pierced with a hole at one end, through which, it is presumed, would have been threaded a cord, and it is covered with red ochre. It is possible, of course, that this artefact might have served some other purpose than being a bullroarer; it may have been a very large pendant, for example. The size of the artefact and the fact that it is covered with red ochre would mitigate against it being used as a fishing weight, however. It could certainly have been extremely

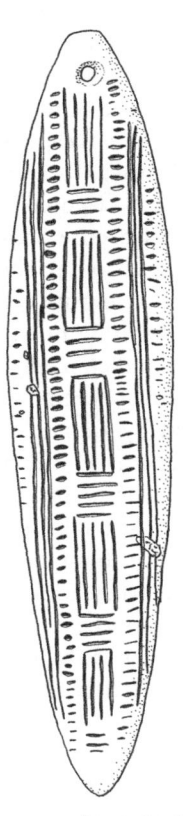

Fig. 4.3. Possible bullroarer from La Roche, Lalinde, France; reindeer antler, covered with red ochre. Magdalenian; 180 mm long, 40 mm wide at widest point. (Redrawn from Bahn and Vertut, 1997, p. 85.)

effective as a sound-producer, if that was its purpose. This was amply illustrated by Dauvois (1989) who produced a replica of the artefact out of reindeer antler using a silex tool, and found that it hummed (*vrombit*) at frequencies between 130 and 174 Hz. As a comparison, a known bullroarer from New Guinea produced frequencies between 90 and 125 Hz (Dauvois, 1989); this was a particularly large artefact, at 370 mm being more than twice the length of the La Roche example, and produced a deeper tone as a consequence. The smaller the artefact, the higher the frequency (and tone) of the sound produced by it. According to Dauvois, the La Roche example produces a particularly impressive sound when twirled in a cave, and even more so when it is played in the total darkness inside the cave, 'vraiment un *son volant*' (Dauvois, 1989, p. 10).

Dauvois (1999) states that spectrograms of the sound made by ancient bullroarers can be superimposed over those of modern ones from Amazonia and Australasia (Australia and New Guinea), showing their capability of producing sounds equivalent to modern examples. In total, Dauvois (1989) describes or illustrates seven likely bullroarers associated with Solutrean and Magdalenian contexts. His figure (Dauvois, 1989, p. 10) represents the La Roche de Lalinde example plus three others: an engraved example, 90 mm long, from Upper Solutrean levels at Lespugue, Haute-Garonne; one from Laugerie Basse, Les-Eyzies-De-Tayac, Dordogne, 107 mm long; and a 190-mm-long Solutrean example from Badegoule, Bersac, Dordogne. It is perhaps worth noting that both the Haute-Garonne area and Laugerie Basse/Les Eyzies have produced several examples of other sound-producers (see Appendix Table 1). A further, rather smaller, example from Magdalenian contexts is illustrated by Harding (1973), from Saint-Marcel, Indre, France, engraved with an elaborate design of lines and concentric circles.

In addition there are three further examples from Upper and Final Magdalenian layers on bone described by Dauvois (1989). These are particularly interesting in that they are all engraved with images of bovids. An example from abri Fontalès, Tarn-et-Garonne is engraved on one surface, and examples from abri Morin (Pessac-sur-Dordogne, Gironde) and Lortet (Tarn-et-Garonne) are engraved on both sides with bovid images. Dauvois does, however, caution against the temptation to equate the sound of a *rhombe* with the mooing of a cow, lest we impose our own single interpretation on the evidence and fail to consider other possibilities (1989, p. 11). It must be said that this is a particularly great temptation given the English name for the items in question.

It is clear that many of these examples can function very well as bullroarers, but how might we attempt to rule out other intended uses, such as being large pendants? It is possible that future use-wear analysis of the areas around the perforations may help to identify the way in which such items were used, and remove ambiguity from their interpretation. In addition to Dauvois's work

(1989), other experimental reconstructions (Alebo, 1986) have tested the sound-making capabilities of a reputed bullroarer, as well as the process of manufacture of tendon string for the purpose, but there seems to have been no use-wear analysis carried out. Future experimental reconstruction and use of bone bullroarers may allow patterns of abrasion around the perforation to be identified, caused by the twine or tendon when used as a bullroarer, which can then be compared with any on Palaeolithic perforated bone fragments. One might expect such wear to be different from that caused by the threading of the object as a pendant or as a fishing weight.

Such research could be extremely valuable in adding confidence to the interpretation of pierced artefacts, especially more ambiguous ones. In the case of the Solutrean and Magdalenian examples described here, however, interpretation of them as bullroarers would seem to be by far the most reasonable attribution of their purpose.

As with pipes, rasps, and percussive instruments (see later in the chapter), any surviving bone, antler, and ivory examples are likely to represent only the tip of the iceberg as regards the original proliferation of such items, as all could be (more) easily made from wood (as is commonly the case in a variety of contemporary cultures, as discussed in Chapter 2), which would not survive to the present day.

PERCUSSIVE INSTRUMENTS

It is possible too, that with our preoccupation with tonal and melodic instruments, we may be neglecting to identify a whole class of sound-producers which is critical in most cultures' musical activities. Ethnographic evidence shows that among many modern hunter-gatherer groups instrumentation is dominated by percussive rather than melodic instruments, melody instead being provided by vocalization (see, for example, Johnston, 1989; Nettl, 1992; Breen, 1994; McAllester, 1996; Locke, 1996; Chapter 2). If this were also the case amongst early humans we would be left with very little evidence of it, as these instruments are usually made from wood or skin, which would leave no archaeological trace. Furthermore, percussive sounds are often made by striking body-parts, which requires no instrumentation at all.

Rasps (scraped idiophones)

One type of percussive sound-producer that might have been used by early humans is the scraped idiophone, or rasp (Huyge, 1990, 1991; Kuhn and Stiner, 1998). A rasp can be a piece of wood, bone, or stone with grooves cut into it perpendicular to its length, which are then rubbed with another object to create a staccato vibration. In the case of this type of sound-producer, there are not only possible artefacts from Palaeolithic contexts, but there is also a representation of what may be a rasp in a bas-relief engraving.

Huyge (1990) has compared properties of reputed rasps from a variety of contexts, in particular suggesting that twelve quite evenly-spaced striations on a Middle Palaeolithic mammoth bone from Schulen (Belgium) might indicate function as an idiophone. It has been suggested by d'Errico (1991) that these are more likely to be due to carnivore action. Although they are strikingly more even in spacing and depth than those on other suggested Mousterian rasps (Capitan and Peyrony, 1912, and Vincent, 1988, quoted in Huyge, 1990), which almost certainly are due to carnivore activity, their anthropogenesis does remain an open question. Microscopic analysis for characteristic traces of lithic tool marks would be valuable (it was found in context with eighteen Mousterian Levallois cores and flakes). This would not, of course, indicate whether or not the object functioned as an idiophone, which could only be suggested by characteristic use-wear.

The anthropogenesis of some later objects is not in question. Examples include a rasp-like artefact from Aurignacian layers at Riparo Mochi (Liguria, Italy) (Kuhn and Stiner, 1998, fig. 5g), a Magdalenian example from Pekárna cave in the Czech Republic, which is made on reindeer antler, as well as very similar Magdalenian examples from Abri Lafaye Bruniquel and Mas d'Azil (Dauvois, 1989) (see Figure 4.4), and an incised rib from La Riera cave, Spain, from a layer dated to 19,000–18,310 ± 260 BP (reported in González Morales, 1986—although no speculation is offered there as to its purpose). For comparison, see also a Bronze Age example from Syria (Huyge, 1990). The example from Riparo Mochi was found in Stratum G, dated to 32,000–35,700 BP and associated with classic Aurignacian split bone points and tools. This artefact, made on a section of ungulate long-bone (probably metapodial, according to Kuhn and Stiner, 1998) is approximately 6.75 cm long, with broad, rounded lobes carved onto one surface (about 4.25 cm of the 'lobed' surface is preserved, on the basis of measurements of the scale drawing). Although the function of this object is not known, Kuhn and Stiner point out its similarity to rasps known from more recent cultures.

Marking bone, antler, and stone artefacts with series of incised lines perpendicular to their length seems to have been a particular preoccupation of populations using Gravettian technologies too (c.27,000–22,000 years ago), with numerous examples of bone and antler shafts and stone plaquettes modified in this way. Whilst this certainly should not be taken to indicate an obsession with rasps—the symbolic and practical functions of these objects could be many and varied—it does clearly reiterate that the technology and competence required to create them was widespread, and highlights a need to include rasps as a possibility in future analyses of their uses.

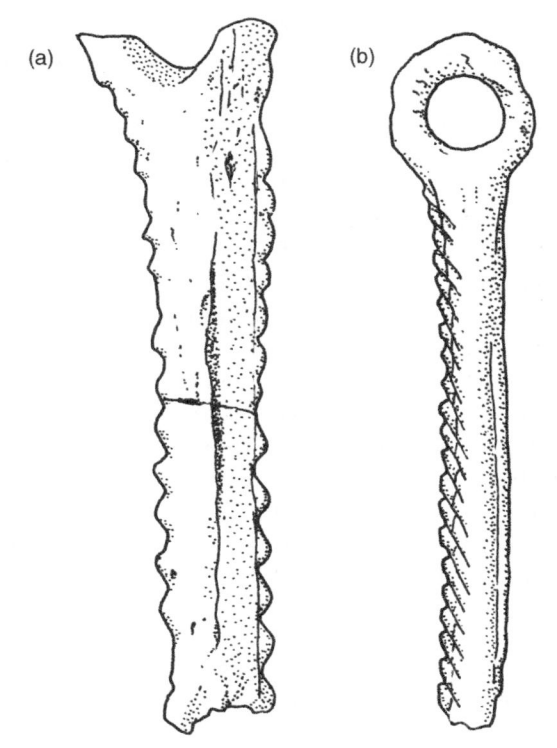

Fig. 4.4. Possible rasps.

(a) Mas d'Azil, Ariege, France. Magdalenian. Length, c.115 mm. (Redrawn from a photograph by Kathy King of the original artefact in the display at Musée d'Archéologie Nationale, Saint-Germain-en-Laye, France (http://donsmaps.com/laugeriebasse.html), and Dauvois 1989, p. 11, image 3.)
(b) Pekarna, Moravia, Czech Republic. Magdalenian. Length: 205 mm. (Redrawn from Dauvois 1989, p. 11, image 1, and Lázničková-Gonysěvová 2002, Fig. 7(b), p. 542.)

Whilst the human agency in the construction of these artefacts is not in question, their use is very much open to debate. They clearly have the potential to function very well as rasps, but there is always the possibility that the grooves on these objects serve some other purpose. Marshack (e.g. 1972) has suggested that incised objects may serve as tallies or calendrical notation, they may function as other Artificial Memory Systems (Lawson and d'Errico, 2002), or it may be that incisions serve to improve grip on the object, which was then used for something else.

In the instance of the Aurignacian Riparo Mochi example, and the Magdalenian examples from Pekarna, Abri Lafaye Bruniquel, and Mas d'Azil, an explanation of them as calendrical notation is not adequate; the grooves on each artefact are deep, evenly sized, and spaced notches and not mere incised lines on the surface of the antler (which is all that would have been required for notation). The cuts are also such that they actually form a serrated edge to the artefact, and are so consistent in form on each object that it is difficult to imagine that the incisions were not all made on the same occasion.

The Magdalenian artefacts also share a feature of a loop (or what appears to be the remains of one) at one end. It is possible that the serrations were created to improve grip on the tool while the loop at the end was used for some other purpose. Conversely, if they were used as rasps as is suggested by Dauvois (1989) and Huyge (1990), then they could have been held using this loop, which would have avoided damping of the sound produced by the serrated edge as it provided the idiophone surface. It is possible that held like this, the serrated edge might also have served some other purpose, but it is hard to imagine what purpose a serrated-edged antler might have fulfilled that a lithic tool could not do better. It is conspicuous too that Mas d'Azil, Bruniquel, and Pekarna have each also produced several examples of other artefacts interpreted as flutes, pipes, and phalangeal whistles made on reindeer and bird bone, from Gravettian through to Magdalenian contexts (Buchner, n.d.; Absolon, 1936; Pequart and Pequart, 1960; Harrison, 1978; Dauvois, 1989; Scothern, 1992; Svoboda et al., 1994). If these sites really did form a focus of activity for large groups of people, incorporating musical activity, then this may add to the plausibility of the interpretation of the antler artefacts as scraped idiophones.

An invaluable contribution to the study of these artefacts would be the identification of characteristic use-wear patterns from the use of bone objects as rasps. The use of experimental reproductions of grooved bones for sound production in this way, using wood, lithic, and bone scrapers, would then allow the identification of characteristic micro- and macroscopic use-wear. Incised objects from archaeological contexts, such as those described here, could then be analysed for equivalent traces. At the very least one would expect to find fine polish and striations running perpendicular to the main incisions (i.e. along the axis of the bone), and the identification of such characteristic marks would open the door to the reanalysis of a potentially large number of hitherto ambiguous or even ignored artefacts. Indeed, Kuhn and Stiner (1998) also report the presence in Riparo Mochi Layer G of a midsection of worked bone shaft with axially-orientated abrasion striae, and it would be valuable to ascertain whether the possible rasp also features such marks at a macro- or microscopic level.

Representations of human artefacts are rare in Palaeolithic archaeology, but one example that is of particular interest is the bas-relief known as the 'Venus of Laussel' or ('Dame à la corne') from Laussel, France, dated to 25,000 years ago (Huyge, 1991). This image consists of a corpulent woman holding an incised horn (generally accepted as that of a bison) (Huyge, 1991). The horn bears thirteen regularly-placed incisions along its length (see Figure 4.5) and is being held up at shoulder height by the female figure. It has been proposed (by the original discoverer of the relief, G. Lalanne, quoted in Huyge, 1991) that

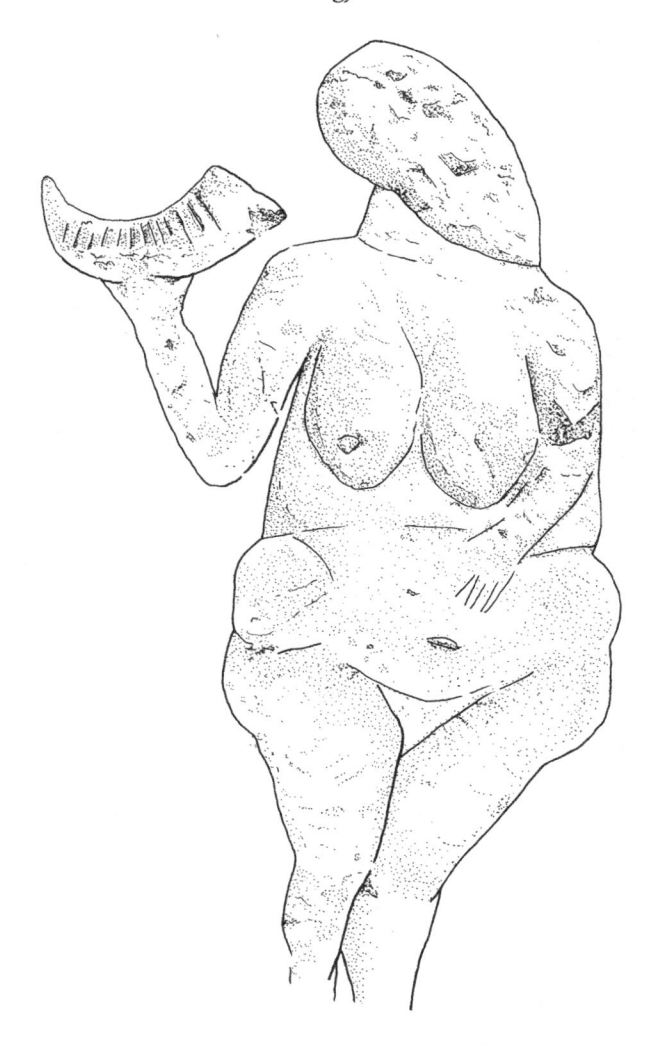

Fig. 4.5. 'Venus of Lausel' or 'Dame a la Corne', Laussel, Dordogne, France. Height 440 mm. She holds an incised bison horn, proposed as an idiophone by Huyge (1990). (Redrawn from Bahn and Vertut 1997, p. 113.)

this may represent a tally of hunting kills, given that the content of another part of the bas-relief implies a possible association with hunting. As well as three females, the relief features a mutilated image of a masculine figure who appears to be about to throw a lance at a deer (Huyge, 1991; Stockman, 1986). It has also been suggested that the thirteen lines may represent a record of the number of moons in a year (Marshack, 1972), with the crescent shape of the horn being associated with the moon. This is a neat explanation, and it is certainly strange that there are not more representations of astronomical phenomena (recognized, at least) in Palaeolithic representation, though it should be noted that there are actually twenty-six crescent moons in a year if both waxing and waning are counted.

Huyge (1991) points out that the horn very closely resembles idiophones made on bovine horn, such as examples from Mexico (Huyge, 1991, fig. 3) and Bonaire (Huyge, 1991, fig. 4). The possible hunting context, if correctly interpreted as such, need not detract from the musicological interpretation of the object, given the many associations between music and hunting represented in the ethnographic literature (see, for example, Kehoe, 1999; Turino, 1992; Bahuchet, 1999; Johnston, 1989). Although the purpose of the horn is impossible to identify with certainty, the relative simplicity of the rasp idiophone and the close resemblance of the image to existing examples makes an idiophone a very plausible—and perhaps even most parsimonious—explanation.

Struck percussion

Percussive instruments (struck idiophones) play a central role in many musical traditions, and are frequently much more prolific than melodic instruments. The identification of them archaeologically presents several problems, however. First, almost any material may be struck to produce sound, and identifying an object that was designed to be used in this way may be difficult, because such objects need have no particular consistent form (unlike a pipe, bullroarer, or rasp). Second, if traces of percussive damage can be identified it could be difficult to demonstrate that they were produced in sound-production activities, as percussive damage may be created by many processes whose primary purpose was not sound production. Third, such instruments are very frequently (and often most effectively) made from biodegradable matter such as split or hollow logs, for example, with or without the addition of skins, none of which would normally preserve archaeologically. Indeed, the human body itself constitutes an excellent percussion instrument.

In spite of these difficulties, some possible evidence for the use of percussive instruments by prehistoric humans does exist. A set of six mammoth bones from Mezin, Ukraine, dated to 20,000 years ago, appears to have been deliberately and repeatedly struck, and the bones were found in context with two beaters and a variety of ivory 'rattles' (Bibikov, 1978). Also found in the same settlement were piles of red and yellow ochre, and other mammoth bones which had been incised and painted. The 'orchestra' bones (or 'osseophones'), including a shoulder blade, thigh bone, jaw bones, and skull fragments, produce a selection of tones when struck with beaters (Bibikov, 1978), and a recording of Soviet archaeologists using them as sound-producers in this way was made in the late 1970s (Hadingham, 1980). A somewhat similar object, a decorated mammoth skull which also shows evidence of having been repeatedly struck, was found at a second Ukrainian site, Mezhirich, along with a set of beaters. Dating to around 15,000 years ago it was also found in the entrance to a hut.

While Bibikov's interpretation of the Mezin objects as a collection of musical instruments was seconded by a selection of Soviet archaeologists, forensic scientists, and medical experts, some other archaeologists and musicologists have remained cautious or sceptical. While the artefacts do show wear from percussion, Scothern (1992) and Lawson et al. (1998) point out that this does not necessarily merit a musicological explanation, as many daily activities involve percussive actions.

The fact that these bones occurred in context with a reindeer-antler mallet and beaters, as well as what appear to be rattles, and in a central and apparently non-residential hut, counts in favour of the musicological interpretation, though. Again, more detailed experimental and use-wear studies are required for the interpretation of objects of this type to be progressed further. Establishing the nature of characteristic impact damage associated with different types of beaters and percussive activity will allow the reassessment of artefacts already in collections and, perhaps, the identification of the traces of deliberate percussive use on future discoveries. Until that happens it is possible that numerous sound-producing artefacts will remain unidentified.

CAVES AND LITHOPHONES

Whilst there is no convincing representation of musical instrumentation in Upper Palaeolithic 'art', with the possible exception of the Venus of Laussel already discussed, rock art has been connected with music and the use of sound in less direct ways. As Boivin (2004) says, 'Links between percussion and rituals related to rock art have been ethnographically documented or archaeologically postulated in a number of cases' (Hedges 1993; Ouzman 1998, 2001), 'and it may be that percussion and/or other sounds contribute to the creation of an appropriate spiritual or emotional state for the viewing or creation of rock art in ritual contexts' (Goldhahn 2002; Ouzman 1998; Scarre 1989) (p. 48).

Reznikoff and Dauvois (1988) carried out an extensive analysis of the acoustic properties of three Palaeolithic caves featuring Palaeolithic painting, in the Ariège area of the French Pyrenees: Grotte du Portel, Niaux, and Fontanet. They made vocalizations throughout the caves, identifying 'points of resonance', places that resonated to particular notes. They then compared the locations of these points with the locations of the cave paintings, which have traditionally been dated to around 20,000 years ago (Scarre, 1989). They found that there was remarkably consistent correlation between the locations of the paintings and the places of particularly great resonance. Most of the cave paintings were within a metre of a point of resonance, and most of the points of resonance were accompanied by a painting of some sort. Further, they conclude that the choice of location of some of the paintings can *only* be explained by their relationship with a point of resonance, as they are often not easily visible or accessible. In fact, some of the locations are marked only with red dots, where there was no room for a full figure (Reznikoff and Dauvois, 1988).

It would seem that the acoustic properties of the cave were at least as significant to the painters as the art itself, as the position of the art seems to have been dictated by the resonance. It would seem that the painting activities could only have been carried out alongside deliberate sound-producing activities, which were part of the process itself of decorating those spaces. The authors suggest that the full potential of the caves' resonance would only be elicited by the great range of the human voice, rather than the comparatively limited ranges of the flutes and whistles. It is not possible to be certain of the nature of the sound production of the peoples responsible for the art in these caves, but what does seem to be true is that they were very aware of the acoustic potential of features of the world around them, and considered such properties to be highly significant.

Reznikoff (2008) has since suggested that the use of sound (and in particular the voice) was actually critical for the exploration of these environments under limited light conditions, using the acoustic response of the environment to detect passages, holes, and surfaces. On the basis of preliminary research in the French caves of Rouffignac, Arcy-sur-Cure, Niaux, and Le Portel, Reznikoff reports that, using sound to proceed through the caves, locating passages and chambers, one is led to the painted areas. When an optimum frequency is achieved in a particular area the sound appears to be amplified by the resonance, and in these places in particular, decoration appears to occur, including pictures and red dots, the latter of which, Reznikoff suggests, are used as 'landmarks' in exploration. He says that the human voice (and male voice especially) is best for this process, because in some places the playing of an instrument is not practical, and the voice is extremely flexible as to pitch.

Reznikoff (2008) outlines three principles, which should certainly be testable by future studies:

> Principle 1: The density of paintings in a location is proportional to the intensity of the resonance in this location;
>
> Principle 2: Most ideal resonant locations are adorned with paintings or signs;
>
> Principle 3: Certain signs are accounted for only in relation to sound.

The idea of privileging auditory cues over visual ones can seem quite alien to us, and a peculiar notion. We live in a very visually dominated society. We rely on our vision to tell us about modern dangers, to decide what to buy, to gain information from images and the written word. But there are two observations to be made about this. First, it is easy to underestimate the extent to which we still use our hearing to gain information about the world—for example, we often need to hear traffic before we can see it, we recognize the footfalls of familiar people before we can see them, and much advertising relies more on auditory 'jingles' than on the content (such as it is) of the visual stimulus they accompany. Second, our seemingly instinctive bias towards the visual is not, in fact, instinctive, and this is evidenced by numerous other societies living in different environments, who process their interactions with the world and entities beyond the world in ways that give far more precedence to the auditory medium. This point is articulately explored in several studies, including Feld (1982), Peek (1994), and Lewis (2009); see also Devereux (2006) for further discussion of the importance of auditory stimuli to past societies.

In the case of the Palaeolithic cave environments their acoustic properties are not limited only to their abilities to resonate sounds made within them, but in some cases they contain features which may themselves be used to create sound. Dauvois (1989) also studied the acoustical properties of features within the caves which bear the marks of being deliberately struck. Several of the stalactites and other natural calcareous depositions bear chips and impact marks where they have been deliberately struck. Dauvois reports that these features are capable of producing several distinct tones, depending on where they are struck, and that the damage marks correspond with these tone-producing locations (Dauvois, 1989; Lawson et al., 1998; Dauvois, 1999). These 'lithophones', like the resonant areas, also tend to correspond with locations of artwork; for example, the Larribau Gallery at Isturitz features some pictures and engravings which, whilst not in resonant locations, are adjacent to stalactitic draperies which can produce sound (Reznikoff, 2008).

The work of Reznikoff and Dauvois was not the first to observe and analyse correlations between Upper Palaeolithic art and the acoustical properties of spaces and features in caves. Lya Dams reported on similar features in the cave of Nerja, in Malaga, Spain (Dams, 1984); Escoural, in Portugal; and Roucador, Cougnac, Pech-Merle, and Les Fieux in France (Dams, 1985). This work built upon observations by Abbé Glory in the 1960s (Glory, 1964, 1965; Glory et al., 1965).

The cave of Nerja contains at least nineteen paintings so far reported, and has excellent acoustics, such that it is now sometimes used as a venue for concerts and dance in the summer months. It also contains a feature known as 'the organ', which consists of an area in an alcove where the calcareous deposits have formed into a fluted, concertina-like feature. The average height of these folds is 3–4 m, and each fold is separated from the next by a gap of 8–15 cm. The total length of the feature is about 5 m (Dams, 1984). In total there are over 200 folds, and they are nearly all decorated with abstract signs, as well as an ibex and a hind. Lines and dots in red and black are the most common decorations, some at ground level and others over 3 m up. Most are not visible from a distance, many are impossible to see unless the observer puts themselves in awkward positions to peer in between the creases; aesthetic considerations do not seem to have been a priority in the creation of the signs. All the decorations use the same colours and are equally faded, so the whole feature was probably decorated over a relatively short period. These examples are not amongst those that have yet been subjected to direct dating (see Bahn and Vertut, 1997), but the style of the ibex and hind illustrated suggest a Solutrean age ($c.$22,000–17,000 years ago) for the illustrations according to Dams (1984). Layers of calcareous deposition overlaying the illustrations confirm a very great time has elapsed since they were painted.

The folds are also worn along the edges and are chipped, apparently deliberately, at different heights. This damage is, like the paintings, covered by more recent layers of calcite deposits, which dates it to the same age as the artwork. Dams (and Glory, 1964) consider this damage to have been caused by deliberate striking of the folds with hard objects. Indeed, when they are struck with a stick or blunt flint they produce clear, harp-like tones (Dams, 1984). The deliberately damaged sections seem to have been broken with the aim of altering the tone produced, as they effectively alter the length of the ridge (Dams, 1984); the shorter the ridge, the higher the tone produced (Dams, 1985). Dams believes that this entire feature was used as a lithophone during the Upper Palaeolithic, and that the artwork on it relates directly to that function, be it abstract or as a type of mnemonic aid dictating striking-places on the rock. Similar features in other caves are also compared with the 'organ' in Nerja (Dams, 1985). The cave at Roucador, France contains a very similar ridged lithophone feature, which has, like the Nerja one, been decorated with black dots and lines, and has had fragments broken off many of the ridges. If there were any doubt of ancient human agency in the damage, it is dispelled by the fact that in one instance, these broken fragments have been wedged back into the wall between two ridges, where they have since been cemented in with a new layer of calcite (Dams, 1985). The painted marks have likewise been covered by calcite layers since their creation, as have soot marks on the rock, presumably from the torches held by the artist/performer or observers. These stalactites, like those of Nerja, are resonant, and produce a bell-like tone when struck. Roucador also contains an unpublished frieze of painted and engraved bison, horses, a mammoth, circular signs, stencilled hands, and other symbols, all located 6 m above the current floor level, and near to the lithophone (Dams, 1985).

Black and red dots and lines mark a similar feature of stalactite ridges at Cougnac cave (also famous for its Upper Palaeolithic art), and a free-standing stalagmite, all covered by layers of newer calcite deposits. These again give forth 'pure and clear' vibrations, and are located near an engraving of the head of a bear. The Pech-Merle and Escoural caves feature lithophones too, marked with red and black lines and dots. These both also exhibit recalcified fractures, and the ground around the former is littered with bone remains and chips from the stalagmite itself. These are, again, situated close to other art and representations of animals. The Pech-Merle example is next to engraved images of ibex and a possible mammoth, as well as cup marks and other engravings (Dams, 1985). The art in all of these caves has been dated (stylistically) to the Solutrean (Dams, 1985). It will be a valuable addition to the study of these features when they are added to the gradually growing catalogue of examples of cave art that have been directly dated by AMS/radiocarbon dating. Whilst confirming most stylistically based attributions of age, there have been notable exceptions; for example, Chauvet cave, some of the cave art of which turned out to be 10,000 years older than expected when dated by AMS (Bahn and Vertut, 1997).

There are several things that can be asserted with some confidence from these studies. Firstly, these calcite formations have been covered with abstract decorations, which can only be viewed through close observation; the formations had specific significance for the people decorating them, but were not decorated for general viewing or aesthetic reasons. Secondly, they occur in areas established as places of considerable artistic, and possibly social, significance. Thirdly, they produce clear tones when struck with wood or stone. Fourthly, they have been chipped and worn, probably through being struck, and this probably occurred around the same time as the painting was carried out, around 20,000 years ago. Finally, they each appear to be part of a network of similar features in southern France, Spain, and Portugal that were all treated and used in the same way.

Considering the level and nature of the attention these calcite formations were given by the Upper Palaeolithic users of the caves, it is far easier to believe that these features *were* used as lithophones, as Glory, Dams, and Reznikoff and Dauvois suggest, than it is to believe that this acoustic quality of the features escaped the notice of these early people.

The idea that rocks and rock formations were used as sound-producing devices is not one without a considerable history, and an extensive global precedent; the lithophonic use of natural rock formations and large stones is also known from more recent prehistory and history throughout the world, and Fagg (1997) carries out an extensive review. There is a number of assemblages of ringing stones in the Canary Islands, which appear to have been used by the native inhabitants of the islands prior to the arrival of the Spanish, and possibly up until the last century (Álvarez and Siemens, 1988). These stones have obviously been struck, being chipped along the edges, and are situated variously in circles and lines in natural contexts which amplify their sound acoustically. Natural lithic features variously known as 'Rock gongs' and 'ringing stones' are also found in India (Boivin, 2004; Boivin et al. 2007), Bolivia, South-East Asia, Australia, and Africa (Montagu, 2007; Laura Basell, *pers. comm.*; Fagg, 1997). Several of the dolerite boulders at Kupgal Hill, India, published by Boivin (2004), produce loud, clear ringing ('bell- or gong-like', p. 47) tones when struck with granite stones. This only occurs when the boulders are struck in particular places, namely on human-created rounded impressions on the surface of the boulders. Other, similarly marked and similarly effective stones were found at other sites in the area, and the Kupgal Hill examples are in close proximity to rock art which shows, amongst other things (apparent hunting scenes, for example), human figures arranged in chains, which have been interpreted in the past as dancing.

Further examples of 'singing stones' traditionally believed to have magical properties in the production of sound are provided in Swedish historical records. These are often cup-marked Neolithic or Bronze Age stones which have continued to be known of as sound-producers over many millennia, even to the present day (Henschen-Nyman, 1988). Britain is not excluded from such traditions of sound-producing with rock; apart from the possibility of Bronze and Iron Age cup-marked stones fulfilling this purpose, several examples of 'geological pianos' were constructed in the nineteenth century from Cumbrian slate (Cross, n.d.), and the systematic use of stone slabs as chime bars is also known from ancient China (Montagu, 2007). As would be expected of a tradition with such a massive geographical and temporal span, the nature and uses of lithophones has varied greatly. Some of the African 'rock gongs' produce a single pitch, others several pitches depending upon where they are struck. 'Uses varied: many for ritual; some for signalling, with sounds that carried for several miles; many also for normal musical purposes, accompanying singing and dancing. Musical practices also varied, some being played by a single player at a time, others by groups as percussion ensembles' (Montagu, 2007, p. 8).

The use of stone to produce sound is not limited to large rocks in the landscape and slabs of stone; smaller, portable stones are also known to have been used as sound-producers (Montagu, 2007). Ian Cross has carried out preliminary work testing the acoustic properties of flint and other stones as potential intentional sound-producers (Cross, 1999b; Zubrow, Cross, and Cowan, 2001; Cross, Zubrow, and Cowan, 2002), and this work has been pursued further by Elizabeth Blake with particular reference to the sound-producing potential of Upper Palaeolithic blades and large Solutrean leaf points. Anybody who has carried out flint knapping will know that it requires a great awareness of the sound that the stone produces as it is struck. A skilled knapper has to be aware of the subtlest variations in the sound produced by the stone being struck, as this provides important clues as to where and how hard to strike the stone next (John Lord, *pers. comm.*). Cross proposes that the necessary awareness of the different sounds produced by stone during tool knapping may have led to its use for percussive or even melodic sound production in other contexts at an early stage.

In any case it seems clear that Upper Palaeolithic humans were especially aware of the acoustic properties of their environment and raw materials with which they worked, and we should take this into account when interpreting the activity sites and material culture of this time.

MUSIC AND DANCE IN LATER PREHISTORY

The relationship between sound production and representation ('art') is one that seems to become more explicit in the later periods of prehistory. Whilst detailed consideration of archaeological evidence for musical behaviours amongst early post-glacial hunter-gatherers and agriculturalists is beyond the scope of this book, it is interesting to note that the great increase in representation of humans amongst these populations in Europe, South Asia, and the Near East includes frequent representations of people apparently engaged in dance activities (Garfinkel, 2003).

Indeed, even when associated with Neolithic populations, representations of dance activities may be focused not on agricultural practices but hunting activities, which remained an extremely important subsistence resource alongside often sporadic agriculture. Excavation of the town of Çatalhöyük in Turkey (occupied from about 9,400 years ago until 8,000 years ago) has produced some wall paintings representing what appear to be dancing and hunting activities, accompanied by horn-shaped, bow-shaped, and drum-shaped objects (Stockman, 1986). The occupants of the city subsisted through a combination of hunting and agriculture for 700 years or more, from at least the middle of the seventh millennium BC. Stockman observes that the material artefacts and wall paintings from Çatalhöyük indicate that the activities of the hunter-peasants of the city were 'deeply rooted in existing traditions, known already from Upper Palaeolithic and Mesolithic times' (Stockman, 1986, p. 12), but that they also were quite creative, trying out new possibilities of tools and materials.

As regards music, the images so far excavated are dominated by what may be idiophones and percussion instruments, most of which, Stockman says, can be traced back to Palaeolithic times. These include figures holding what appear to be horns, bows (often two such items, one in each hand, which questions their use as hunting bows), beaters, and sticks (Stockman, 1986). These paintings occur on the walls of a room known as 'the Hunting Shrine' and the many figures depicted are almost always in poses of dynamic movement, surrounding a large representation of an animal; in one instance this is a bison, another shows a stag and a boar. As the paintings feature no perspective, it is unclear whether the animals are supposed to be standing or lying down on their sides. This allows for a couple of interpretations of the pictures: the people could be running, participating in an active hunt, or they may be dancing around a kill.

The latter interpretation, although initially perhaps seeming less intuitive, could be more likely for a couple of reasons. Firstly, as mentioned, several of the figures are carrying *two* bows or *two* sticks or beaters, which suggests that they were not being actively employed in hunting at the time represented in the image. Secondly, the animals depicted are not in dynamic poses of movement; in contrast to the humans, the positions of their legs are those of an animal that is still. This interpretation may be reinforced by the fact that both the bison and the stag seem to have their tongues hanging out, an image evocative of a dead animal.

If these interpretations are correct, of the objects as instruments and the activity as dancing, then this provides further evidence for the prehistoric use of music in association with celebratory activities, in particular related to hunting, as well as a prehistoric musical tradition dominated not by melodic instruments but by percussive ones.

The relationship of the proliferation of dance imagery with the spread of early agriculture is explored in detail by Garfinkel (2003). He identifies and discusses over 400 representations of dance activity from parietal art and pottery contexts associated with early agriculturalists (10,000–6,000 years ago) from Mesopotamia, the Levant, Iran, Egypt, Anatolia, the Balkans, Greece, and the Danube basin. In fact, according to his analysis, 'dancing is the oldest and one of the most persistent themes in Near Eastern prehistoric art, and this theme spreads with agriculture into surrounding regions of Europe and Africa' (Garfinkel, 2003, p. 3). Garfinkel concludes that the scenes show dance as an egalitarian communal activity associated with ceremony.

The representation of these activities—all of which appear to be representations of actual, rather than supernatural, events—occurs at precisely a time when social stratification was increasing, yet egalitarianism and communality appear to be emphasized in the dance activities, rather than hierarchical relationships. To Garfinkel this betrays the main purpose of the activity: he suggests that the music and dance activities illustrated betray a change in the importance and role of these activities attendant upon the changing social and cosmological structure of society associated with settlement and seasonal agricultural activities. He suggests that the egalitarianism and communality illustrated in the dance scenes acted as a foil to the increasing social stratification that was occurring in other aspects of life and economy, and that 'The importance of these ceremonies is also borne out by ethnographic observations of pre-state communities, in which dance is indeed the most important component in religious ceremonies. Dancing together creates unity, provides education, and transmits cultural messages from one generation to the next' (Garfinkel, 2003, p. 100).

Garfinkel, however, sees a disjuncture between the practices of hunter-gatherers and the early agriculturalists, attributing to the latter the development of new and more 'sophisticated' types of performance. He goes on to state that 'Hunter-and-gatherer societies are characterized by ad hoc rituals, mainly rites of passage, dictating a set of actions. Calendrical ritual, which became dominant in agricultural societies, has a different set of priorities and thus is represented by different types of performances. In urban societies, with a complex hierarchy and sharp specialization, new types of performances developed, such as theatre and sophisticated music' (p. 102).

Whether such a disjuncture between hunter-gatherer and early agricultural practice is justified requires further consideration. First, it is a common misconception to imagine that hunter-gatherer subsistence is not dictated by seasonal considerations. Much of hunter-gatherer subsistence is reliant upon critical annual changes, movements of animals and flowering and fruiting of plants, for example; i.e. seasonality is extremely important to hunter-gatherers, as it is to agriculturalists, and this is reflected in their rituals. In addition to personal-social rituals such as, amongst others, rites of passage, there are other essential rituals related to subsistence, seasonality, cooperation, and knowledge about the world, which are far from being ad hoc (Chapter 2 and references therein; Williamson and Farrer, 1992, and contributions therein). Second, many hunter-gatherer societies also feature hierarchies and craft or task specializations—these are not unique to urban agriculturalists. Third, performance of narratives ('theatre') and social mediation are common in hunter-gatherers and are not unique to agriculturalists (see Chapter 2). Fourth, there is plentiful evidence from the study of musical activity in hunter-gatherer societies that it is sophisticated in both form and intended function, and there thus seems little reason to reserve the term 'sophisticated music' for urban agriculturalists.

For the reasons Garfinkel describes, music and dance activities indeed have a great potential to facilitate communality, unity, and the transmission of knowledge; however, this is by no means unique to agricultural settled contexts—these are precisely the roles that it fulfils in many traditional hunter-gatherer societies (see Chapter 2; Williamson and Farrer, 1992, and papers therein; also Morley, 2007). Through his analysis, Garfinkel (2003) has successfully identified music and dance activities with characteristics and roles that are well known from a great diversity of traditional societies; in fulfilling these functions in early agricultural contexts music and dance activities were perpetuating social roles with a long history. This is a notable finding, and what is particular about these early agricultural populations is that they chose to illustrate the activities, where their predecessors had not (or had even avoided it). Equally interesting is that they apparently perpetuated the communality of the activities in the face of changing social structures, using them to help offset social tensions emerging with stratified social organization. What these representations may be doing for us, then, is providing us with the earliest pictorial record of activities which had been going on through thousands of years, thanks to a change not in performance practice, but in representational practice.

ARCHAEOLOGY CONCLUSIONS

The archaeological record of Upper Palaeolithic Europe has provided us with numerous examples of early sound-producers, some of them dating to the earliest occupation of the continent by modern humans. Research in this area has hitherto been rather disparate, and many of the finds have been open to debate as regards their anthropic (human) origin or use as sound-producers. The oldest-known confidently-identified musical instruments (bone and ivory flutes from the Swabian Alb, Germany) (Hahn and Münzel, 1995; Richter et al., 2000; Conard et al., 2004; Conard et al., 2009; Higham et al. 2012) actually pre-date any of the currently known rock art from the European Upper Palaeolithic, although the musical record is not discussed anything like as often as the record of rock art. It appears that, alongside producing figurative representation and other characteristic behaviours, modern humans were making and using sound-producing pipes within (in relative terms) a very short time of being in Europe, if not upon arrival.

After 40,000 or so years ago we are presented with a diversity of sound-producers, with bone flutes, possible rasps and percussion, and the acoustic use of rocks and caves, in the case of flutes from Aurignacian, Gravettian, and Magdalenian contexts, in the latter cases particularly during the eras associated with the Solutrean and Magdalenian techno-complexes. The majority of past authors on the subject have taken this apparently sudden advent of the use of sound-producers to indicate the beginning of the use of music. While the interpretation of some earlier bone objects as instruments seems unlikely to be correct, it is worth treating the majority conclusion with caution too, for a number of reasons.

First, the use of instrumentation and the use of music are often considered synonymous in the literature, an assumption which is unjustified, as the ethnographic evidence considered in Chapter 2 illustrated; there are many types of music which do not rely on instrumentation, and many instruments which would leave no archaeological trace.

Second, the first evidence that we see of instruments in Europe (the Swabian Alb and earliest Isturitz pipes) is highly sophisticated in design. As d'Errico et al. (2003) put it,

The sophistication of the pipes' various design elements (technological and ergonomic), both at Isturitz and Geissenklösterle, suggests that such instruments must, even at around 35,000 years, be several conceptual stages removed from the earliest origins, even of instrumental musical expression, to say nothing of those universal vocal, manual-percussive and dance forms which must have existed independently of—and before—any need for such tools. (d'Errico, Henshilwood, Lawson, et al., 2003, p. 46.)

The redating of the Geissenklösterle finds to in excess of 40,000 years ago further reinforces this position.

Third, it is possible that the determining factor in the increase in the production of bone flutes was not that flutes were not used before that point, but that the necessary raw materials for manufacturing *bone* flutes were not available—either that the lithic technologies available were unsuitable for working bone finely, or that suitable bone resources were not available. Wood and cane may have been used long before, both being easier to work than bone. Alternatively, the advent of bone flute manufacture may be as a consequence of the increase in the use of avian fauna for subsistence during the Aurignacian, providing the raw material necessary for them.

Fourth, all the evidence so far has been accrued from Europe. There is increasing evidence (e.g. McBrearty and Brooks 2000; Mellars, 2000; Henshilwood et al., 2001; d'Errico et al., 2001; d'Errico et al., 2003) that many of the cultural behaviours, such as bone-working, personal adornment, and abstract representation, hitherto associated with the 'Upper Palaeolithic revolution', were being evidenced in Africa during the Middle Stone Age by anatomically modern humans. Vast areas of Africa still require archaeological exploration, and continued excavation of African evidence may well show that musical instrumentation was also used there prior to the European Upper Palaeolithic.

Fifth, musical behaviours have tended to be equated with a 'package' of symbolic behaviours, and it has been thought that music could not have been performed prior to this emerging. The view that music need be symbolic is rather Western-centric, however; the issue of the requirements for symbolism (or lack of it) for musical behaviours is touched upon in the ethnographic evidence considered in Chapter 2, as well as being discussed in Chapter 11; for musical activities to be efficacious does not rely on the presence of a symbolizing capacity (although when such a capacity is in place it can clearly contribute a great deal to the variety of ways in which music is experienced).

Finally, as becomes clear in the following chapters, the evidence for the nature and evolution of the capacities underlying musical behaviours shows that these have an extremely long evolutionary heritage and are rooted in cognitive and physiological abilities that pre-date Upper Palaeolithic modern humans by many millennia. So the archaeological record for the earliest sound-producers seems unlikely to represent the nascence of musical behaviours amongst modern humans, but instead to constitute evidence for a particular form of a well-established behaviour with a long lineage.

So is there anything that we can say about the nature and possible roles of these behaviours in the Upper Palaeolithic with the evidence as it stands? Despite the fact that a large amount of evidence has undoubtedly been lost to the archaeological record, and that shortcomings in the recording of the excavation of many of the early finds interpreted as sound-producers limit our understanding of the contexts of their use, it has been possible to ascertain some clues as to the nature of music in the Upper Palaeolithic. It is difficult to be certain of the human origin of any of the artefacts dating to earlier than around 45,000 years ago, but after that date there are several examples of sound-producing objects, mostly bone, in which we can be confident of human agency. The contexts in which these objects occur may provide some clues as to the nature and use of music produced.

The cave site of Isturitz, which has produced many of the Upper Palaeolithic bone flutes, was an important focal point for large groups of people throughout the Upper Palaeolithic (Bahn, 1983), and seems to have been a focus for major gatherings in spring and autumn in particular. Similarities in the stone and bone artefacts with those from the Dordogne area, and the transferral of flint and other resources over distances exceeding 100 km suggest wide social contact (Gamble, 1983; Scothern, 1992; Normand et al., 2007). Mas d'Azil shows similar evidence of large aggregations of people, and Magdalenian finds from Le Placard and Marcamps show parallels with Isturitz in the bone flute design and engravings (Scothern, 1992). The flutes are commonly found inside the decorated caves of these sites, suggesting either that acoustics were particularly important, or their relation to the cave art, or both.

It is difficult to demonstrate whether the production of music in the caves was a group activity indulged by all or an activity performed only by a select few, an issue which also applies to the production of cave art. In her analysis of the Solutrean lithophones (described earlier) Dams (1984, 1985) notes that the lithophones occur in caves which generally also allow a number of people to congregate in nearby chambers, up to twenty to thirty people at Roucador, for example (Dams, 1985). It is worth noting, however, that the locations of the lithophones are dictated by natural cave features rather than human agency so their position in relation to other large chambers is not premeditated. What would be particularly valuable as a comparison would be to know of caves with equivalent calcified structures in smaller chambers that *haven't* been decorated and used percussively. If such examples exist, it might suggest that the ones in the caves described by Dams were chosen selectively because they could accommodate many people.

The similarities between the features and decoration of the lithophones Dams describes suggests wide social contact between the people responsible, so the possibility exists that large groups of people congregated to experience (and perhaps participate in) the sound produced by them. Whether the activities involved many or few people, it is clear from the work of Glory (1964, 1965), Dams (1984, 1985), and Reznikoff and Dauvois (1988) that the acoustics of the caves were highly significant, and that sound production bore an important relationship with both abstract and representational art.

As with rock art and with other more general aspects of daily life we need to be careful when considering sound-producing activities that our understanding is not distorted by the fact that the majority of our evidence comes from cave environments. Whilst cave environments have acoustic properties that could clearly play an important role in the nature and effects of sound production, it would be wrong to assume, firstly, that these environments formed the only locations for these activities, and secondly, that other environments did not also have significant acoustic properties (see, for example, Scarre and Lawson, 2006). Musical production may also have formed an important solitary activity in the wider environment, for example, but artefacts betraying this would be rarely if ever found.

A rare example of possible Palaeolithic sound-producers being preserved in an open-air context is provided by the Mezin bones, reported by Bibikov (1978). These, as it happens, may also provide more clues towards communal use of music than solitary activity. The 20,000-year-old mammoth-bone 'orchestra' was found in a large, open-fronted, communal hut in the Mezin settlement, rather than one of the smaller dwellings. This does suggest, if it is indeed a collection of instruments, that their use was supposed to be experienced by the community; it would have been impossible to use them privately in these circumstances. Again, it is difficult to draw any conclusions regarding the use of the music, other than as a social activity. It may have been significant to the mammoth-hunters that the reputed instruments were made of mammoth bone; on the other hand, it may simply be that this was their most ready raw material for such artefacts. There is evidence for the trafficking of raw materials and visually related goods between European populations and those of the Russian plains of the Magdalenian (Scothern, 1992), so it is quite possible that sound-production traditions were also shared. Certainly it is clear from the preceding and following chapters that activities incorporating musicality draw upon aspects of cognition that are fundamentally related to social interaction, and that such activities can be particularly efficacious in social contexts.

There are two principal dangers of which we must beware in interpreting objects as sound-producers, however. One is of falsely accepting artefacts as sound-producers when they are not, and the other is of falsely failing to identify objects as sound-producers when they are. Both of these errors have almost certainly been made in the past.

We need to be very careful that objects are not identified as instruments solely on the basis of their appearance; they must be subject to detailed analysis which tests whether the features that can lead to the identification of the object as an instrument have genuinely been created and used in the way that is supposed. The analyses to which the bear-bone femur from Divje babe I have been subjected (by Turk and colleagues, and d'Errico and colleagues, for example) provide an excellent example of the level of detailed examination that can be required. The excavation strategies and recording of Nicholas Conard and Maria Malina at Geissenklösterle and Hohle Fels represent exemplary cases of the level of detail of information that can be retrieved from Palaeolithic sites, and will ultimately prove invaluable in understanding the circumstances under which the instruments discovered there were deposited. The contrast between the level of contextual evidence pertaining to the artefacts from the Swabian Alb sites and Isturitz will have been obvious, and we can hope that future discoveries will also furnish us with an increasing level of information about their deposition—as well as the discovery of artefacts that would otherwise never have been identified. But the more recent analyses of the Isturitz examples (e.g. by Buisson, 1990, and Lawson and d'Errico, 2002) also provide great encouragement with regard to the amount of information that might still be recoverable from artefacts excavated already, the true use of some of which might not yet have been identified.

There is much further work that can be carried out. With regard to rasps, bullroarers, and bone and stone percussion, use-wear analysis of replica examples will allow comparison with artefacts reputed to be instruments in order to determine whether they were used in the ways suggested. Many existing artefacts could be reassessed in this way, and newly discovered artefacts will be able to be assessed in light of such avenues of interpretation. In the excavation of sites, the potential existence of artefacts such as these must be borne in mind lest they be overlooked.

The evidence we have has the *potential* to allow us to draw conclusions at two very different levels. One pertains to capabilities of humans at this time at a general level; the other pertains to the very specific interpretation of individual artefacts in individual circumstances at particular sites at particular times. Bridging the gap between these two levels of interpretation may prove extremely difficult. The fact that some populations in Europe in some locations were clearly carrying out instrumental sound production does not mean that all were carrying out the same behaviours—whilst the capabilities for musical activities clearly have a far older heritage and were present in all these populations, we must be careful not to make generalizations about the precise nature of behaviours across a span of time in excess of 35,000 years. The same applies, of course, to rock art and other activities. On the other hand, the capabilities, technologies, and motivations to do so were evidently in place at the time of the arrival of modern humans in Europe. Or more particularly, the capabilities, technologies, and motivations to use materials which happen to preserve archaeologically were in place at that time.

It is a valuable truism in archaeology that absence of evidence does not represent evidence of absence, but neither can absence of evidence be reliably asserted to represent evidence of loss. At least, this is certainly true in the total absence of evidence. If we had no evidence whatsoever of musical activity from the Upper Palaeolithic then, whilst perhaps being surprising, we would be unable to make any assertions about its importance or otherwise in Upper Palaeolithic societies. However, having *some* evidence, as we do, and not just from one region or period but from many regions and all periods, we can be pretty confident that what we have represents a tiny fraction of what was actually produced during this long period. If we are to ensure that we look in the right places for further relevant evidence, we need to maintain a very open conception of the types of contexts in which musical sound production might have occurred and what artefacts might have been used in that process. What we do have clearly represents an important activity amongst these populations, and one which there is no doubt that we would recognize as music.

5

The Palaeoanthropology of Vocalization 1: Vocal Anatomy

> Music is our oldest form of communication, older than language or art; it begins with the voice, and with our overwhelming need to reach out to each other.
>
> Yehudi Menuhin (in Menuhin and Davis, 1979)

INTRODUCTION

This chapter explores the vocal apparatus and fossil evidence for its evolution. At their most basic level musical stimuli consist of a combination of tonal and temporal information. As explored in Chapter 1, and in more depth later, of course musical experience is derived from far more than just auditory information, and not every temporally organized sound will necessarily fulfil a conception of music. However, it would be true to say that in order to experience auditory stimuli as musical we certainly require the ability to extract and process tonal and temporal information from them. We require this ability when we are producing such stimuli ourselves too, of course, which relies also on the capacity to plan and control the muscular movements required to produce varying pitch tones and sounds with a temporal structure. This, in turn, relies on the existence of musculature that is not only controllable to that level of precision, but capable in the first place of producing the sounds required.

Western music, for the past few hundred years at least, has tended to be dominated by instrumental sound production and, notwithstanding choral music, we have a tendency to think of music as being an 'instrumental' activity. In fact, this is not representative of the diversity of musical forms that exist in other societies (some of which are discussed in Chapter 2), and we have no reason to believe that instruments predominated in musical sound production in the past, either. Rhythmic sound production, which always involves precisely controlled rhythmic muscular movements, can be created using instruments or the human body. The principal tonal sound-producing apparatus possessed by all primates is the vocal tract, and in humans, over the course of our evolution, this has become an instrument *par excellence*, with the potential to produce a great diversity of sounds, and to communicate information in a variety of ways. Indeed, it is this *biological* instrument, possessed by all of us, which constitutes the principal tonal sound-producer in many musical traditions. We often think of our auditory abilities as being rather modest in comparison to some other animals, but these too, in humans, have evolved alongside our vocal abilities to be specialized at what they do.

Artefactual instruments do indeed feature in the prehistoric archaeological record, and form the subject matter of Chapters 3 and 4, but instruments constitute an accessory to existing human capacities; the origins of musical behaviour would not have relied upon the invention of instruments. The study of the origins of the capacities for musical behaviours must therefore examine the evolution of the biological features that are used in such activities. This chapter and the next examine the evidence for the evolution and development of the various physiological features responsible for the production and perception of tonal vocalizations (rhythmic muscular control is dealt with in Chapter 9).

This chapter examines the fossil evidence for the evolution of a variety of features of the vocal tract, with a view to mapping the evolution of the capacity to produce extended tonally varied vocalizations. This is followed in Chapter 6 by an examination of the fossil evidence for the evolution of neurological structures used in primate (including human) vocalizations, and for the evolution of the inner ear and sound-perception capabilities. Ultimately, the developments of the physiology and neurological control required to carry out behaviours that we would recognize as musical need to be understood in light of each other and, after further investigations, should feed into an understanding of the path by which such features evolved, and how these capacities might interrelate.

THE VOCAL APPARATUS AND FOSSIL EVIDENCE FOR ITS EVOLUTION

The evolution of the physiology necessary for vocalizations has received much attention in the last thirty years or so, in particular since interest in language evolution has blossomed (e.g. Pinker, 1994; Deacon, 1997; Jablonski and Aiello, 1998). A consequence of language evolution being the driving interest behind most of this research has been that the results have generally been presented in terms of the development of linguistic speech capabilities in hominins. Dealing as it does with the fundamental physiological features required for *controlled* and *varied* vocalizations, however, the research is relevant to hominin capabilities to produce *any* vocalizations, linguistic, melodic, or otherwise. Note that 'vocalizations' are *not* equivalent to 'verbalizations', despite the fact that these concepts have been used interchangeably in some literature; the ability to vocalize has often been spuriously equated with linguistic speech ability, with little consideration of other roles of vocalization although, more recently, some attention has been given to these features with specific reference to music as well (e.g. Frayer and Nicolay, 2000).

As is the case with bipedal locomotion, vocal capabilities are a product of both physiological anatomy and neurological control of that anatomy. In reconstructing the evolution of bipedalism in fossil hominins, we seek to understand how the locomotor physiology and control that we exhibit today was derived from an ancestral form in our last common ancestor with chimpanzees; the locomotor form of each fossil hominin on the lineage between that ancestral form and us will to a greater or lesser extent resemble either that of the ancestral form or that of modern humans, and has the potential to illustrate how the changes from one form to the other occurred. The evolution of vocal physiology and control represents an equivalent situation, and the same rationale applies.

All of our hominin ancestors were, of course, higher primates, like the apes and humans of today. As a consequence our most appropriate comparators with regard to vocal tract anatomy and capability are those higher primates which survive today—great apes and humans. The vocal anatomy and capabilities of past hominins will have had more in common with those of the great apes and those of humans than any other animals, resembling one or the other to a greater or lesser extent, depending upon the hominin in question—with the caveat that the higher primates of today may also have undergone evolutionary changes of their own over the time since we shared an ancestor with them.

Vocalizations obviously do not fossilize, so what we have to look for is evidence in the surviving bone material for the changes that must have taken place from the vocal tract and capabilities present in our common ancestor with *Pan* (common chimpanzees and bonobo chimpanzees), around 6 million years ago (Stringer and Andrews, 2005), to those present in modern humans. We can derive some idea of the ancestral form by examining the vocal anatomy and capabilities of modern African apes: the characteristics that are shared today by bonobos (*Pan paniscus*), chimpanzees (*Pan troglodytes*), and gorillas (*Gorilla gorilla*) are likely to have been shared by their (and our) common ancestor around 8–10 million years ago, and thus also by our common ancestor with *Pan* around 6 million years ago. We can thus be confident that the vocal anatomy and capabilities of our ancestral form shared the features that African ape vocal anatomy and capabilities share with each other today.[1] This approach addresses the caveat mentioned above.

Further, to continue this analogy, in the study of the evolution of bipedal locomotion in hominins, cases of other animals that have developed bipedal locomotion (for example, birds and kangaroos) are of interest in illustrating ways in which bipedalism can manifest itself, and possible reasons for it doing so; however, they are not treated as homologues for bipedalism in humans, though they may represent analogues. In these cases, the capability, its mechanisms and control have emerged over very different timescales, from very different starting points, under very different ecological and social circumstances—and, whilst superficially similar, even the end point is significantly different. They are thus limited in their applicability to understanding the evolution of bipedalism in another zoological family. Again, the evolution of vocal physiology and control is an equivalent situation and, whilst the vocal physiology and behaviour of other animals have the potential to add to our understanding of the potential capabilities of vocal physiology, they generally do not provide the most appropriate model for understanding the capabilities, and transformations in capabilities, that occurred between our ancestral condition and humans today.

In talking about vocal 'capabilities', there are two principal changes to consider: one is the development of vocal physiology and its *potential* for the production of sounds (the focus of the present chapter), and the other is the development of the neurological control of that physiology; the development of each ultimately has to be understood in light of the other, and they also have to be understood in the context of the selective environment(s) in which they took place. The first question to address is whether we can detect changes in the vocal tract anatomy of fossil hominins away from the ancestral form and towards that exhibited by humans today, and if so, what their implications are for vocal capabilities at the times of those changes.

Vocalizations are produced in two stages, in the supralaryngeal tract (the throat above the larynx) and the oral cavity (the mouth) (see Figure 5.1). The acoustic energy of vocal sound is generated when air exhaled from the lungs passes through the larynx, which modulates the airflow in the supralaryngeal tract. This sound is then adjusted by the action of the tongue, lips, teeth, and palate in the oral cavity (Lieberman, 1992). So the principal evidence with the potential to inform us about vocal capabilities relates to control of exhalation

---

[1] This is more likely than the alternative explanation, which is that the features of modern African ape vocal physiology that are shared by the different apes have each emerged independently from a different precursor form by convergent evolution.

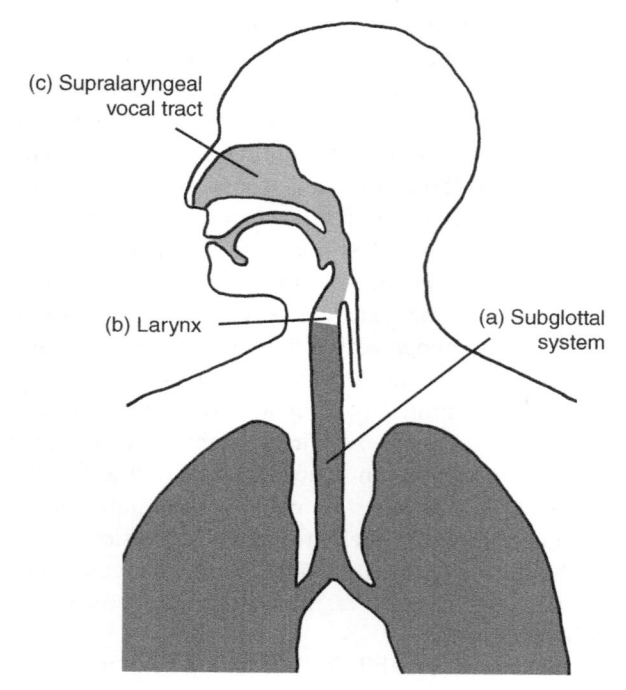

Fig. 5.1. Key components of human vocalization anatomy:
(a) Subglottal system (lungs and trachea (windpipe))
(b) Larynx
(c) Supralaryngeal vocal tract (pharynx, oral cavity, nasal cavity)
(Redrawn from Lieberman, 1992, p. 134.)

from the lungs, the form and control of the larynx, the position of the larynx (and thus the form of the supralaryngeal space), and the physiology relating to the control of the tongue and other parts of the oral cavity.

Unfortunately, the vast majority of the physiology responsible for vocal sound production (for example, the tongue, larynx, and supralaryngeal tract) consists of cartilaginous and soft tissue that is not preserved in the fossil record. Consequently, attempts to reconstruct the evolution of vocal anatomy and ability in our ancestors have relied upon drawing conclusions about the nature of these features from the bony structures supporting this soft-tissue anatomy, which do survive. These include the jaw (mandible), bottom of the skull (basicranium), in some rare instances the hyoid bone, and the neural canals in the bones through which some of the nerves responsible for vocal control passed. The extent to which we can be confident about such reconstructions has been questioned by some, however, on the basis either of the evidence used in the reconstructions themselves, or of conclusions drawn from comparisons with the anatomy of other animals.

The larynx and basicranial flexion

Vocal ability is determined by the physiological form of the vocal anatomy (such as the size and position of the larynx and tongue, and the form of vocal tract and the oral cavity), and by control (precision and integration) of these anatomical features. We know already that non-human primates and other mammals are capable of creating vocalizations, and our understanding of mammalian vocalization capabilities is improving all of the time, as Fitch (2009a) elaborates. Humans, like other animals, use vocal anatomy to produce formant frequencies, the tonal sounds which constitute a critical element of vocal communication. Other mammals can also use their anatomy to produce formants, but our ability to make rapid and precisely controlled changes between them seems to be unique to our species today (Fitch, 2009a). The range of frequencies that we produce, and move between, also seems to be unusual.

What is it that allows us to produce this wide range of frequencies, not just low and high, but all of the pitches in between? One of the influences is the morphology of the pharyngeal cavity, the area between the epiglottis and soft palate. In most animals the resting position of the larynx is near the top of the throat (Duchin, 1990; Budil, 1994). In this position the larynx allows the animal to breathe and swallow simultaneously without choking, as it separates the upper respiratory and digestive tracts (Laitman, 1984; Budil, 1994); it also restricts the resonance (and thus variety of vowel sounds) produced by the pharyngeal cavity (Lieberman, 1984; Budil, 1994). It was traditionally thought that the larynx was fixed in this position during vocalization, but as Fitch (2000a, 2009a) explains, more recent research observing living animals (dogs and red deer, in particular) has shown that various mammals in fact have the ability to lower their larynx temporarily during vocalization. This lowering enlarges the supralaryngeal space in which resonance occurs, thus allowing additional, lower frequency sounds to be produced.

At birth, the larynx occupies the top of the throat in humans, as in other mammals, with the same important benefit with regard to feeding and breathing. However, as the child grows up, particularly after the first year of life, the larynx descends towards its adult position (Figure 5.2). This process is only completed during adolescence (Magriples and Laitman, 1987). In adult humans, therefore, the resting position of the larynx is further down the throat, meaning that it does not need to be temporarily lowered in order to produce additional frequencies, and the pharyngeal resonance cavity is permanently larger in size. This allows the production of the full wide range of vowel sounds (such as [a], [u], and [i], the 'vowel triangle') that are universals in all human languages (Lieberman, 1984) at any time during the process of vocalization. Other animals are now known to have permanently lowered larynges (Fitch, 2009a); however, humans remain unique amongst higher

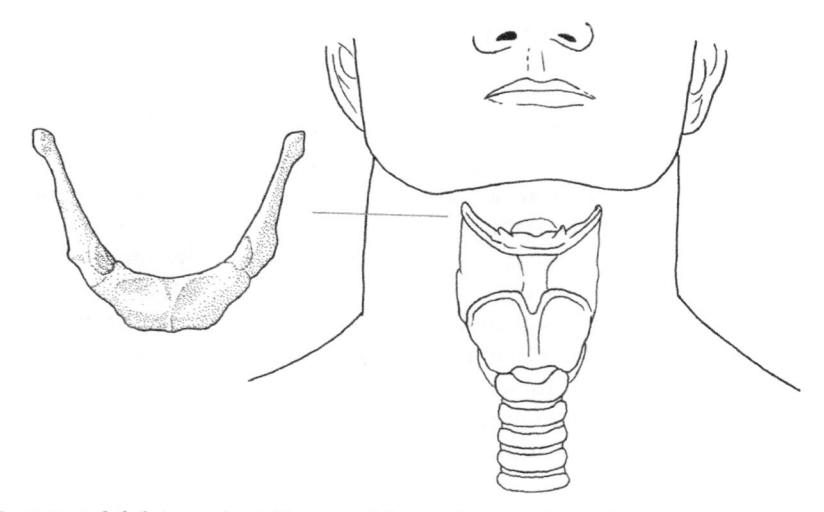

Fig. 5.2. Adult human hyoid bone and larynx location, frontal view.

primates in having a permanently lowered larynx, and this appears to have an impact upon the vocal potential of that anatomy. This capacity to produce such a wide variety of sustained vowel sounds, and the control to be able to shift between them at will, do seem to be unique to humans, and are also fundamental to the ability to produce vocal melody.

Analysis of chimpanzee vocal anatomy (Laitman and Heimbuch, 1982; Duchin, 1990) and other non-human mammals including baboons (Laitman and Reidenberg, 1988) shows that non-human primates have a larynx which, in its resting position (cf. Fitch, 2009a), is high in the throat; in contrast, of primates only humans (Duchin, 1990, p. 688) have a larynx that is permanently located low in the throat (see Figure 5.3).

An examination of position of the larynx in the fossil record of human ancestors should shed light on the extent to which their vocal anatomy at different stages resembles that of chimpanzees or that of modern humans, and may allow us to draw conclusions regarding how and when the anatomy that we see in modern humans developed—and the extent of their vocal abilities at different stages, in comparison with modern humans and non-human primates. Whilst in some mammals the position of the larynx in death is not necessarily a good indicator of its position during vocalization (Fitch, 2009a, discussed further later), we do know that in humans and other higher primates the resting position of the larynx is indeed indicative of the range of vocal sounds producible.

Laitman has asserted that there is a direct relationship between laryngeal position and the degree of curvature of the underside of the base of the skull, or *basicranial flexion* (Laitman et al., 1979; Laitman and Reidenberg, 1988) (Figure 5.4).

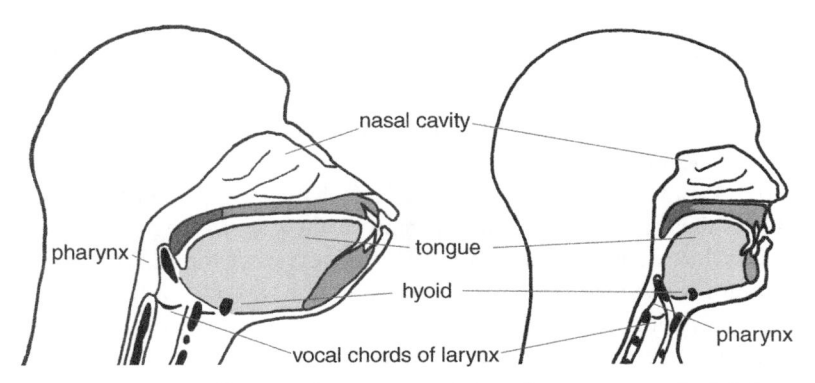

Fig. 5.3. Comparison of key vocalization anatomy in chimpanzees and adult humans, showing positions of larynx, hyoid bone and tongue. (Redrawn from Lieberman, 1992, p. 135.)

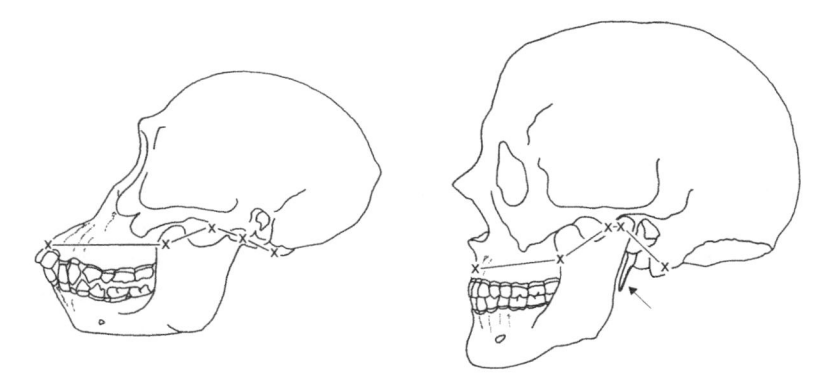

Fig. 5.4. Basicranial flexion in the skulls of chimpanzee (left) and human (right), showing the change in degree of flexion in these two species. Crosses mark the relative locations of the five principal points on the centre-line of the base of the skull for which coordinates are taken, linked by the basicranial line. The arrow indicates the styloid process on the human skull. (Chimpanzee skull and basicranial lines redrawn after Laitman & Heimbuch 1982, Fig. 2, p. 325 and Fig. 4, p. 330.)

This curvature is observable in varying degrees in extant pongids (*Pan*, bonobos, and chimpanzees) and humans, and has been argued to have a direct influence on the position of the tongue and laryngeal muscular attachments to the base of the skull. The basicranial flexion is taken as an indication of the inclination of the styloid process, a bony protrusion from the temporal bone at the base of the skull, which is rarely preserved (Budil, 1994). This, in turn, has been used to reconstruct the position of the hyoid bone and the larynx, as the muscles which support these structures attach to the styloid process (Budil,

1994). Thus study of the degree of basicranial flexion of hominin fossils has been used to extrapolate the form of the vocal tract in those specimens.

Laitman and Heimbuch (1982) carried out examinations of Plio-Pleistocene specimens such as Sts 5 (*Australopithecus africanus*, from Sterkfontein, South Africa) and KNM-ER 406 (*Paranthropus boisei*, from Koobi Fora, Kenya), 2.5 and 1.5 million years old respectively (Johanson and Edgar, 2006), and found that these australopithecine specimens show no flexing of the basicranium, as is the case with modern apes. The first evidence for any arching of the basicranium is displayed by *Homo ergaster* skull KNM-ER 3733 (from Koobi Fora, Kenya), approximately 1.75 million years ago (Laitman and Heimbuch, 1982). A fully flexed, modern-like, basicranium is first seen in species that are the descendants of *Homo ergaster* and *erectus*, in specimens such as Petralona (*Homo heidelbergensis*, Greece, 300,000–400,000 years ago), Broken Hill 1 (*Homo heidelbergensis*, Kabwe, Zambia, *c*.300,000 years ago), and Steinheim (*Homo heidelbergensis*, Germany, *c*.250,000 years ago) (Laitman, 1984; Johanson and Edgar, 2006). Laitman suggests that *Homo ergaster* would have been able to produce the vowel sounds found in 'boot', 'father', and 'feet', whereas *Homo heidelbergensis* and other contemporaneous descendants of *Homo erectus* would have been able to produce a full modern repertoire of vowel sounds (Laitman, 1984).

Other researchers have argued that Neanderthals would not have been able to produce this full repertoire. Lieberman and Crelin (1971) created computer simulations of the vocal tract based on the anatomy of the La Chapelle-aux-Saints Neanderthal specimen. This led them to attribute only limited vocal ability to Neanderthals, not comparable to that of modern humans. Although groundbreaking at the time, this original study has since been severely criticized on a number of fronts. First and foremost, the La Chapelle-aux-Saints cranium is that of an aged and crippled Neanderthal, who was also suffering from the bone-altering disease osteoarthritis, and so is not a good approximation for the species as a whole. Second, Lieberman and Crelin's simulation was based partly on the original reconstruction of the skull by Marcellin Boule from 1911–12, which has since been substantially revised by other researchers, including Heim (1985), who showed that the skull in fact had a basicranium very similar to that of modern humans (Heim, 1985). This illustrates the extent to which variations in methods and quality of reconstruction of crania can make such research difficult, and that the choice of sample for analysis must be made with care. Despite the criticisms levelled at the original reconstruction, however, it does seem that Neanderthal basicrania are, on average, longer and flatter than those of modern humans and of both European (Steinheim) and African (Broken Hill) examples of *Homo heidelbergensis* (Laitman et al., 1979), and this is often still used as a basis for suggesting incomplete speech production amongst Neanderthals.

Lieberman has since modified the severity of his original conclusions in the light of the criticisms of the 1971 simulations (Lieberman, 1989), shifting his emphasis towards the connections between physiology and neurology. He still maintains that 'classic Neanderthal hominids appear to be deficient with respect to their linguistic and cognitive ability... they probably communicated slowly at extremely low rates and were unable to comprehend complex sentences' (Lieberman, 1989, p. 391). He makes this assertion on the basis that there are important neural connections between vocal physiology and the cognitive substrates responsible for language production and perception. He hypothesizes that since the physiology appears to be underdeveloped in Neanderthals, so too was the neurology.

There are still problems with these assertions, however. Whilst a physiological system is of no use without the neurology to control it and process information from it, in contrast a neurological system can exist and later come to be used by a different combination of physiological systems than that with which it was originally involved. Once such a link is established, of course, the relationship between the physiological and neurological development must be symbiotic. But this means that the level of development of a physiological system is an indicator of the *minimum* level of neurological capability, not the *maximum* as Lieberman (1989) suggests. Kuhl (1988) concludes from her studies of the perceptual mechanisms and abilities of non-human mammals and prelinguistic infants (see also Chapters 7 and 8) that the mechanisms relied upon to process speech sounds were in place long before humans began to use speech. She concludes instead, that speech was shaped by our perceptual abilities. In relation to Neanderthals Kuhl's (1988) studies show that the ability and perceptual mechanisms to perceive the complexities of speech sounds could have been in place, even if the speech sounds that they were able to produce were different from those that we produce ourselves.

Fidelholtz (1991) suggests that the brain capacity for language production was present in the 'generalized Neanderthals' (now usually classified with *Homo heidelbergensis*, although considered to be of a type ancestral to Neanderthals) such as Swanscombe and Steinheim, around 250,000 years ago, and that the subsequent developments in the vocal anatomy (including the loss of jaw prognathism) were responses to the use of the speech ability.

Furthermore, research regarding the vocal abilities of individuals with Down's syndrome, who can possess similar features of abnormally high laryngeal position and large tongue, as well as lower than average intellectual capacities (Lieberman originally used their linguistic ability as an analogy for Neanderthals'), suggests that such individuals can be capable of producing virtually all of the sounds present in modern speech as well as a high level of singing ability (Scothern, 1987, cited in Scothern, 1992). It would seem that these anatomical features are not in themselves such a barrier to vocal production as has sometimes been asserted.

Other supralaryngeal reconstructions have been attempted. Budil (1994) compared the basicrania of 109 modern adult humans (from a total of seventeen disparate ethnic groups) with those of 103 non-human primates (from eighteen genera), twenty-eight fossil hominins, and the measurements from a further sixteen fossil hominins as quoted in Laitman's publications. He used the evidence from all of the modern specimens (human and primate) to devise a formula relating basicranium, styloid process, hyoid, and laryngeal position. Based on this, he reconstructed the supralaryngeal tract of the Petralona specimen, a *Homo heidelbergensis* 300,000–400,000 years old. His results indicate that a modern-like supralaryngeal resonance cavity was present in the Petralona specimen, and others with similarly flexed basicrania, including other *Homo heidelbergensis* such as Steinheim (Germany) and Broken Hill 1 (Zambia), and *Homo sapiens* such as Taforalt 12, Cro-Magnon 1, Predmosti 3 and 4, Afalou 5, and Skhul 5. His conclusions thus agree with those of the earlier work by Laitman and Reidenberg (1988) and Arensberg et al. (1990, see later), that an upper respiratory tract similar to that of modern humans (and thus similar vocal sound-production ability) had appeared by *Homo heidelbergensis*, and further concludes that the laryngeal physiology of the australopithecines resembled that of modern apes. The results for Neanderthal specimens—in contrast to *Homo heidelbergensis*—again fall outside the range of any modern humans, despite post-dating the other specimens by many millennia, agreeing with Lieberman's analysis of the physiology. To Budil, this finding 'remains enigmatic'.

However, what does seem to be the case is that the late 'classic' Neanderthals, who possessed this anomalously low level of flexion, were not 'throwbacks' perpetuating an earlier basicranial form, but had developed this basicranial shape as a result of specific adaptations of their own. Earlier Neanderthals closer in time to *Homo heidelbergensis*, such as Saccopastore 1 and 2 (c.120,000 years ago), are closer in form to *Homo heidelbergensis* in that they retain a more flexed cranial base than the later, so-called 'classic' examples (Johanson and Edgar, 2006). Whilst it is clear that having modern-like vocal anatomy has the potential to allow modern-like capabilities, it is not clear to what extent the divergence from this norm in later Neanderthal physiology would actually limit their capability—especially given that their immediate predecessors appear to have had such a capability through their physiology and, presumably, control of it (see also Clegg, 2012).

There are several influences on cranial base anatomy. Arensberg et al. (1990) also cite evidence of craniofacial reconstructions of hominins (Solow, 1966) that basicranial flexion is negatively correlated with maxillary width (width of the upper jaw, the maxilla) and posterior height; i.e. as flexion increases, maxillary width and posterior height decrease. This has the consequence of overall nasopharyngeal volume apparently remaining consistent, irrespective of the degree of flexion. Thus, although the flexing indicates a change in the position of the larynx and other elements of the vocal apparatus, the nasopharyngeal soundspace may not be much affected by the process.

It would also seem that direct parallels that have been drawn between the flat basicranium of apes and that of Neanderthals are mistaken. In apes and humans, the hyoid and larynx rest in the same position *relative to the individual vertebrae of the neck*, but due to the forward posture (kyphosis of the vertebral column) of apes the hyoid and larynx are much nearer the cranial base. In hominins with a fully upright posture the hyoid and larynx, although in the same position relative to the vertebrae, rest further from the cranial base. Consequently, in fully upright hominins (including, of course, Neanderthals), a lack of basicranial flexion is of far less significance with regard to the supralaryngeal volume and range of vocalizations than it is in apes.

The main selective influence on the change in the flexing of the basicranium seems to have been that it was a way of accommodating increasingly large brain size along with a relatively short cranial base (Spoor and Zonneveld, 1998); in extant primates the degree of basicranial flexion is positively correlated with brain size relative to basicranial length (Ross and Ravosa, 1993; Spoor and Zonneveld, 1998). As usual with physiological evolution, there are numerous factors at play. Whilst brain size increased in absolute terms, the cranial base was shortened, both relatively and absolutely, by the reduction of kyphosis of the spine with bipedalism (the shift to an upright spine) and the consequential midsaggital location of the foramen magnum (the hole in the base of the skull where the spinal cord enters), as well as the reduction in facial prognathism (Aiello, 1996; Spoor and Zonneveld, 1998). All of these processes would have had effects upon the positions of the vocal tract and larynx relative to the skull, and more recent studies have begun to take these factors into account, with several writers (e.g. Arensberg et al., 1990; Aiello, 1996; Clegg, 2012) now suggesting that the basicranium is only one of many complementary features which could indicate vocalization quality and enunciation ability. Likewise, supralaryngeal space is only one of several physiological factors implicated in sound production that would be affected by laryngeal lowering. This does imply that a dismissal of Neanderthal vocal sophistication solely on the basis of basicranial morphology and supralaryngeal volume is unsustainable; basicranial flexion can be a good indicator of the extent to which vocal tract morphology resembles that of modern humans, but it is not *the* determining factor in vocal ability. Other factors are discussed further later.

It is worth noting at this stage that the potential individual cost of the lowering of the larynx in the throat (namely, that the individual is increasingly subject to choking on liquid) could be a great one; for a progressive lowering of the larynx to have occurred over the course of hominin evolution, the selective advantage to be gained from this process must have outweighed this potential cost. However, Clegg and Aiello (2000), having looked at death certificates from 1894 to 1994, do point out that the incidence of mortality from choking in recent history has been small, and that the cost of laryngeal lowering may have been somewhat exaggerated. Of course, it may be that the incidence of fatal choking is now low as a consequence of laryngeal lowering being offset by increased control of breathing (itself intimately related to increased vocal control, discussed later), allowing humans (and other hominins) to control against inhaling liquid where before the high larynx was required to fulfil this role. Additionally, it is possible that the potential cost of the lowered larynx has been offset by increasingly sophisticated social relations, which are able to act in aid of individuals who are choking—thus reducing deaths from choking, although not necessarily the incidence of choking. Modern human food preparation also contributes to ease of eating, and possibly accordingly reduced risk of choking. A second potential cost of increased basicranial flexion and decreased prognathism, which has only recently been mitigated, is that the shorter palate and jaw often results in impacted teeth (especially M3 'wisdom teeth'), which until the nineteenth century commonly led to fatal infections (Callaghan, 1994).

The body of research relating fossil basicranial anatomy to vocal sound-production range has more recently been criticized by Fitch (2009a), on the basis of observations of vocal anatomy in other animals. Fitch (2009a) observes that, because it is demonstrably the case that some animals are capable of adjusting the position of their larynx from its resting position during vocalization, the resting position of the larynx (i.e. as occurs when not vocalizing, when the animal is dead, and as is reconstructed from fossil specimens) is not representative of the vocal sound-producing potential of the animal in life. In particular, the bony protrusions that indicate laryngeal muscular angles, and thus laryngeal position, do not necessarily indicate laryngeal position during vocalization, only at rest. He thus argues that any attempt to reconstruct vocal sound-producing capability from fossil specimens is doomed to failure (citing, as examples, the studies by Lieberman and Crelin, 1971; Crelin, 1987; and Duchin, 1990). This is certainly a salient word of caution about the conclusions that we draw from dead (and long-dead) specimens; we perhaps need not be as pessimistic as Fitch, however. We *do* know that there is a correlation between laryngeal position in the throat and the frequencies that it has the potential to produce at the time of vocalizing (and observations of temporary laryngeal lowering in animals reinforce this). The larger the supralaryngeal space, the lower the formant frequencies that can be produced. What this means is not that the laryngeal muscular angles are no indicator of vocal tract sound-producing potential, but that they do not indicate its limitations; i.e. a reconstruction of resting laryngeal position on the basis of the angles of laryngeal muscular attachments is indicative of the *minimum* vocal range capability of the hominin in question. Fitch's cautionary words are particularly applicable to cases where the laryngeal muscular attachment angles indicate a resting laryngeal position that is high in the throat, as this does not rule out the potential to temporarily lower it during vocalization, to add some additional frequencies to the range of the resting larynx (Fitch's (2000b) proposals about the possible role of this sort of mechanism in human evolution are discussed in Chapter 8). The question under these circumstances is of which of the living mammals observable today constitutes the best model for reconstructing the sound-producing potential of a fossil hominin possessing a larynx that is located high in the throat in its resting position. The answer must surely be those mammals with which we share our most recent evolutionary heritage, namely, the African great apes, and in particular, chimpanzees.

So we cannot confidently attribute specific *limits* to the vocal sound-producing capabilities to those specimens who *do not* possess a human-like arrangement. It is certainly essential, however, that we fully understand the functioning of great ape vocal tract anatomy in life, because this is likely to represent the minimal capabilities of those hominins least divergent from our ancestral state—that of the last common ancestor with chimpanzees. We need also, as Fitch (2009a) elaborates, to be aware of the vocal possibilities that we cannot rule out, on the basis of the capabilities of other mammals; though we should assess the relative likelihood of those different potentials with reference to the capabilities that we do know exist in the higher primate lineage, namely, those possessed by higher primates (including humans) today, as well as in light of our understanding of hominin environment, ecology, and social organization.

But what are the minimum vocal capabilities of a hominin with a laryngeal anatomy that is low in the throat, like that of modern humans? Fitch (2009a) observes that the big cats of the *Panthera* genus (lions, tigers, jaguars, and leopards) have a laryngeal and tongue root position around halfway down the throat (Weissengruber et al., 2002), which is similar to that of humans, yet these animals do not have the ability to produce the range of formant frequencies that we can. This leads to the important conclusion that laryngeal position alone is not enough to indicate the range of formants that could be produced by a particular vocal anatomy. Of course, *Panthera* do not share other features of vocal anatomy with humans, such as the form of the tongue muscle itself, palate shape, dentition, lips, and other orofacial musculature. Furthermore, the neurology for the fine control of that whole system must also be in place to be able to produce the range of sounds that we can.

In summary, the two main implications of the zoological evidence cited by Fitch (2009a) are that

1) physiological indicators that laryngeal position was high in the throat in its resting position need not indicate the *limit* of vocal frequency producible, because some animals can produce some lower frequencies by temporarily lowering the larynx;

2) physiological indicators that laryngeal position was low in the throat in its resting position need not indicate the capability to produce a wide range of frequencies because some animals that have a low laryngeal position cannot produce a wide range of sounds, because they do not possess other human-like features of vocal anatomy.

The former point means that a high resting laryngeal position indicates only the minimum level of vocal tract sound-production capability, and the latter reinforces the point that, in the absence of other primate-like vocal tract features, non-primate mammals make poor models for primate laryngeal sound-production potential. Laryngeal anatomy needs to be understood in the context of other vocal sound-producing anatomy and control mechanisms, and our starting point for modelling reconstructions of these in extinct hominin species should be the higher primates alive today, namely, the great apes and humans.

The hyoid bone and mandible

In any case, laryngeal position alone may not be enough to draw precise conclusions about the extent of vocal ability possessed by a specimen. Fortunately, in assessing the extent to which the vocal anatomy of an individual hominin resembles that of humans or apes, other physiological features can be considered. One example is the hyoid bone, the small bone to which the tongue base attaches and which supports the larynx in the throat. This is a rare find archaeologically, since it is very fragile, and is attached to other bones only by soft tissues. Five examples have been published from the fossil record to date, one from an *Australopithecus afarensis*, two from European *Homo heidelbergensis*, and two from *Homo neanderthalensis*.

The oldest-known example comes from specimen DIK-1-1, a 3.3 million-year-old juvenile *Australopithecus afarensis*, also nicknamed 'Lucy's baby' (after the well-known adult of the same species), from Dikika in Ethiopia (Alemseged et al., 2006). The individual is thought to have been around 3 years old at death, and appears to have been rapidly buried in sediments very soon after death, perhaps in a flood event. As a consequence many fragile skeletal elements which are often lost, including the hyoid bone, were preserved in the sandstone matrix. The hyoid has a form virtually identical to that of modern chimpanzees and gorillas, leading Alemseged et al. (2006) to conclude that this and the form exhibited by modern chimpanzees and gorillas is representative of the ancestral form of the hyoid, rather than the more bar-like form possessed by modern humans. They also suggest it may relate to the possession of laryngeal air sacs, like those of the African apes. It seems that the hyoid form possessed by modern humans has arisen in the human lineage since *Australopithecus afarensis*.

The first fossil hyoid bone to be published was found in Kebara Cave, Israel (Arensberg, 1989; Arensberg et al., 1990); dating to 60,000 years ago, it is remarkable in its completeness. The cranium of the Kebara 2 specimen does not survive, but the postcranial skeleton indicates that it is of a robust Neanderthal (Arensberg, 1989). Arensberg et al. (1990) carried out a reconstruction of the vocal tract of the hominin using evidence from the muscular attachments on the hyoid, spine, and mandible, and compared the hyoid bone with those of modern apes and humans. They conclude that the hyoid and vocal tract morphology of the Kebara 2 Neanderthal is virtually indistinguishable from that of modern humans, and is in contrast to that of great apes. A further Neanderthal hyoid bone from El Sidrón cave in Asturias, Spain, dated to around 43,000 years ago has been published by Rodríguez et al. (2003), and this shows a morphology consistent with that of the Kebara 2 example, also falling within the range of modern humans' hyoid morphology (Martinez et al., 2008).

So these examples strongly suggest that the modern human hyoid anatomy emerged from an ape-like form at some time subsequent to *Australopithecus afarensis* and by the time of our common ancestor with *Homo neanderthalensis*. The publication of two more hyoid bones, from *Homo heidelbergensis* specimens at least 530,000 years old, has strongly reinforced this conclusion. These two specimens, AT-1500 and AT-2000, were excavated from the Sima de los Huesos, Atapuerca, Spain, and represent the oldest-known hyoid bones from the *Homo* genus (Martinez et al., 2008). Critically these, like the later Neanderthal examples, are modern-human-like in morphology, very unlike those of great apes and the australopithecine example, and confirm that modern-human-like hyoid anatomy has existed in the *Homo* genus since *at least* 530,000 years ago (Martinez et al., 2008). In fact, modern-human-like hyoid anatomy was almost certainly possessed by our last common ancestor with *Homo heidelbergensis*. Tracing how the morphological change from an ape-like to a human-like hyoid anatomy took place will rely on further hyoid bone finds from earlier specimens, but our present evidence collectively pinpoints the process to having occurred in the earlier stages of the *Homo* lineage, between the australopithecines and late *Homo erectus*.

In the course of their reconstructions Arensberg et al. (1990) conclude that rather than basicranial flexion, features of the mandible are a far better indicator of the position of the hyoid (and thus the larynx) and consequent vocal ability. The positions of the muscular systems supporting the hyoid bone and larynx are indicated by the angle of a groove on the inside of the mandibles, called the *mylohyoid groove* (Figure 5.5).

Evidence from Middle Palaeolithic hominins, including Neanderthals (Kebara 2, Amud 1, Tabun I and II, Krapina J and C, Regourdou) and early modern humans (Qafzeh 9, Skhul V) shows that the mylohyoid groove is identical in position and orientation to that of modern humans 'suggesting a

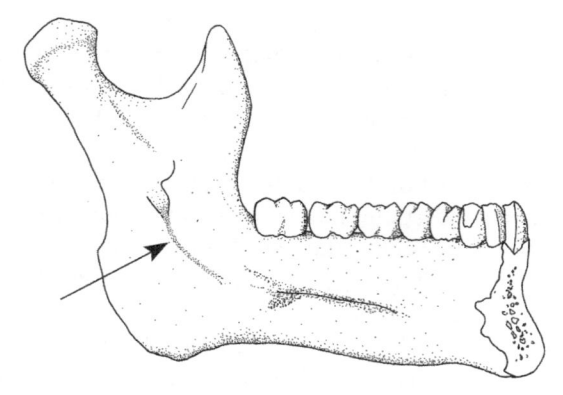

Fig. 5.5. Location of the mylohyoid groove in the human mandible (indicated by arrow).

suite of muscular relations similar to modern humans must have been present' (Arensberg et al., 1990, p. 141). The conclusion drawn from all this is that, regardless of basicranial flexion, the hyoid and larynx of Neanderthals must have been low relative to the cranial base; they should thus have had supralaryngeal spaces approximately the same size as those of modern humans, and no significant difficulties in articulation.

So the hyoid bones and mylohyoid groove reinforce the conclusions from the other studies discussed so far, that australopithecines' vocal anatomy was virtually indistinguishable from that of the great apes, and Neanderthals' vocal anatomy was similar to that of modern humans. As Fitch (2009a) has pointed out, this does not necessarily justify an attribution to Neanderthals (and now *Homo heidelbergensis*) of modern-human-like articulatory ability—in particular, articulate speech—but it does indicate that the form of our vocal anatomy has a very ancient provenance, and that critical foundations of our sound-producing ability were in place more than half a million years ago. To gain a better understanding of vocal function we must view this in light of the development of the neurological control of these physical systems.

The hypoglossal canal and tongue

Another hard tissue feature that has been used as a source of information regarding the nature of the vocal soft tissue of human ancestors is the *hypoglossal canal*, which transmits the nerve that supplies the muscles of the tongue (Figure 5.6). Kay, Cartmill, and Balow (1998) hypothesized that the relative size of this canal correlates with its degree of 'innervation' (i.e. the size of the hypoglossal nerve that occupies it in life), and that this in turn may reflect the degree of motor coordination of the tongue.

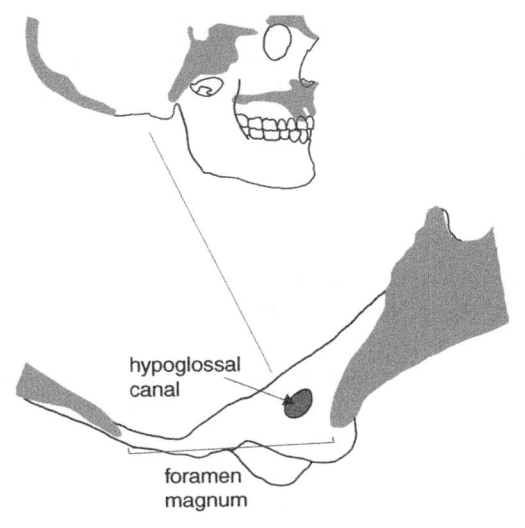

Fig. 5.6. Location of the hypoglossal canal in the human cranial base. (Redrawn from Kay, Cartmill & Balow 1998, p. 5418.)

Kay et al. (1998) found that the relative sizes of the canal in *Australopithecus africanus* (Sterkfontein specimens Stw 19 and Stw 187) and what is probably *Homo habilis* (Stw 53) (Curnoe and Tobias, 2006) are within the range of modern *Pan troglodytes* and *Gorilla gorilla*, whereas those of *Homo heidelbergensis* (Kabwe and Swanscombe), *Homo neanderthalensis* (La Chapelle-aux-Saints and La Ferrassie 1), and early *Homo sapiens* (Skhul 5) are well within the range of modern humans and significantly larger than *Pan troglodytes*. Given the estimated age of Swanscombe and Kabwe, Kay et al. suggested that the modern human pattern of greater innervation allowing increased motor control of the tongue, alongside other vocalization abilities, had evolved by around 300,000 years ago, in *Homo heidelbergensis*, and was present in subsequent species including Neanderthals.

These findings were subsequently questioned by DeGusta, Gilbert, and Turner (1999), following a thorough analysis of hypoglossal canal size in a large sample of non-human primates (prosimians, New World and Old World monkeys, and apes), modern humans, and fossil hominin skulls. They found that many non-human primates have hypoglossal canal dimensions within the range of modern humans, both in absolute terms and relative to the volume of the oral cavity, and that the same applied to fossil samples from *Australopithicus africanus* (the same three Sterkfontein specimens used by Kay et al., 1998), *A. afarensis* (A.L.333-45, A.L.333-105, and A.L.333-114), and *A. Boisei* (Omo L.338-y-6). This difference in the findings of DeGusta et al. (1999) and Kay et al. (1998) seems largely to be due to the greater sample size and precision of measurement in the more recent analysis. It is also evident that there is a large range of hypoglossal canal size in modern humans; this is more conspicuous in DeGusta et al.'s sample of 104 specimens than in Kay et al.'s sample of forty-four. Furthermore, a study of five modern human cadavers by DeGusta et al. (1999) revealed no confident relationship between hypoglossal canal volume and the size or number of axons in the hypoglossal nerve. The authors of the original paper have concurred that the findings of DeGusta et al. (1999) constitute a more accurate assessment of the available evidence (Jungers et al., 2003).

However, the issue of hypoglossal control of the tongue is not entirely a moot one; Jürgens and Alipour (2002) identified stronger and more direct neurological connections in the brain from the motor-cortical tongue area to the hypoglossal nucleus in non-human primates than in non-primate mammals. They suggest that this indicates a phylogenetic trend toward stronger cortico-motoneuronal connection for tongue control in humans. They also suggest that 'this might be one reason for the superior role the tongue plays in human vocal behaviour in contrast to non-human vocalisation' (p. 245). A similar comparison between these neurological connections in human and non-human primates would be particularly useful. The latter stages of such a phylogenetic development may also be *a consequence of* the increased importance and complexity of tongue movement in vocalizations in human ancestors. Certainly it suggests the laying of an important foundation for greater control of the tongue musculature at an early stage in primate evolution. Whilst this does not facilitate the attempt to date the occurrence of such a development from the *fossil* record of hominins, the identification of such phylogenetic trends, apart from being useful in its own right, does give a context in which to understand other physiological changes that are identifiable, and may also suggest that with refinement of technique there may yet be hope for identifying fossil physiological correlates with such a neurological development. This has indeed proven to be the case with regard to innervation through the vertebrae—the spinal cord.

Vertebral innervation, intercostal musculature, and breathing control

Another very important consideration pertaining to vocal ability is the fine control of the rib (intercostal) muscles; these muscles are responsible for aspects of breathing, through control of the ribcage, and thus are also implicated in the fine breath control required for extended and modulated vocalizations. MacLarnon and Hewitt (1999) have carried out a detailed analysis of the evidence for intercostal muscular control in primates as well as some fossil hominins.

Palaeoanthropology 1: Vocal Anatomy 149

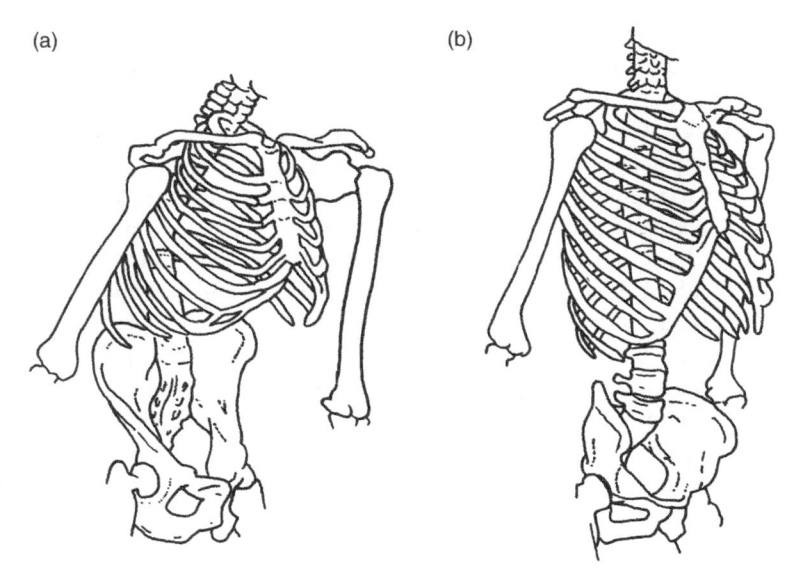

Fig. 5.7. Comparison of the rib-cage shapes of chimpanzees (a) and humans (b); the thoracic cage of great apes and australopithecines is funnel-shaped, whereas that of *Homo ergaster* and subsequent hominids, including humans, is more barrel-shaped. (Redrawn from Stringer & Andrews 2005, p. 18.)

The physiology and innervation of the thorax of extant primates and humans are quite different: the thoracic cage of great apes and australopithecines is funnel-shaped, whereas that of *Homo ergaster* and subsequent hominins, including humans, is barrel-shaped (Figure 5.7). In both cases, though, vertebral canal dimensions are quite well correlated with the dimensions of the spinal cord they contain, particularly in the middle and upper vertebrae, so their dimensions can be interpreted as an analogue for the degree of spinal cord innervation.

The thoracic spinal nerves which innervate the intercostal and subcostal muscles of the thorax are known as 'grey matter' (in contrast to the spinal nerves which control the hindlimbs, known as 'white matter'). The former have several functions: as well as maintaining posture, and controlling movements such as body-rotation, they are also involved in coughing, parturition, and defecation, and the control of breathing. MacLarnon and Hewitt's direct comparison of modern primates and humans indicated that it is 'only the grey matter of the modern human thoracic spinal cord that has expanded beyond the typical relative size for the order'; in contrast, 'the bulk of white matter containing nerve fibers passing through the thoracic region, to and from the hindlimbs, is of expected relative dimensions for a primate' (p. 347). This means that any increase in thoracic vertebral canal dimensions visible in fossil specimens is not a consequence of disproportionately increased innervation of the hind limbs.

What is the implication of this greater innervation of grey matter for vocal control? The respiratory muscles are responsible for controlling the air pressure from the lungs ('subglottal pressure') that reaches the larynx and upper respiratory tract. Of course, this is the case in all primates, but in humans these muscles are capable of maintaining a constant air pressure throughout vocalizations of varying length, preventing the reduction in volume and intensity (and tonal control) that would otherwise occur over the course of a long vocalization. Not only does this allow humans to 'speak fluently in long sentences, without disruptive pauses for inspirations, and with the necessary pauses placed at meaningful linguistic boundaries', but it also allows fine control of intensity, syllabic or phonemic emphasis, and pitch and intonation patterns (MacLarnon and Hewitt, 1999, p. 350). This latter function would certainly be essential in making full use of the potential range of the modern vocal tract in both speech and melodic vocalization, and MacLarnon and Hewitt go so far as to suggest that it is more important than the length and shape of the upper vocal tract structures in modifying volume and emphasis.

MacLarnon and Hewitt highlight the importance of subglottal pressure control for the production of human speech 'At both the larger scale of breath cycles and the finer scale of detailed features within phrases and words' (MacLarnon and Hewitt, 1999, p. 350). It is important to note that human speech *makes use of* this capability, rather than the capability being 'designed' to allow the production of linguistic speech, in the form of the features of phrases and words. Rather than working from *language* backwards towards the physiology, for the purposes of this study we need to consider how the development of these physiological capabilities could form the basis for the complex vocalizations. Although MacLarnon and Hewitt do not say so, the function of fine control of intensity, syllabic or phonemic emphasis, and pitch and intonation patterns would almost certainly precede that of controlling long expirations. The evidence from primatological and neurological studies discussed in Chapters 6 and 8 shows that control of tonal quality, pitch, and amplitude of vocalizations seems to have a far older provenance than lengthy vocal sequences. Such capabilities would have been of importance before 'long sentences', 'linguistic boundaries', and 'the features of phrases and words', linguistic structures which would surely have evolved *in response to* or at least *in tandem with* physiological capabilities, rather than such capabilities being shaped by linguistic demands. There would almost certainly be 'bootstrapping' (i.e. mutual reinforcement) between the changes in physiological features and the *vocal communicative* demands, but *linguistic* structures should emerge out of such already extant vocalization capabilities—the physiological capabilities would not have been shaped by linguistic demands, but linguistic form would have been shaped by the physiological capabilities present.

At what stage in the fossil record do we see the earliest evidence of this increased vertebral innervation? MacLarnon and Hewitt (1999) found from their analysis of primate and fossil hominin skeletal remains that the relative thoracic vertebral column diameter is increased in modern humans and Neanderthals relative to body mass, in comparison with extant higher primates, but that of australopithecines and *Homo ergaster* is not. Australopithecines appear to have had thoracic vertebral canal dimensions within the range of great apes (on the basis of measurements of AL288-1, Sts 14 and Stw 431), as did *Homo ergaster* relative to its body size (KNM-WT 15000, the 'Nariokotome Boy'), whereas Neanderthals (La Chapelle, Shanidar 2, Shanidar 3, and Kebara 2) and early modern humans (Skhul 4) had dimensions within the range of modern humans. Unless this is a case of convergence, this would suggest that the common ancestor of Neanderthals and modern humans (and thus also *Homo heidelbergensis*) also had vertebral canal dimensions equivalent to those of modern humans.

At what point such increase occurred is more contentious, as the case of *Homo ergaster* (in particular, KNM-WT 15000) is the focus of some debate. Whilst MacLarnon and Hewitt (1999) assert that the thoracic canal arches in *Homo ergaster* fall within the range of modern primates in relation to estimated overall body mass, Frayer and Nicolay (2000) contend that they fall within the range of modern humans, if considered relative to cranial capacity. It would seem in this case that MacLarnon and Hewitt's interpretation of the spinal innervation relative to body mass makes for a better grounded comparison, for two reasons. First, since cranial capacity has also increased disproportionately relative to body mass over the course of hominin evolution, surely human thoracic canal size *relative to cranial capacity* is *less* than that of other higher primates. Second, there is no reason to believe that innervation of the spinal cord should be related to neocortical development in the case of a fundamental physiological process such as breath control, unless you *presuppose* neocortical involvement, implying voluntary control, which is actually what we are hoping to identify. The null hypothesis should instead be that vertebral innervation correlates with body size and mass, in which case it is disproportionate growth in these terms that is interesting, as it does imply that other processes are at work.

Frayer and Nicolay (2000) also question whether thoracic canal measurement is relevant as an indicator of breath control at all. Whilst MacLarnon and Hewitt (1999), MacLarnon (1993), and Walker (1993) assert that the thoracic nerves are most significant for fine control of breathing for vocalization, Frayer and Nicolay (2000) state in contrast that thoracic diameter is of little relevance: 'most of the muscles involved in speech respiration are innervated by either cranial or cervical nerves' (p. 226). The only *cervical* vertebra preserved from KNM-WT 15000 is C7, which does not have dimensions lower than those expected in modern humans (citing MacLarnon, 1993).

In fact, though, the authors' views are artificially polarized; both thoracic and cervical innervation is implicated in breathing control, but thoracic innervation is more important than Frayer and Nicolay (2000) acknowledge and less fundamental than MacLarnon and Hewitt (1999) assume. Whilst the thoracic intercostal nerves are indeed responsible for *some* control of breathing (Palastanga et al., 2002), they seem to be implicated in additional fine control of a function that is principally controlled through cervical innervation. Cases of spinal injury in modern humans confirm this. Spinal injury below C6 still allows control of breathing without assistance, and vocalization capabilities indistinguishable from normal, except that the patient is limited in the depth of inhalation. What the thoracic nerves seem to allow is control of especially prolonged vocalizations. The duration of vocalizations can be affected in patients with spinal injuries below the cervical vertebrae, but the *quality* of such vocalizations is not affected by the loss of thoracic control (the PoinTIS Spinal Cord Occupational Therapy site of the SCI Manuals for Providers; *pers. comm.* G. Price, clinical physiotherapist).

If the cervical vertebrae of KNM-WT 15000 are indeed of modern dimensions, as Frayer and Nicolay (2000) and MacLarnon (1993) assert, this indicates that there is no reason why we should not attribute to *Homo ergaster* the ability to voluntarily moderate its breathing sufficiently well to produce vocalizations controlled for pitch, intensity, and contour, in coordination with its laryngeal and orofacial anatomy. If its thoracic vertebrae are indeed within the range of higher primates rather than humans, relative to body size, then what it may have lacked was the ability to produce *extended* voluntarily controlled vocalizations. It seems that this *is* an ability which has developed over the course of human evolution, as indicated by thoracic canal innervation, between *Homo ergaster* and the common ancestor of *Homo neanderthalensis* and *Homo sapiens*.

Other conclusions of Frayer and Nicolay (2000) should also be subject to further scrutiny. In line with their interpretation of the evidence regarding breathing control, they conclude from their analyses that by 1.5 million years ago, in *Homo ergaster* 'both the articulatory capacity to form vowels and the respiratory capacity to maintain high-volume airflow were present' (Frayer and Nicolay, 2000, p. 233). Whilst the latter may well be the case (albeit limited in duration, as discussed earlier), they base their conclusions regarding the former, articulatory capacity, on the evidence from the hyoid bone and basicranial flexion. The hyoid evidence they cite comes from the Neanderthal Kebara specimen, dating to 60,000 years ago (as mentioned earlier), so conclusions based on this cannot be extended to *Homo ergaster*. This is also the case for the *Homo heidelbergensis* hyoids found subsequently to Frayer and Nicolay's study. The basicranium, whilst showing the first evidence of flexion in *Homo ergaster*, does not achieve modern proportions until *Homo heidelbergensis*, 300,000–400,000 years ago (as already mentioned) and, in any case, is only one of a suite of features that may influence vocalization ability. It should be possible to attribute to *Homo ergaster* some articulatory versatility, but not modern abilities; greater value, in developmental terms, is gained from the evidence by identifying that it was with *Homo ergaster* that the first major changes towards modern vocalization abilities occurred.

What is especially significant is that both early anatomically modern humans and Neanderthals *do* show modern thoracic vertebral canal development, suggesting that such control over extended-duration vocalization has an ancient provenance and is not a feature solely of modern humans. It is likely that the common ancestor of *Homo neanderthalensis* and *Homo sapiens* also exhibited this development, which had occurred between *Homo ergaster* and that point; hopefully fossil specimens that can be confidently identified as common ancestors of Neanderthals and modern humans will confirm or refute this when they are subjected to analysis.

SOME PREVIOUS EXPLANATIONS FOR INCREASED TONAL RANGE

Several explanations have been advanced for the development of the versatility of the modern vocal tract, from research in a variety of disciplines, including primatology and zoology, developmental psychology (child development), palaeontology, linguistics, and neurobiology. Lieberman et al. (1969, 1972) proposed that the modern form of the vocal tract was selected for to support articulate speech, and this proposal remains very influential. On the basis of developmental research, Dissanayake (2000) has suggested that vocal versatility and perception arose to support infant-directed (ID) speech; Fernald (1992b) has also suggested that modern vocal capabilities are derived from those originally used for ID speech (these ideas are discussed further in Chapter 8). Following studies of primate and other mammalian vocalization, Fitch (1999, 2000a, 2000b, 2009a) has pointed out several important areas of comparison between the vocalizations of humans and other animals, including an ability in some animals to lower their larynx (temporarily) during vocalization. This has led him to suggest that the lengthening of the vocal tract through laryngeal lowering might have taken place to lower the formant frequencies produced.

In Fitch's opinion (2000a), the best explanation for the permanently lowered larynx in humans is still provided by the hypothesis of Lieberman et al. (1969, 1972) (and many others since, it should be noted): that it is used for producing articulate speech. He proposes incorporating into this hypothesis the ability, at an early evolutionary stage, to temporarily lower the larynx to produce lower formant frequencies and greater tongue versatility; he then suggests that in a species that spends a lot of time vocalizing, it may be more energy-efficient to have a permanently lowered larynx than to have to lower it before each vocalization (using the strap muscles, sternothyroid, and sternohyoid). A low larynx has implications for tongue control (a fact relevant to all discussions of vocal tract development, not just this particular scenario). The low position of the tongue root concomitant with a low larynx allows for greater movement and control of the tongue body to produce sounds in the oral cavity. In Fitch's words, 'having the skeletal support for the tongue body in a permanently low, stabilised position may provide significant advantages for producing the rapid, precisely controlled vocal tract movements that characterise modern human speech' (Fitch, 2000a, p. 215). Formant frequency, on the other hand, determines the characteristic pitch of vowel sounds; the sound quality of a vowel is normally determined by a combination of formant frequencies. In humans, monkeys, and dogs, vocal tract length is positively correlated with body size, and in turn, formant frequencies are closely related to vocal tract length (Fitch, 2000b). Low formant frequencies are produced by large animals with long vocal tracts, and formants play a role in individual identification, being a sort of vocal signature; certain animals have the ability to temporarily lower their larynx during vocalization to produce lower formant frequencies (discussed later).

In other words, Fitch suggests that a capacity to lower the larynx temporarily to produce lower formant frequencies and increase tongue control could have been an important early component of articulate vocalization, and if the larynx became permanently lowered this would have been more efficient in a vocally very active species. This rationale requires that low-formant-frequency vocalizations with a temporarily lowered larynx constituted a selectively important component of the vocalizations of a vocally active species, before sophisticated control of pitch contour and vocal control in the supralaryngeal tract developed. What it is important to remember is that the temporary lowering of the larynx by the strap muscles does not allow a greater *range* of controlled vocalizations to occur—it simply allows a *different, additional* vocal sound to be made. In both cases this sound could be moderated by tongue and orofacial movements, but this is not the same as control over the entirety of an increased range of tonal vocalizations. This rationale requires the existence of a hominin ancestor which was very vocally active using this facility, prior to the development of a wide vocal range, versatility, and control. So in this scenario a lowered larynx essentially formed a pre-adaptation (exaptation) to complex controlled vocalizations in a creature that was already very vocally active.

All this leads Fitch (2000a) to propose a three-stage process for this development:

> In stage one, some early ancestor used a standard mammalian vocal gesture to produce calls, but introduced tongue body perturbations during larynx lowering to produce a wider range of formant patterns (and hence a greater diversity of

discriminable 'calls' or phones)... In stage two, the dual use of degrees of freedom of the tongue body was consolidated into the communication system, with a variety of vowel-like sounds and formant transitions being produced. However, these would be made with a temporarily lowered larynx, and the larynx would be returned to the nasopharynx during resting breathing. Finally, in stage three, the larynx would have assumed a permanent low resting position during ontogeny, as it does today, giving these hominids less effortful speech and perhaps more vocal control. (Fitch, 2000a, p. 216)

This seems a very plausible scenario for the beginnings of the process of vocal tract evolution, and the early stages of complex vocalization. Into stage one of this model Fitch has subsequently integrated his 'size exaggeration' hypothesis (Fitch 2000b), discussed later, although it is by no means necessary to the model, with the other arguments for increased tongue control and tonal range. The vocalizations outlined in stage two would have been of limited duration; this fits with the fossil and neurological evidence described earlier which suggests that the ability to plan and control vocalizations of extended duration is something that arose after the beginnings of vocal versatility in *Homo ergaster*. However, as already pointed out, the model provides little explanation of the development of the fine tonal and pitch control, muscular control, and complex vocalizations that use the whole of the supralaryngeal tract; these are the subject matter of the subsequent chapters here. These elements of vocal sophistication would have to have developed during and subsequent to stage three of Fitch's proposed scheme; in this respect the model accounts for the beginnings of some articulatory control but can provide an account only of the earliest stages of vocal tract evolution and pitch control. It is at stage three of the model that fine control of vocal tone could become important and selected for, and although Fitch's model seems to imply that the lowering of the larynx occurred in one fell swoop, the fossil evidence we have considered would seem to suggest that this had several stages; further lowering of the larynx would also have occurred subsequently, with concomitant increasing control over the necessary musculature.

As noted, Fitch (2002b) has elaborated on the first stage of his model by suggesting that the selective force for the initial lowering of the larynx may have been an advantage associated with creating an exaggerated impression of size through lower frequency vocal formants. A lowered larynx increases vocal tract length and decreases the formant frequencies produced, allowing an animal with a lowered larynx to duplicate the vocalizations of a larger creature, thus exaggerating the impression of size conveyed by its vocalization. Fitch outlines several precedents for the size-exaggeration role of vocalizations elsewhere in the animal kingdom. Many birds have elongated tracheae, which lower the formant frequencies that they produce. Fitch (1999, 2000b) suggests that this might be highly effective at exaggerating apparent size 'at night or from dense foliage' (Fitch, 2000b, p. 264). Another example, this time in mammals, is provided by red deer and fallow deer males. These have the ability to lower their larynx temporarily to produce their loud roar with low formant frequencies, used to intimidate rivals and impress females.

These observations lead Fitch to suggest that this ability might have been the principal selective influence on the evolution of a lowered larynx in hominins, and that our primate ancestors might have used formant frequencies as an estimate of body size from vocalizations. He then suggests that this might have formed a pre-adaptation for 'vocal tract normalisation . . . a crucial feature of speech perception whereby sounds from different sized speakers are normalised to yield equivalent percepts' (Fitch, 2000b, p. 263). It is not clear, however, how a physiological feature that he suggests was selected for in order to *distinguish* and *differentiate* individuals might have formed a pre-adaptation for a crucial feature of human speech perception which apparently acts to the opposite purpose, *standardizing* a percept.

Whilst illustrating that size-exaggeration vocalization has precedents in the animal kingdom, the examples given would seem to provide little direct analogy with human vocalization, or for how the elements of vocalization which are exclusive to humans must have developed. In the case of the birds (and perhaps with the size-exaggeration hypothesis in general), Fitch's suggestion that size exaggeration might be especially effective at night or in dense foliage should perhaps read that it might *only* be effective at night or in dense foliage (i.e. in situations of limited visibility, when the vocalizer cannot be clearly seen). The first evidence for permanent lowering of the larynx in hominins occurs with *Homo ergaster*, at a time when hominins were living predominantly in more open, less arboreal habitats than their ancestors (Klein, 2009). If size exaggeration was the selective force behind laryngeal lowering one would expect to see it at a time when hominins were living in habitats of limited visibility, not moving away from living in such environments, so it would have to have had its role amongst earlier hominins than *Homo ergaster*, such as australopithecines or, perhaps, *Homo habilis/rudolfensis* (and it is with the latter that we first see possible evidence for increased manual and orofacial muscular control, in the form of a disproportionate increase in size of the left temporal region). But even in the case of these species the evidence suggests that they lived in more open environments than higher primates do today. An increasingly open habitat has often been argued to have been the driving force for the decreased arboreality and increased bipedalism of the australopithecines, who show no anatomical or cerebral difference in vocal control ability relative to higher primates of today. It is conceivable that size exaggeration might have been useful at night in intimidating predatory animals on the savannah, but all the examples of size exaggeration in nature given by Fitch (1999, 2000b) instead confer sexual selection advantages in mating and contest with rivals.

In the case of the laryngeal lowering of the red and fallow deer bulls in their roar to intimidate rivals and impress females, this illustrates a further potential use of lowered formant frequency. But it is worth reiterating that, in this form, the mechanism is not a direct analogy for the human vocalization ability—the deer can only lower its larynx temporarily to roar, whereupon it returns to its original position. It has no control over the intervening frequency ranges. In contrast, we have very fine control over the range of frequencies it is possible for our permanently lowered larynx to produce.

It may be that Fitch is correct in suggesting that the ability to temporarily lower the larynx formed a *precursor* to a permanently lowered larynx, and there is a variety of evidence to suggest that formant frequency is still used as an indicator of body size by humans (Fitch, 2000a). In this sense, the size-exaggeration hypothesis provides an interesting model for vocal tract use in our ancestors which has received less attention than linguistic hypotheses, and describes a potential *additional* role of the vocal tract, but its likelihood needs to be judged in light of the ecological circumstances in which these hominins operated.

The relationship between vocalization frequency and body size does provide a possible explanation for the second lowering of the vocal tract at puberty that is especially prominent in adolescent human males, an effect of hormonal change at sexual maturity (Fitch, 2000a, 2000b). This is a secondary sexual characteristic which might be selected for. Australopithecines and *Homo habilis/rudolfensis* exhibited high levels of sexual dimorphism—males were up to 50 per cent larger than females (Klein, 2009)—which would result in distinct differences in formant frequency production between the sexes. It is possible that as sexual dimorphism decreased to the level seen in *Homo ergaster*, which is little different from that in modern humans (Klein, 2009), the sex-difference in vocalization formant frequency was maintained through the secondary lowering of the larynx in males at puberty. In this circumstance, however, it should be viewed as an exaggerated signal of body size, rather than a signal exaggerating body size; i.e. it functions to remove ambiguity about body size (and sex), rather than to mislead about body size.

As an aside, rather than asking why the male larynx lowers at puberty, it might instead be appropriate to ask why the female larynx *does not* lower at puberty, at a time when body size increases almost as much as it does in males. Instead, females retain a vocal range much closer to that of children. Possible explanations include this being as a consequence of sexual selection for characteristics in females which are associated with youthfulness, and thus fertility, or because of advantages associated with retaining a vocal register to which young infants' auditory systems are more sensitive. Further investigation from this angle may provide valuable insights. The value of infant-directed vocalization is discussed further in Chapter 8.

While providing an interesting additional set of possible reasons for the beginnings of vocal variability, the size-exaggeration hypothesis seems to have little to tell us about the evolution of the properties of the vocal tract that are particular to humans. It provides no explanation, or path, for the evolution of fine vocal control of the wide diversity of frequencies that the human larynx is able to produce, or how the larynx came to be permanently lowered; the latter is accounted for by the three-stage model proposed by Fitch (2000a) (discussed earlier), but the former is not. Nor does it address the question of why we are so sensitive to the prosodic content (i.e. frequency variations) of utterances. It is these properties of the vocal tract (and their evolution) with which an investigation into the development of the capacities for musical behaviour must be principally concerned.

If increasing versatility and control over complex vocalizations is taken as the main selective force for the development of the vocal capabilities of melody and speech that we now exhibit, it follows that the muscular control (and cognitive capacity) necessary must have evolved in tandem with the vocal tract. Chapter 6 examines what the driving forces for the development of such control could have been.

CONCLUSIONS

In summary, the australopithecines studied so far show characteristics of anatomy related to vocalization that are little different from African apes (gorillas and chimpanzees) today. Changes in the resting position of the larynx away from an ape-like form are first identifiable in *Homo ergaster*, which shows the first indications of a lower resting laryngeal position and increased supralaryngeal soundspace, changes which were probably initially instigated by a shift to fully upright human-like bipedal posture, as well as changes in brain size. While this has the potential to allow the production of a larger range of formant frequencies than the ancestral (and australopithecine) form, the true formant-producing potential of the *Homo ergaster* anatomy is difficult to model. However, this development appears to have been coupled with an increase in cervical vertebral innervation, which allows for the control of airflow through the larynx, thus permitting some control of the pitch, intensity, and contour of sounds produced by the larynx. *Homo ergaster* appears not to have undergone any increase in thoracic canal innervation over the ape-like condition, however, so although able to produce a greater variety of sounds, it would have been limited in the control of the duration of the sounds it could produce, as are other higher primates today.

By the time of the last common ancestor of Neanderthals and modern humans, probably around 600,000 years ago, human-like thoracic innervation had emerged, allowing control over utterances of extended duration, alongside a modern-human-like hyoid anatomy and position, and thus supralaryngeal soundspace. Certainly European *Homo heidelbergensis* specimens ancestral to Neanderthals, and African *Homo heidelbergensis* specimens ancestral to modern humans, as well as Neanderthals and modern humans themselves, all possessed all of these features. So a rearrangement of laryngeal anatomy into a form essentially indistinguishable from that of modern humans, along with the neurological control over pitch, intensity, contour, and duration of sounds produced by it, appears to have taken place at some point(s) over the 1-million-year or so period of the evolutionary development of *Homo erectus*, from *Homo ergaster* to the common ancestor of Neanderthals and ourselves.

Note that none of this evidence is able to directly indicate the possession of control over the tongue and orofacial musculature that we possess, which allows us the further fine control of the sounds that are being produced in the vocal tract (such as is required for articulate speech), but it does indicate the possession of a vocal tract anatomy capable of producing sounds of variable pitch and extended duration, considerably developed from the form and capabilities of higher primates. It also indicates that the laryngeal anatomy that we possess, and the control over our breathing necessary to make use of it to produce a wide range of sounds, have a very ancient evolutionary heritage. This evidence also does not necessarily indicate that these anatomical capabilities were selected for specifically for the purposes of making complex extended vocalizations; it has been argued that the thoracic innervation allowing control over deep breathing may have emerged to support long-distance running (Bramble and Lieberman, 2004), and the changes in basicranial and laryngeal anatomy could conceivably have been a by-product of brain size increase and reduced facial prognathism (though these developments have themselves often been argued to be associated with complexity of communication and social organization). The emergence of such capabilities could clearly subsequently confer considerable advantages as *exaptations*, however, whatever their initial causes. To better understand possible reasons for these changes we will need to look at other sources of evidence, in the following chapters.

The lowering of the larynx allowed greater versatility of vocal tone to be produced (and lower formant frequencies) and also allowed greater potential for control and movement of the tongue due to the lower position of the tongue root. Rationales for the evolution of the human vocal tract have to account for the fact that it has developed in such a way that it allows us not only to produce a greater range of sound frequencies, but also to have very fine control over the entire range of those frequencies. They also have to account for why we are so sensitive to these frequency variations in utterances (their prosodic content). These elements of our vocalization capabilities have very important communicative roles. It is suggested by Fitch (2000a) that the ability to *temporarily* lower the larynx in order to produce more complex vocalizations may have led to a preliminary permanent lowering of the larynx. This would subsequently have continued to lower, with concomitant increasing control over the vocal and orofacial sound producing capabilities possible.

That an increase in control over pitch, intensity, and contour seems to have occurred before the ability to produce vocal sounds of extended duration is interesting. As MacLarnon and Hewitt (1999) point out, many primates vocalize in the form of discrete units of sound created with single air movements, but are limited in the duration of these and the order in which certain sounds can be made in the breathing cycle. They are also limited in the diversity of such sounds that they can make. An evolutionary path in which the ability to produce long sequences of controlled vocalizations developed out of an initial ability to make discrete vocalizations which were controlled for pitch and tone would seem to be consistent with the foundations for these capabilities which are already evident in higher primates.

On the basis of all of the evidence considered in this chapter, the most parsimonious scenario would seem to be that increasing control of intensity, pitch, and intonation patterns of discrete vocalizations occurred initially, to date first exhibited by *Homo ergaster*; pitch and intonation control increased subsequently with the development of a greater supralaryngeal soundspace, and control over maintaining long sequences of such utterances also followed, until these levels of control over vocal range and duration were essentially modern-like in *Homo heidelbergensis*. It is possible that the ability to control extended sequences increased at the same time as vocal range increased, but the resolution of the record does not, at present, allow us to identify intermediate phases of either development—only where they both start (with *Homo ergaster* at least c.1.7 million years ago, or an as yet undiscovered predecessor) and where they both appear complete (with *Homo heidelbergensis*-like hominins 300,000–600,000 years ago). In fact, this sequence of the emergence of control, as suggested by the fossil evidence, makes more sense than the reverse—it is difficult to imagine how long sequences of vocalizations with little control over pitch, contour, and intensity could be as meaningful as short sequences of vocalizations controlled for pitch, contour, and intensity. The latter could be communicative in their own right, and as control increased, the length of sequences of such pitched and contoured utterances could also increase; subsequently, the order in which the expressive vocalizations occurred could assume importance.

But at this stage this is just a 'best explanation' for the fossil evidence; a fuller understanding of the emergence of vocalization capabilities requires that we look at other sources of evidence too, and establish what their mutual implications are. What other evidence is there for control over the vocal system? And what is the use of vocalizations which are controlled for pitch and contour?

6

The Palaeoanthropology of Vocalization 2: The Brain and Hearing

INTRODUCTION

The use of the physiological features described in Chapter 5 would, of course, require controlling input from the brain. How far back in our evolutionary history do we have to go to see evidence of the brain systems responsible for the control of vocalizations? To address this question we can look at a different aspect of the fossil record—evidence of the shape and proportions of the brain—and we can examine the neurological structures and capabilities possessed by other primates. The final section of this chapter examines the evolution of the inner ear and sound-perception capabilities, and how these relate to sound-production capabilities and the evolution of other physiological features.

EVIDENCE FOR THE EVOLUTION OF VOCAL CONTROL IN THE BRAIN

Fossil endocasts

There are several difficulties associated with making inferences regarding the development of hominin neuroanatomy from the archaeological evidence. Not least of these is that there is no *direct* fossil evidence for the interior neurological structure of early human brains, because the brain structures do not fossilize. Nevertheless, the fossil record can provide some valuable information. Some light can be shed on the anatomy of early brains by the study of *endocasts*. These are casts of the inside of fossil skulls, which can form naturally or be artificially created, and show the shape and features of the brain which occupied the cranium (e.g. Kochetkova, 1978; Falk, 2004a; Johanson and Edgar, 2006). Because, in life, there is very little space between the brain and the interior surface of the cranium (cranial vault), the form of the cranial vault actually reflects very closely the shape and features of the brain which occupied it. If fine sediment enters and fills a cranium during fossilization then it can form a near-perfect cast of the inside of the skull, and thus of the brain that occupied it. The same result can be achieved by making a cast of the inside of a complete skull today, or through the use of scanning technology to build a 3-D model of the cranial vault.

The extent to which one may extrapolate brain function from the brain physiology revealed by an endocast is debated, however. An endocast only shows the outer surface of the brain, as it is a cast of the imprint left by the brain on the inner surface of the skull; it can tell us nothing about the internal brain structure and little about connections between different areas (Leakey, 1994). Endocasts can show the *shape* of the brain, though, and the relative proportions of different parts of it. For example, it has been possible to identify development in Broca's area (a part of the brain which in humans has important roles in language production; see Figure 6.1) of hominin brains (Holloway, 1983).

This identification has led to the assertion that *Homo habilis* had some language abilities, as it shows evidence of development in the left temporal region where Broca's area is located (Holloway, 1983). This, however, was on the basis of analysis of skull KNM-ER 1470, from East Turkana, which is now commonly classified as *Homo rudolfensis*, and may not be closely related to *Homo habilis* or the direct human lineage (see, for example, Johanson and Edgar, 2006).

There are some other problems with this inference. As discussed later here and in Chapter 7, many other areas of the brain are also directly involved in language production and processing (e.g. Borchgrevink, 1982; Benson, 1985;

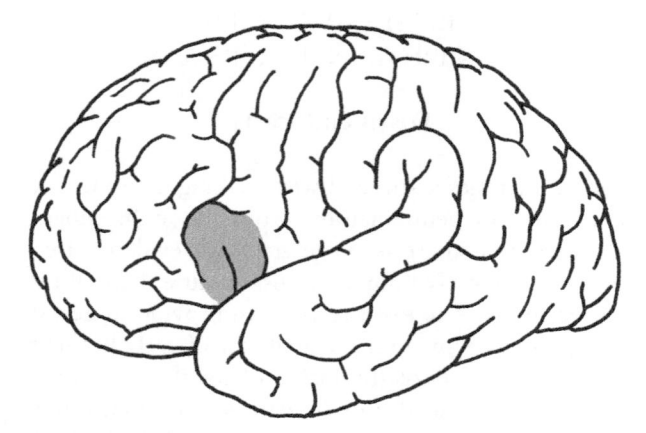

Fig. 6.1. The location of Broca's Area in the left hemisphere of the human brain. (Redrawn from Carlson 1994, p. 6.)

Kimura, 1993; Poeppel and Hickock, 2004; Patel, 2008). Furthermore, the areas around Broca's area are also used for processes other than language; for example, complex sequences of muscular movements and fine motor muscular coordination of oral and facial muscles are controlled in and around this region, including musculature necessary for production of facial expressions and vocalizations (Ojemann et al., 1989; Calvin, 1996; Duffau et al., 2003; Nishitani et al., 2005; Petrides et al., 2005). This seems to be an evolutionarily very ancient function of this region, as the analogous area in macaque monkeys is also involved in orofacial musculature control (Petrides et al., 2005). The great apes also show some asymmetrical development of this area in the left hemisphere, confirming an ancient development of these substrates prior to the development of language (Cantalupo and Hopkins, 2001). In humans Broca's area is also involved in executing sequences of movements for rhythm production, and has been shown to be involved in playing music and when musicians were carrying out a rhythmic task (Sergent et al., 1992; Platel et al., 1997; Besson and Schön, 2003).

Note that *vocalizations* are not the same thing as *language*. Thus development in Broca's area and the surrounding areas may indicate the development of the ability to produce complex muscular sequences and finely controlled vocalizations, and not have any bearing on language ability in the sense of syntax and grammar. In fact, Peterson et al. (1988) conclude from PET studies of the brain that the classic Broca's area (Brodmann's areas 44 and 45) is not concerned with semantic associations of words at all, but exclusively with motor functions (see also Mohr et al., 1978; Poeppel and Hickock, 2004); it seems to be the nearby Brodmann's area 47 that is responsible for semantic associations. Later studies (Brown et al. 2006a; 2006b) have found that BA 44 and 45 in the left hemisphere are activated in the generation of improvised sentences, and that BA 45 is also involved in the generation of improvised melodic phrases, suggesting that overlaps exist here in the production of the complex sound structures of melody and speech (see Chapter 7). Uncertainty about the precise functions of Broca's area may therefore have led to the attribution of language abilities to hominins who probably had none; Broca's area and the areas around it have important roles in fine vocal, orofacial, and manual control, and the identification of development in this area in endocasts gives important information regarding the development of these capabilities essential to complex vocalization, rather than necessarily indicating *linguistic* function.

Before about 2 million years ago, fossil hominin brains show no development in the area of endocasts corresponding to Broca's area (Falk, 2004a). As noted, the first evidence for any growth in this area this comes from KNM-ER 1470, *Homo habilis sensu lato/Homo rudolfensis* (Falk, 2004a). Between these specimens and *Homo ergaster* there is then a disproportionate level of growth in this area, in comparison to the rest of the brain's growth (Kochetkova,

1978). While there is a general increase in brain size of around 50 per cent between *Homo habilis* specimen OH7 (at around 674 cc; Johanson and Edgar, 2006) and *Homo ergaster* KNM-WT 15000 (estimated adult cranial capacity 909 cc; Johanson and Edgar, 2006), there is an 'especially intense' growth of the lateral tuba, corresponding to Broca's area (Kochetkova, 1978, p. 200). This suggests that there was a significant development of fine motor control of vocalization and/or gestural sequences as part of the emergence of the new species. Was this an increase in control over complex manual sequences related to the development of stone tool production? This is possible, but the recent publication of animal bones cut-marked by stone tools and apparently dating to around 3.4 million years ago (McPherron et al., 2010) now makes this seem less likely, preceding as they do these neurological developments by over a million years. Was this an increase in control of vocal and orofacial musculature related to the development of the vocal tract and interspecific communication? Or does it represent interrelated developments in both of these sets of abilities—for example, the integration of vocal communicative behaviours with manual grooming activities? The relationships between manual and vocal functions are explored further in Chapter 9. By early *Homo sapiens* this area is 'very protuberent' (Holloway, 1981, p. 387, on the basis of Djebel Ihroud I, 1,305 cc; Holloway, 1981—although at the time of Holloway's publication Djebel Ihroud I was identified as *Homo neanderthalensis*).

Neurology of vocal production in primates and humans

Other methods must be used to investigate the history—or prehistory—of connections between different areas and functions within the brain, as these will not show up in endocasts. Comparison between human and other primate vocalization neurology can inform us as to which structures in the human brain are unique to humans, having probably emerged in the hominin lineage, and which are shared with other primates (and thus, probably, shared with our ancient common ancestors and present in the lineage of hominins leading to humans).

Jürgens (2002) carries out a comprehensive review of the physiology and neurology underlying vocalization, in a variety of mammals, including primates and humans. In terms of the neurological areas involved in the production of vocalizations, there is a big difference between the level of input required for involuntary (instinctive) and voluntary utterances. Instinctive vocalizations do not rely on any input from the forebrain, but voluntarily controlled vocalizations do. Control over the acoustic form of utterances depends on input from the motor cortex, of course, and the motor planning of extended, purposeful utterances relies on input to the motor cortex from the ventral premotor and prefrontal cortex, including Broca's area; the control of transitions between elements of the vocalization requires input from the cerebellum (Jürgens, 2002).

Voluntary control of emotional content

The integration of emotional content into vocalizations also requires specific neurological involvement when this is voluntary as opposed to instinctive communication of emotion. Neuropathological studies have identified the anterior cingulate cortex (anterior limbic cortex) as being responsible for the production of voluntary emotional vocal expressions in all primates (Jürgens, 1992) (Figure 6.2).

Destruction of this area in humans does not affect the ability to produce *involuntary* vocal reactions to external stimuli but does result in an inability to *voluntarily* produce joyful exclamations, angry curses, or pain outbursts. Without the use of the anterior cingulate cortex voluntary vocal utterances, which would normally be highly emotionally communicative, sound more or less monotonous, with very flat intonation (Jürgens and von Cramon, 1982; Jürgens, 1992); the anterior cingulate cortex is responsible for the inclusion of emotional content in voluntary utterances. This is an evolutionarily ancient area of the brain, fulfilling the same function in rhesus monkeys (and almost certainly all primate and hominin species on the evolutionary line between

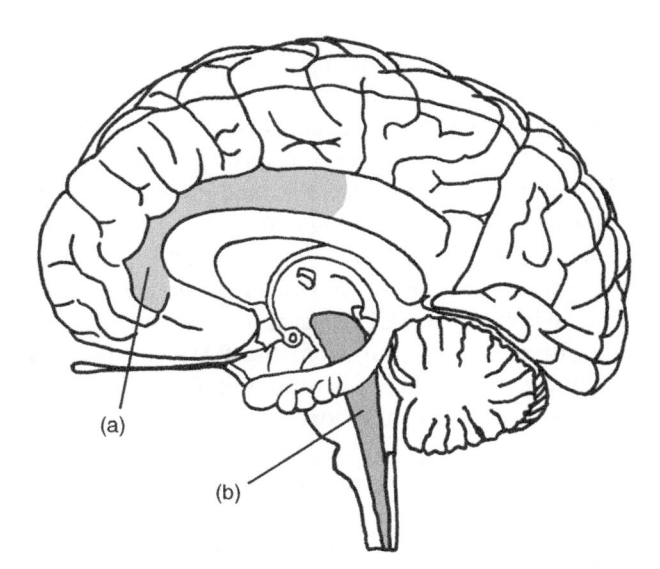

Fig. 6.2. Section through the human brain showing the location of (a) the anterior cingulate (limbic) cortex and (b) the periaqueductal grey matter (PAG).

rhesus monkeys and humans), with the same effects when destroyed (Aitken, 1981; Kirzinger and Jürgens, 1982, reported in Jürgens, 1992).

The role of the periaqueductal grey matter (PAG)

It seems therefore that the biological basis of non-verbal vocal emotional utterances is the same in humans as other primates. Jürgens (1998) concludes, for example, on the basis of many years researching the neurological roots of vocalization, that vocalizations carried out by squirrel monkeys 'can be considered as a suitable model for the study in humans of the biological basis of nonverbal emotional vocal utterances, such as laughing, crying and groaning' (p. 376). Critical in the production of non-verbal vocal emotional utterances is the periaqueductal grey matter (PAG) of the midbrain (grey matter that surrounds the cerebral aqueduct at the very heart of the brain), which acts as a critical relay station for such utterances (Jürgens and Zwirner, 1996; Schulz et al., 2005) (see Figure 6.2). The PAG is responsible for mediating emotional processes in all mammals; it and the tegmentum of the midbrain that borders it collect the various auditory, visual, and somatosensory stimuli that trigger vocalization, as well as motivational controlling inputs and volitional impulses from the limbic structures, including the anterior cingulate cortex mentioned earlier (Jürgens, 1998). In lower primate species (and other mammals) it is responsible for the production of species-specific calls, communicating information about emotional state and arousal (Schulz et al., 2005).

On the perceptual side the PAG also contains opiate receptors that reinforce positive emotional experiences, including social attachments. This includes atunement to the vocal characteristics of conspecifics and, in humans, 'our atunements to voices of those we love, and hence, by parallel reasoning, to certain types of music (Panksepp, 1995)' (Panksepp and Trevarthen, 2009, p. 117). The significance of opiate receptors and emotional involvement in perception is discussed in detail in Chapter 10.

Muscle control for both voiced 'vowel' (laryngeal) and voiceless 'consonant' (orofacial) vocal sounds seems to be generated in the lateral column of the mid-brain PAG (Davis et al., 1996). Laryngeal muscle control is thought to be mediated by a neurological pathway from the PAG via the nucleus retroambiguus to the nucleus ambiguus, where monosynaptic connections to the laryngeal motoneurons exist (Vanderhorst et al., 2001, on the basis of the study of rhesus monkeys; but see also Jürgens, 2002, pp. 240–1). This pathway is absolutely critical for the production of vocalizations; as well as being responsible for integrating vocal fold control, the nucleus ambiguus is also crucial for expiratory control, orofacial muscular control, and overall control of the laryngeal system (Jürgens, 1998)—i.e. all of the elements of fine muscular control that must be integrated in order to carry out precise and varied vocalization. This alone indicates how closely interrelated these functions are—and reiterates the potential significance of the development of this area of the brain identified in endocasts. As Davis et al. propose, 'the PAG is a crucial brain site for mammalian voice production, not only in the production of emotional or involuntary sounds, but also as a generator of specific respiratory and laryngeal motor patterns essential for human speech and song' (1996, p. 34).

So it seems that these foundations for fine laryngeal and orofacial control necessary for vocalizations have a long phylogenetic history, and that they have since been built upon during the *Homo* lineage. So what do humans have that other primates do not? Humans have a direct monosynaptic connection between the primary motor cortex (M_1, responsible for planned motor control) and the phonatory, articulatory, and respiratory areas of the medulla in the brain stem, but chimpanzees do not (Jürgens, 2002; Okanoya and Merker, 2007). Jürgens also reports that in contrast to the primates studied, modern humans have a direct connection between M_1 and the nucleus ambiguus (site of the laryngeal motoneurons) (Jürgens, 1976, 1992). This connection is *not* present in monkeys or chimpanzees, and probably not in any non-human mammals (Jürgens, 1992; Schulz et al. 2005), and would seem to be what allows humans to carry out planned structured vocal utterances that rely on laryngeal control. In contrast, control of the species-specific utterances communicating information about emotional state and arousal is largely involuntary in the lower mammals. Amongst non-human primates, although there are direct neurological connections from the neocortex to nuclei that control lip, tongue, and jaw articulators, allowing voluntary control of them, there is no such direct connection to the nucleus ambiguus that allows fine control of the larynx too (Schulz et al., 2005). As Lieberman (1994) reiterates, although higher primates can produce many vowel sounds and phonetic features of human speech, these are 'bound' into relatively narrow range of stereotyped calls based on affect (emotion), and they seem not to be able to produce novel voluntary vocal motoric sequences (see also Ploog, 2002). So, of the higher primates alive today, only humans possess a direct monosynaptic connection from the primary motor cortex to the nucleus ambiguus controlling the larynx (Jürgens, 1976, 2002; Schulz et al., 2005), allowing us to regulate the sound produced by the larynx itself, in combination with the use of our orofacial articulators and respiratory control, which are also controlled via a unique monosynaptic connection with the primary motor cortex (Okanoya and Merker, 2007).

Nevertheless, humans do not bypass the involvement of the visceromotor (instinctive emotional motor) control elements of the PAG in carrying out voluntary vocalizations; PET scan research by Schulz et al. (2005) suggests that the visceromotor mechanisms that organize reflex-like vocalizations in other mammals are still activated in humans during propositional speech vocalization, even though the instigation of the vocalization comes voluntarily from the neocortical primary motor cortex. This strongly reinforces the hypothesis that human vocal behaviour, whilst being unique today amongst higher primates in its degree of voluntary control, built upon the existing system for vocalizations communicating emotional state and arousal—a system which has become 'exapted' from that original purpose.

So, it is highly likely that the ability to perform emotional vocal expression involving orofacial control and laryngeal *activation*, in response to external stimuli and internal affective state, was present in all members of the *Homo* genus, and probably in all primates on the lineage between rhesus monkeys and humans. What separates us from other primates, however, is vocal behaviour which involves voluntary *control* of the larynx, and voluntary control and planning of the *structure* and *complexity* of vocal utterances. Although they have a separate pathway for controlling learned vocalizations, without the connection between the primary motor cortex and the laryngeal motoneurons in the nucleus ambiguus, other primates cannot show the human capacity for learning complex vocal patterns by imitation *and by invention*, and adapting them to novel situations.

The fossil evidence considered in Chapter 5 concerning cervical and thoracic vertebral innervation suggested that voluntary control over intensity, pitch, and contour of vocalization, through the cervical canal innervation of breath flow through the larynx, emerged first, being present in *Homo ergaster*. The monosynaptic connections allowing voluntary control over the larynx, articulation, and breathing, present in humans but absent in other higher primates, would have been of little use unless accompanied by this cervical innervation, so both would seem likely to have arisen around the same time. The later development of the thoracic neurology for voluntary extended breathing control would have had to have been accompanied by, or preceded by, the development of the monosynaptic connection between the primary motor cortex and the respiratory areas of the medulla in the brain stem—the area of the medulla which is also responsible for phonatory and articulatory control—allowing the planning and control of these aspects of vocalization. It would seem that these neurological connections, now unique to humans, must have developed during the period between the emergence of human-like cervical and thoracic canal innervation; i.e. between *Homo ergaster* and our last common ancestor with Neanderthals.

The FoxP2 genetic mutations

Recent genetic studies have shed some light on the emergence of voluntary control over vocalizations in the human lineage, by identifying two mutations on the Forkhead Box Protein P2 ('FoxP2') coding gene that are possessed by humans, defects in which are related to speech and language dysfunction

(Lai et al., 2001). The FoxP2 gene is possessed by all mammals so far analysed, as well as various other vertebrates, and in most animals is responsible for regulating gene expression, including for lung development and, in some cases, neural plasticity (Krause et al., 2007). In *Homo sapiens*, however, the FoxP2 gene features two mutations, at points 911 and 977, which are not possessed by our nearest relatives, chimpanzees (Lai et al., 2001; Krause et al., 2007). Studies of the KE Family, who show an inherited speech and language disorder, have shown that in humans *inactivation* of one FoxP2 copy leads to deficits in voluntary orofacial movement control and linguistic processing that are comparable to Broca's aphasia. Without both of these mutations functioning correctly, affected individuals show reduced orofacial motor coordination ability for speech, as well as difficulties in some elements of language comprehension. In other words, it is clear that in humans, the FoxP2 mutations have a key role in regulating some other genes which have specific roles in aspects of language and speech, including for fine orofacial muscular control (Krause et al., 2007).

These mutations on the human FoxP2 gene clearly arose through a selective sweep at some point in the human lineage. Are they really unique to *Homo sapiens*, however? Recent sequencing of this part of the genome of Neanderthals (two individuals from El Sidron cave, Spain) by Krause et al. (2007) has shown that these vocal-control and language-related mutations of FoxP2 were also possessed by Neanderthals—their results show that the Neanderthals carried a FoxP2 protein that was identical to that of present-day humans in the only two positions that differ between human and chimpanzee, and that the selective sweep that led to these mutations must have pre-dated our last common ancestor with Neanderthals. This finding accords well with the previously discussed physiological and neurological evidence regarding the development of increased voluntary control over vocalizations in the *Homo* lineage.

THE EAR, SOUND PERCEPTION, AND EVOLUTION

The production and perception of sound are inextricably linked; in most mammalian species the frequency range of their vocalizations (mainly determined by their laryngeal system; see Chapter 5) largely coincides with that most effectively perceived by their auditory system (Wind, 1990). The human auditory meatus (ear canal), at 25–30 mm in length, resonates at about 3,500 Hz, resulting in the ear being progressively more sensitive to frequencies up to this range; this is also the upper range of normal human vocalization frequencies (Benade, 1990; Campbell and Greated, 1987), meaning that the whole of the human vocal range is very effectively perceived. A number of important changes appear to have occurred over the course of the evolution of the human auditory system with specific relation to its use for perceiving vocalizations. It would seem that sometimes the development of the auditory system has been spurred by changes in vocalization capabilities and at other points vice versa.

Liberman et al. (1967) identified certain *articulatory invariants* across all human cultures, regardless of language. Articulatory invariants are the sounds produced by certain mechanisms of articulation that are human universals (Liberman et al., 1967; Wind, 1990); because the mechanisms are universal, so are the sounds they produce (this is not to say that the *meaning* attributed to these sounds is universal, only the sounds themselves). The conventional view has been that the auditory system in humans is a special perceptual mechanism which evolved to detect these articulatory invariants.

Over the course of mammalian evolution, laryngeal and respiratory differentiation made more variation in pitch and amplitude possible, as well as more frequent vocalizations, while increasing auditory sophistication enabled a greater variety of sounds to be perceived and analysed. As Wind puts it, 'Mammalian vocalisations are mainly used for intraspecific communication. Therefore auditory function must have been shaped by selective pressures gearing them to vocalisation capacities' (Wind, 1990, p. 185).

However, Kuhl (1988) suggests that in fact the evolutionary process may have been the reverse of this: that instead, the acoustical properties of speech were selected for as a consequence of the particular invariant properties that are detected by the auditory system. Kuhl found that non-human animals and prelinguistic human infants (neither of whom are capable of producing articulate speech) show particular tendency to perceive speech sounds. She found that the categories of sound perceived most effectively ('natural auditory categories') conform to the phonetic categories of speech. This set of phonetic sounds is universal across all human languages; despite the great variability of the sounds that it is possible to produce with the human vocal tract, a very limited set of acoustic forms is used in the worlds' languages.

That sounds within these 'natural auditory categories' are more effectively perceived than other sounds by non-human mammals and prelinguistic humans suggests that the perceptual mechanisms relied on for the perception of speech sounds are very ancient, pre-dating humans by millions of years. As hominins developed the ability to control their vocalizations in order to communicate, there would have been strong selective pressure to be able to vocalize using those sounds that were most easily perceived by others. So, it would seem that audition, specifically the existence of 'natural auditory categories', was responsible for the shaping of the phonetic categories in speech, and that the mechanisms for perceiving such categories were in place in our hominin ancestors long before they were capable of actually producing articulate speech. This has important implications for the debate concerning

Neanderthals' information-processing abilities with respect to their supposed vocal abilities (*contra* Lieberman, 1989; Chapter 5)—we have no reason to believe that Neanderthals lacked the ability to perceive and process those sounds which form the core of human vocalization.

However, whilst the phonetic categories of speech appear to have been shaped by some fundamental auditory system characteristics, it also seems to be the case that the human auditory system has become fine-tuned to be especially sensitive to the range of human vocalization. The discrimination of pitch has a fundamental role in the recognition of sources of sounds in the environment, clearly important for all primates. Studies of monkeys show that tonotopy (systematic mapping of tones in the brain) is a universal feature of the primate brain (Morel et al., 1993; Brugge, 1985; Recanzone et al., 1993; cited in Turner and Ioannides, 2009). Liégois-Chauvel et al. (2003) describe the tonotopic organization of human auditory cortex responses to different frequencies, i.e. which parts of the primary auditory cortex respond most to which frequencies. They found that lateral areas of the primary auditory cortex produce the greatest electrophysiological response to sounds of around 400 Hz (low frequency sounds; their 'Best Frequencies'), whilst medial areas of the primary auditory cortex respond most to 4 KHz (high frequency sounds; their 'Best Frequencies'), with intervening areas of the primary auditory cortex responding most to intervening frequencies of sounds.

This range of sounds (400 Hz–4 KHz) corresponds to the frequency range most useful for perception of human speech. Studies of non-human primates (owl monkey and marmoset, Imig et al., 1977; Aitkin et al., 1986) have shown a similar lateral–medial/low–high frequency organization, but in their case the 'Best Frequency' range covers almost the entire range of frequencies of hearing of non-human primates, from <100 Hz–*c*.32 KHz, rather than just the range of their vocalizations. This constitutes further evidence that the human auditory perceptual system has become specifically fine-tuned to the perception of human vocalization as its principal concern.

A further piece of evidence indicating that the characteristics of the auditory perceptual systems are intimately linked with the development of vocalization is that the stapedius muscles of humans (which regulate the intensity of the operation of the stapes bones on the eardrums) involuntarily contract before the onset of vocalization. This reduces the intensity—to ourselves—of the sound of our own voice and prevents our own speech from masking other sounds from the surrounding environment, allowing us to hear other sounds even while we speak (Borg and Counter, 1989). The value of this function would increase as the length of vocalizations, and the diversity of their frequencies, increased; vocalizations obscuring environmental sounds would be less of a handicap when such utterances were short, but the more extended they became, the greater the proportion of the 'auditory scene' that would be lost. The same goes for the frequency range that constitutes the vocalizations:

the greater it is, the greater the number and variety of environmental sounds that would be obscured during an utterance. The existence of this system suggests that the auditory system, over the course of its evolution in human ancestors, also faced significant selective pressure as a consequence of our vocalization abilities. It is clear that regardless of the order in which the major evolutionary developments occurred, the vocal and auditory systems must have evolved in response to each others' capabilities, each changing to be 'fine-tuned' to changes in the other.

Another major factor in the development of the human auditory system was bipedalism (Daniel, 1990; Spoor et al., 1994; Spoor and Zonneveld, 1998). The labyrinthine structure of the inner ear is important for the regulation of balance, as well as hearing. Daniel (1990) and Spoor et al. (1994) link the anatomy of this structure with bipedal posture. In all quadrapedal mammals, the labyrinthine capsule of the inner ear (see Figure 6.3) is situated vertically, whereas in anatomically modern humans it is rotated through almost 90 degrees (Daniel, 1990).

Fortunately, the bony labyrinth is often preserved inside fossil hominin skulls, which has allowed Spoor et al. (1994) to carry out detailed computer reconstructions of the labyrinthine system of various fossil hominins, based on computer tomography (CT) scans of the skulls. They then compared the rotation and dimensions of the labyrinths of the fossils with those of modern humans and extant great apes. Whilst there is not the same extreme of difference in rotation of the labyrinth between humans, hominins, and great apes as there is between humans and quadrapedal animals, there are still important differences. They found that the labyrinths of *Australopithicus* and *Paranthropus* closely resemble those of modern great apes, whereas fully modern morphology is first exhibited by *Homo ergaster* (SK 847, *c.*1.5 million years ago), and they take this to imply that the australopithecines and paranthropines were only partially bipedal, whereas *Homo ergaster* had full bipedality.

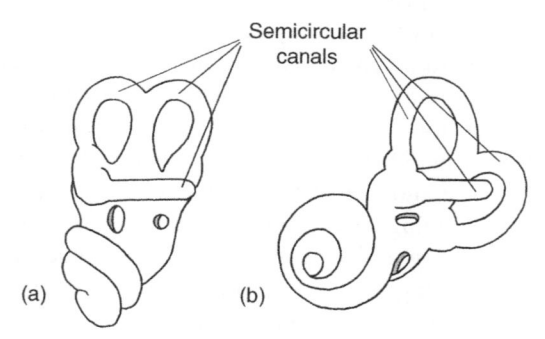

Fig. 6.3. The left labyrinths of the inner ear of (a) a quadruped and (b) *Homo sapiens*. The arrangement has rotated through 90 degrees, apparently as a consequence of bipedalism. (Redrawn from Daniel 1990, p. 260.)

A more comprehensive comparison of the entire labyrinthine structure in humans, great apes, and other primates was subsequently carried out by Spoor and Zonneveld (1998), who showed that there are other important phylogenetic differences in morphology of the labyrinth. Especially pronounced are differences in the relative size and some of the angles of the semicircular canals, and the angle of rotation of the cochlea. Their analysis also showed that all four of the great ape species (*Pan troglodytes* (common chimpanzee), *Pan paniscus* (bonobo), *Gorilla gorilla* (gorilla), and *Pongo pygmaeus* (orangutan)) share the same morphological pattern, as do the South African australopithecines and the hominoid *Dryopithicus brancoi* (RUD 77; Spoor, 1996; Kordos and Begun, 1997) found in 10-million-year-old deposits at Rudabanya, Hungary. These similarities strongly suggest that the traits which are unique to humans have emerged over the course of the development of the genus *Homo* (Spoor and Zonneveld, 1998).

In something of a moderation of the conclusions presented by Spoor et al. (1994), Spoor and Zonneveld acknowledge that the semicircular canals are responsible for a whole suite of functions in addition to balance and the coordination of body movements during locomotion, including stabilization of gaze during movement of the body, through moderation of neck and ocular reflexes. These functions are still related to locomotive patterns, however, and it seems likely that the moderation of head movement with particular regard to locomotion would have been a significant selective force in shaping the morphology of the semicircular canals. These canals may also fulfil a more fundamental role in the regulation of bipedal motion itself. As Spoor and Zonneveld describe, Ito and Hinoki (1991) found that in quadrapeds the vestibulo-spinal tract, which is supplied from the semicircular canals, extends only to the upper thoracic levels of the spinal cord, whereas in humans it seems that neurological impulses from the semicircular canals reach the lumbosacral cord, influencing the activity of lower- as well as upper-extremity movement.

The important question for the present study remains at what point these changes in the morphology of the inner ear occurred. Spoor and Zonneveld (1998) reiterate the findings of Spoor et al. (1994) that *Homo ergaster* displays modern human canal dimensions, whilst the South African australopithecines show great-ape-like morphology. Analysis of the inner-ear structures of 500,000-year-old *Homo heidelbergensis* from Atapuerca Sima de los Huesos, Spain, also confirms the presence of a modern-human-like inner-ear anatomy as a well-established early trait of the *Homo* genus (Martinez et al., 2004).

Interestingly, given that *Homo ergaster* is thought to be ancestral to all later forms of *Homo*, and that the Atapuerca *Homo heidelbergensis* are thought to be directly ancestral to Neanderthals, Neanderthals prove themselves yet again to be the exception to the rule by having smaller *vertical* canal dimensions than modern humans. This highlights the fact that there is clearly more than simply a shift from quadrapedalism to bipedalism at work in determining the morphology of the semicircular canals. Nevertheless, the fundamental changes in inner-ear anatomy to that exhibited by modern humans do appear to have occurred with the emergence of *Homo ergaster* (or in some as-yet undiscovered precursor), and the morphology of the human inner-ear labyrinth is unique amongst extant primates.

At least some of the changes in the features of the labyrinthine anatomy appear to be correlated with another physiological change particularly associated with *Homo*, namely, changes in the shape of the basicranium. The cranial base accommodates the labyrinthine structure, and the shortening and flexing of the basicranium with increased brain size, which occurred in all subsequent hominins to some extent (see Chapter 5), seems to have had distorting effects upon the labyrinth. This means that the changes in the shape of the inner-ear labyrinth would have been occurring at the same time as the changes in vocal tract morphology which are also influenced by basicranial anatomy (see Chapter 5).

Daniel (1990) relates the rotation of the inner-ear anatomy in humans to various other neurological, as well as physiological, changes. In humans, rotation of the labyrinth from vertical to horizontal occurs ontogenically too, during the development from foetus to adult, which reinforces the idea that bipedal locomotion is a major cause of the rotation. Daniel proposes that these adjustments to the labyrinthine system as a consequence of bipedalism had important neurological knock-on effects with regard to language and learning. The vestibular neurological system associated with the labyrinth is also phylogenetically and ontogenically the first to exhibit lateralization tendencies. According to Daniel this, and clinical connections between vestibular dysfunction and language and learning deficits, suggest that the foundations of linguistic functions are first evidenced by *Homo ergaster*, being the first hominin to exhibit full bipedalism, and, as Spoor et al. (1994) have shown, a modern-like inner-ear labyrinth. This would correlate with the other evidence that *Homo ergaster* is the earliest hominin to exhibit signs of vocal ability distinct from those of higher primates (Chapter 5). Ultimately, the conclusion is that regardless of when *language* actually emerged in humans, the beginnings of the neurological predisposition for it, and other planned, extended, complex vocalizations such as singing, coincided with the development of full bipedalism and the associated labyrinthine-vestibular changes, which were in place by at least 1.5–1.9 million years ago, in *Homo ergaster*.

CONCLUSIONS

In discussing the earliest foundations of musical behaviours in the human lineage, one is necessarily investigating the origins of the production and processing of complex vocalizations and muscular movements. Although there are clearly difficulties with carrying out reconstructions and analyses of fossil specimens, the amount, quality, and diversity of evidence available are improving all the time, and the understanding that underlies the reconstructions is accordingly increasingly refined. The major changes in the vocal apparatus that can be tracked with reasonable confidence in the fossil record nevertheless have to be understood in terms of other changes in functionally related neurological systems and behavioural capabilities in great apes and humans.

Fossil endocasts of hominin brains show particular development of regions in the left hemisphere, around Broca's area, that are associated with fine muscular control of sequences of vocalization and manual muscular movements; the earliest disproportionate development of this area occurs with *Homo rudolfensis*, and its development is especially pronounced in *Homo ergaster*.

In primates, including humans, the motor planning of extended, purposeful utterances relies on input to the motor cortex from the ventral premotor and prefrontal cortex, including Broca's area, and the *voluntary* integration of emotional content into vocalizations relies on input from the anterior limbic cortex and the periaqueductal grey matter (PAG). The PAG is also involved in reinforcing positive emotional experiences, including attachments to conspecifics and their vocal characteristics. The nucleus ambiguus is responsible for integrating vocal fold control, expiratory control, orofacial muscular control, and overall control of the laryngeal system. But of the higher primates alive today, only humans possess the neurological connection allowing us to regulate the sound produced by the larynx itself, in combination with the use of our orofacial articulators and respiratory control. In doing this, however, we still rely on input from the mechanisms that organize reflex-like vocalizations in other primates, reinforcing the idea that human vocal behaviour, although unique today amongst higher primates in its degree of voluntary control, built upon the existing system for vocalizations communicating emotional state and arousal.

The ability to perform emotional vocal expression involving orofacial control and laryngeal activation, in response to external stimuli and internal affective state, seems to have been present in all primates on the lineage between rhesus monkeys and humans, but unlike other primates we are capable of vocal behaviour which involves voluntary control of the larynx, voluntary control and planning of the *structure* and *complexity* of vocal utterances, and a capacity for learning complex vocal patterns by imitation *and by invention*. Over the course of our evolution we have developed the monosynaptic neurological pathways necessary for this control, most likely since our divergence from the essentially ape-like australopithecines, and before our last common ancestor with Neanderthals, around 600,000 years ago.

As with the vocal apparatus, the first significant developments of auditory anatomy occur with *Homo ergaster*, by around 1.7 million years ago, and also seem to be related to a shift to a fully upright posture. It seems likely that the moderation of head movement with particular regard to locomotion would have been a significant selective force in shaping the morphology of the semicircular canals of the inner ear, and they may also fulfil a more fundamental role in the regulation of bipedal motion itself. The shortening and flexing of the basicranium with increasing brain size had distorting effects on the shape of the labyrinth, housed as it is in the cranial base. Earlier hominins, on the other hand, exhibit auditory anatomy that is not significantly different from that of other higher primates.

Some features of auditory perception are obviously very ancient, being present in mammalian audition generally. One of these is the preferential perception of the so-called 'natural auditory categories'; these are also universal features of human speech sounds. This suggests that these qualities of vocalization were tailored to the capabilities of the auditory system: as hominins developed the ability to control their vocalizations in order to communicate, there would have been strong selective pressure to be able to vocalize using these sounds that are most easily perceived by others. So, it would seem that audition, specifically the existence of 'natural auditory categories', was initially responsible for the formation of particular vocalization properties, and that the mechanisms for perceiving such phonemic categories were in place in our hominin ancestors long before they were capable of actually producing articulate speech.

By contrast, other features of human auditory function appear to have faced significant selective pressure as a consequence of hominin vocalization capabilities. The value of the function of the stapedius inner-ear muscles in reducing the intensity of our perception of the sound of our own voice would presumably increase as the length of vocalizations, and the diversity of their frequencies, increased. Vocalizations obscuring environmental sounds would be less of a handicap when such utterances were short and limited in pitch and intensity, but the more extended and variable they became, the greater the proportion of the 'auditory scene' that would be lost without the operation of the stapedius muscles in the middle ear. So, it seems that important changes towards modern auditory anatomy occurred in *Homo erectus* with full bipedalism, that the vocal system has formed to produce sounds which are particularly well perceived by the auditory system, and that the human ear contains features which are particularly useful in an animal which carries out extended tonal vocalizations.

Overall, the evidence considered concurs that the emergence of modern-human-like vocal anatomy, and the neurology for control of it, emerged between *Homo ergaster* and our last common ancestor with Neanderthals, being present in both African and European *Homo heidelbergensis*.

The following two chapters look in particular at the nature of the complex vocal capabilities of humans, and how they relate to the capabilities required for musical production and processing.

7

Neurological Relationships Between Music and Speech

> People really do want to touch each other, to the heart. That's why you have music. If you can't say it, sing it.
>
> Keith Richards (2010), p. 56

INTRODUCTION

The fact that elements of music and language use the same physical mechanisms for production and perception—namely, the vocal apparatus and the auditory apparatus—has led to various proposals that language and music are closely related in terms of their origins. Chapter 5 examined the evolution of the physiology that would allow the production and perception of extended vocalizations of varied pitch, volume, and contour—the kind of vocalizations that would form the foundations for melodic behaviours. This was followed in Chapter 6 by an examination of the neurological foundations for the production of vocalizations in primates, and how those foundations are used in carrying out vocalizations in humans. These studies showed that human ancestors evolved the apparatus necessary to make complex, extended, tonally varied vocalizations over a period of around 1.5 million years, between *Homo ergaster* and *Homo heidelbergensis*. By around 600,000 years ago, *Homo heidelbergensis* had a brain of 1,200–1,400 cc (Atapuerca 4, Atapuerca 5, Petralona 1; Johanson and Edgar, 2006) and a vocal apparatus that was essentially modern. It is not generally suggested that *Homo* had modern language capabilities by 600,000 years ago, however, in terms of developed lexical complexity and syntax; in fact, the predominant view has traditionally been that there is little evidence for fully modern language use before the time that symbolic artefacts are first produced by anatomically modern humans (e.g. Mithen, 1998). If this is the case, what were these ancestors to modern humans, and modern humans themselves, using their large brains and developed vocal tracts for? What was the selective pressure responsible for vocal versatility being developed so long before it was used for fully developed language? This is a question that the following chapters seek to address.

The preceding evidence can tell us something about the development of the *capabilities* for vocal versatility, but little about the form of the actual behaviours which resulted from those capabilities. Having physiological mechanisms does not necessarily imply possession of the cognitive capacity to use them in particular ways—for example, for the production or perception of vocalizations, melody, and rhythm—nor does it give insight into why such physical and cognitive abilities evolved over time. To gain an idea of how these cognitive capacities developed, in relation to each other and to the physiological capabilities already explored, it is important to identify the ways that these functions relate to each other—or do not—in the modern human brain and behaviour.

The following chapters (7–10) examine the evidence from a wide range of areas of research, including neurology, music psychology, developmental psychology, and primatology, with a view to addressing these questions. The aim is to determine which of the capacities fundamental to musical behaviour are unique to humans and which share foundations with higher primates, which are innate and which are learned, which appear to be specifically dedicated to music and which are related to other behaviours.

Whilst later chapters deal in more detail with relevant neurological and behavioural evidence from developmental psychology, music psychology, and primate studies, this chapter is limited to evidence from a selection of neurological studies identifying areas of the modern human brain responsible for the production and processing of elements of music, speech, and language, and how they relate to each other. Are there different mechanisms that deal with different elements of sound production and perception? Are there particular mechanisms for music and different mechanisms for speech? Or do they share some mechanisms? This chapter then looks at whether we should expect to be able to find a neurological 'module' especially dedicated to music, or whether music makes use of mechanisms also used for other functions, and what these different possibilities mean for music's biological foundations.

Research relating musical functions (and deficits) to neurological study has a long history, beginning with some of the earliest investigations into brain function (and malfunction) on the basis of pathological studies in the nineteenth century; this body of research is now growing more rapidly than ever, particularly burgeoning with the refinements in brain-scanning technology in recent years. This has resulted in dedicated volumes such as

Peretz and Zatorre's (2003) *The Cognitive Neuroscience of Music*, Patel's (2008) *Music, Language and the Brain*, and Rebuschat et al.'s (2011) *Language and Music as Cognitive Systems*, in addition to hundreds of scientific journal articles. Earlier research and key conclusions are reviewed comprehensively by Borchgrevink (1982), Marin and Perry (1999), and Brust (2003).

Until recently, much of the research dealt predominantly with the production and perception of elements of the Western musical tradition, being concerned with technical instrumental ability, the recognition of pieces of music, and reading and writing of notation. However, some investigations, the more recent in particular, have sought to understand music-related brain functions at a more fundamental level. We are concerned here with musical capabilities at this fundamental level, of course, rather than such culturally generated and relatively recent elements of musicality as written notation and the naming of tunes. Consequently, of greatest interest are those studies exploring prosodic, tonal, rhythmic, and lexical-semantic comprehension and production in music and language. It would be impossible in a single chapter to do justice to the full range of investigations that have been carried out and the explanations that have been offered, but we can at least examine some key findings and draw out some of their major implications for critical relationships (and dissociations) between melodic and speech sound production and perception.

There are two main purposes here of examining the neuropsychological literature regarding the neurological organization of musical and other vocal functions. One is to attempt to relate this information to such evidence as we have of neurological development from the primate and fossil record (discussed in Chapter 6); the other is to identify interrelationships and differences between the neurology responsible for musical and linguistic (and other) functions, in attempting ultimately to propose a path for their development. As such, in some cases the neurological locations responsible for individual aspects of the cognitive functions are relevant, in other cases it is the implications of the associations and disassociations between those functions that are most significant. Areas of the brain referred to in the text are illustrated in Figure 7.1.

As singing and speech make use of the same physiological mechanisms for their production and perception (namely, the vocal tract and ear) one might expect the two to be closely linked, in evolutionary terms and in terms of the neurological mechanisms they use. This is indeed the case in many respects, though neurologically the relationships between the various music and speech functions are rather more complicated.

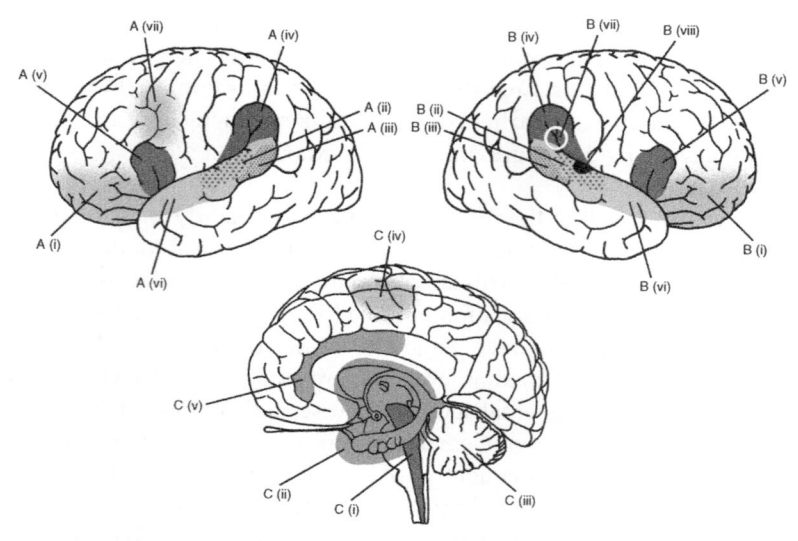

Fig. 7.1. Locations of parts of the human brain referred to in text.
A: *Left hemisphere*:
A(i): Left inferior frontal lobe
A(ii): Left hemisphere primary [BA 41, 42] auditory cortex
A(iii): Left hemisphere secondary [BA 22] auditory cortex
A(iv): Left hemisphere supramarginal gyrus [BA 40]
A(v): Broca's area [Left hemisphere BA 44 and 45]
A(vi): Left hemisphere superior temporal gyrus (STG) (includes A(ii) and A(iii))
A(vii): Premotor area
B: *Right hemisphere*:
B(i): Right inferior frontal lobe
B(ii): Right hemisphere primary [BA 41, 42] auditory cortex
B(iii): Right hemisphere secondary [BA 22] auditory cortex
B(iv): Right hemisphere supramarginal gyrus [BA 40]
B(v): Right hemisphere Brodmann's Areas 44 and 45
(including right frontal operculum [right hemisphere BA44])
B(vi): Right hemisphere superior temporal gyrus (STG) (includes B(ii) and B(iii))
B(vii): Right hemisphere temporo-parietal region
B(viii): Heschl's gyrus (part of primary auditory cortex in BA 41)
C: *Section*:
C(i): PAG (periaqueductal grey matter)
C(ii): Sub-cortical basal ganglia
C(iii): Cerebellum
C(iv): Supplementary motor area
C(v): Anterior cingulate cortex

HEMISPHERIC ORGANIZATION: LANGUAGE IN THE LEFT BRAIN, MUSIC IN THE RIGHT?

Conventional belief for many years was, as Borchgrevink (1982) and Sloboda (1985) observe, that language function is controlled almost exclusively by the left hemisphere, and music function by the right. Whilst there are certainly differences in left- and right-hemisphere involvement in tone, language, and rhythm processing, detailed neurological studies using a variety of techniques have shown that the simple traditional division of language = left and music = right is not accurate. It must be remembered that both music production and perception and speech production and perception activities are highly complex and each inevitably requires the involvement of numerous neurological areas simultaneously, in different aspects of each task.

For example, as Brust (2003) points out, music is a multi-modal activity; it involves the processing of numerous elements, including auditory phenomena, such as pitch, harmony, timbre, interval, and contour, embodied motor action, memory processes, emotion, and other associations. Not all of these functions will be located in the same brain hemisphere, and we cannot even necessarily expect each of these elements to be sharply localized and isolated in individual areas of the brain. Furthermore, musical experience is more than the sum of these parts, and the interaction of these stimuli creates a 'psychological whole' that is widely distributed throughout the brain. Similar observations, unsurprisingly, apply to linguistic communication. Nevertheless, as will be seen, it is indeed the case that there is distinct lateralization for particular aspects of music and speech production and processing, with certain functions occurring in predominantly either the left or right hemisphere (see Brown et al., 2006a, for a useful review of research to date), and these can tell us something about the ways in which these functions are related to each other. They can be investigated using a variety of techniques generating various types of data.

IDENTIFYING FUNCTIONAL NEUROANATOMY: BRAIN SCANNING AND NEUROPATHOLOGY

Attempts to identify which areas of the brain are used for particular activities involve selectively separating the different aspects of these activities and examining which areas of the brain are required for their normal functioning. This can be attempted by examining which neurological areas are used when certain activities are carried out or, conversely, which aspects of those activities are detrimentally affected by the absence of certain neurological input. Brain scanning technology can attempt to identify the former, and studies of neurological pathologies (brain damage resulting from injury, stroke, or birth defects) or selective anaesthetization can help to identify the latter.

Neuropathological studies from the mid-twentieth century onwards have investigated relationships between language and music with respect to hemispheric organization, highlighting not just areas of the brain critically (although not necessarily exclusively) involved in their perception, but also relationships between music and language processing (Brust, 2003). Observations of brain pathologies which result in aphasia (loss of language production/processing abilities) and/or amusia (loss of music production/processing abilities) illustrate that some functions are shared and some are lateralized (Borchgrevink, 1982; Schweiger, 1985; Marin and Perry, 1999; Patel, 2008). The process of identifying precise locations of neurological areas implicated in different tasks is complicated by the fact that pathologies are rarely neatly localized, and cerebral anaesthetization (to induce a temporary 'pathology') is very difficult to achieve selectively; but valuable conclusions about lateralization and localization of functions have nevertheless been achieved.

The neurological localization and organization of these various functions have been further illuminated by the refinement of technologies for brain scanning, showing the functioning of the brain 'in action'—an area of investigation which has massively expanded in recent years, with the generation of huge quantities of new data. Whilst they can provide considerable insight, such investigations are complicated by the necessity of 'factoring out' other cerebral processes that are inevitably occurring at the same time as those under observation; this is a problem especially salient when investigating musical processing, since musical activities by their very nature involve multiple simultaneous cerebral processes, each of which is potentially relevant to the investigation, as are the relationships between them.

Analysis of brain activity using scanning techniques is either *direct* or *indirect*. Direct methods detect the actual electrical activity of individual neurons (via intracellular recordings) or populations of neurons (via electroencephalograms, EEGs), or through changes in the magnetic activity that is associated with the electrical activity (via magnetoencephalograms, MEGs). Indirect methods detect changes in the brain's metabolic activity, through the analysis of blood flow, for example (via positron emission tomography, PET, or functional magnetic resonance imaging, fMRI). In general, direct methods have very good temporal but poor spatial resolution, and indirect methods good spatial but poor temporal resolution (Besson and Schön, 2003).

Turner and Ioannides (2009) review the various brain imaging techniques, and their advantages and disadvantages, and also point out that, whilst the techniques for *identifying* neurological activity have developed very rapidly, the techniques for *analysing* the neurological activity identified are still developing and being refined. They caution that brain imaging studies should be considered to be suggestive rather than definitive. Amongst other things, the differences in temporal resolution of the different techniques, and the variety of thresholds used, mean that it is difficult to representatively combine the finer aspects of the data from different techniques.

Whilst it can be difficult to combine the *data* from different techniques the individual conclusions derived from particular studies can be very usefully examined in light of each other. Functional imaging studies (indirect methods) such as PET and fMRI are very useful in showing which areas of the brain change their level of activity in response to certain task demands. As Zatorre (2003) points out, however, what they cannot show is which parts of those activated areas are critically or predominantly involved in the process; this is where lesion (neuropathology) studies and artificial stimulation/inhibition techniques (such as transcranial magnetic stimulation) provide important complementary evidence. The more we can draw on all these forms of study the fuller our picture of brain function will be. By examining a combination of neuropathological and scanning studies it will be possible to draw some important conclusions regarding the interrelationships (and independence) of various elements of tonal, linguistic, and rhythmic processing in the brain.

SPEECH AND MELODY PRODUCTION

The discussion of primate and human vocalization neurology in Chapter 6 showed that over the course of their divergence from other higher primates, humans not only developed a vocal tract with the ability to produce a greater range of tones, but also built upon the higher primate ability to perform emotional vocal expression involving orofacial control and laryngeal activation, by developing a set of neurological connections that allow for voluntary control of the larynx, and voluntary control and planning of the *duration*, *structure*, and *complexity* of vocal utterances. Humans can learn complex vocal patterns by imitation *and by invention*, and adapt them to novel situations, but they nevertheless still involve activation of deep-rooted and evolutionarily ancient instinctive emotional motor control neurology.

The primatological evidence discussed by Jürgens (2002) and the human PET scan evidence discussed by Schulz et al. (2005) indicates that the visceromotor (instinctive emotional motor) control elements of the periaqueductal grey matter (PAG) are automatically used in carrying out voluntary voiced utterances (Chapter 6). In humans, with regard to the production of emotional expression in the intonation of utterances, Snow (2000) concludes that 'linguistic analysis suggests that even intonation patterns traditionally described as nonemotional have their underpinnings in the speaker's emotions' and goes on to observe that pathological studies 'indicate that intonation patterns described as either linguistic or emotional are mediated by right-hemisphere substrate specialised for emotional experience' (Snow, 2000, p. 1). It appears, then, that whatever the content of speech, its production, and specifically its intonation, unavoidably involves some input from neurological systems concerned with affective (emotional) function. What implications does this have for the ways that we use vocalizations today, in speech and singing? Clearly there are deep-rooted neurological mechanisms used in *all* vocalizations, but other aspects seem to be later-emerging specializations.

Brown, Martinez, and Parsons (2006a) used PET to observe the brain structures activated during the creation of improvised melodies and sentences, to see the extent to which there are overlaps and unique activations for each task. Both melodic phrase creation and sentence creation relied upon nearly identical activation in many areas of the brain, showing significant overlap in cognitive activity, but with some task-specific activation. In particular, they found that the generation of sentences relied upon activation of Brodmann's areas (BA) 44 and 45 in the left hemisphere, corresponding with Broca's area, and that the generation of melodic phrases relied upon activation of BA 44 in the *right* hemisphere along with BA 45 in the left hemisphere. Brown et al. (2006b) also found that right-hemisphere BA 44 (the right frontal operculum) was activated in carrying out dance movements; i.e. right-hemisphere BA 44 is involved in generating both dance movements and improvised vocal melodies. Brown et al. (2006b) proposed that it is involved in general motor sequencing (its level of activation was the same regardless of the movement task required). Their observations and model suggest that cognitive overlaps exist between dance and vocal motor sequencing, as well as between *phonological* aspects of melody and speech production (the generation of complex sound structures), but that melody and speech production rely on distinct, specialized, areas for the generation of their specific information content, or semantics.

Brain pathologies can provide us with some further clues as to which functions are shared between speech and melodic vocalization, and which are specializations. Music and speech production can be simultaneously affected by brain pathologies. As Marin and Perry (1999) observe, those instances in which both speech and musical vocalizations are simultaneously affected by avocalias (impairment of vocalization ability) do 'suggest the possibility of a common neural substrate at a lower level in the hierarchy of vocal expressive functions', again entirely as we would expect on the basis of the neurological evidence discussed in Chapter 6. They go on to observe, however, that 'dissociations in performance between verbal and melodic vocal tasks give support for a contralateral localisation of crucial neural processing at higher levels' (p. 665).

There is an important distinction to note between the terms 'vocal' and 'verbal', and 'speech' and 'language'. *Verbal* necessarily implies linguistic content, whereas *vocal* does not, encompassing non-linguistic utterances too. This distinction is an important one, made carefully by Marin and Perry (1999), but sometimes neglected in other literature where the terms have, on occasion, been used interchangeably. Likewise, for the purposes of this discussion, *language* will be used to refer specifically only to the lexical, syntactic elements of *speech*, while *speech* also encompasses tonal elements of communication.

Marin and Perry (1999) go on to state that examples of aphasia without amusia 'decisively contradict the hypothesis that *language* and music share common neural substrates' (p. 665, emphasis added). Such cases of aphasia without amusia indicate that music and the lexical, syntactic elements of speech do not share common neural substrates. For example, Broca's aphasia, which can result in severe disruption of semantic speech performance, can leave affective-tonal vocal performance quite intact; Marin and Perry (1999) describe one of their own clinical cases of Broca's aphasia, which resulted in severe linguistic expressive disorder, but no impairment of the performance of non-lyrical vocal melodies, highlighting the distinction between vocal tonal control and linguistic function.

A further example of a strong dissociation between linguistic ability and the ability to communicate using vocal tonal contour is provided by another case observed by Marin and Perry (1999), a patient with severe bilateral cortical atrophy. The patient had no spontaneous language at all, and an inability to sing lyrics. However, her spontaneous vocalizations contained appropriate sound contours expressing emotion and intention, consisting of iterative sounds in sentence-like sequences, demonstrating that the prosodic element of speech was intact; preserved along with this was perfect rhythm, intonation, and prosody in singing, lacking only lyrical content. That aphasics often continue to be able to produce and comprehend the *prosodic* components of language, its emotional tonal content, confirms the distinction between the linguistic and prosodic elements of speech function.

Lesions isolating speech areas in the left hemisphere can also result in the condition of echolalia, whereby the patient is able to faithfully repeat any linguistic utterance they have heard, but have no understanding of the linguistic content of the utterance (Geschwind et al., 1965). The fact that the phonological loop (allowing repetition of heard utterances) can function in isolation from the linguistic cognitive processes indicates that the system controlling complex vocalization ability and the system controlling language comprehension are distinct from each other. It appears that the neurology responsible for the comprehension and production of linguistic content is quite distinct from that responsible for the physical act of carrying out complex vocalizations and, furthermore, from that responsible for the emotive prosodic content of such complex vocalizations, be it in speech or in song.

PROCESSING OF TONAL INFORMATION IN MUSIC AND SPEECH

Processing auditory signals at a neurological level involves numerous systems, and has been the subject of extensive research, with regard to auditory signals generally, and more specifically with regard to speech and musical signals and relationships in the ways in which they are processed.

In his review of the history of neurological investigations of musical processing, Brust (2003) highlights that processing of tonal content in both speech and music can be detrimentally affected by right-hemisphere damage. In particular, pathology of the right-hemisphere temporo-parietal region can result in loss of the ability to comprehend music *and* to understand nuances and inflections in speech. Timbral content in both speech and music can also become impossible to process as a consequence of right-hemisphere damage (Brust, 2003). In Peretz's (2003) view, the only consensus regarding specific areas of the brain involved in music processing relates to pitch-contour processing, for which the superior temporal gyrus and frontal areas of the right side of the brain seem to be responsible. But it seems that 'the intonation patterns of speech seem to recruit similarly located, if not identical, brain circuitries (Zatorre et al., 1992; Patel et al., 1998)' (Peretz, 2003, p. 200).

These findings are paralleled by the studies of congenital amusics carried out by Ayotte et al. (2002) and by Patel et al. (2008), who found that both musical pitch perception and speech prosody perception are simultaneously affected in congenital amusics, apparently being processed by the same cognitive mechanism.

Ayotte et al.'s (2002) investigation examined the effect of congenital amusia on pitch processing in speech. In their discussion they elaborate that

> It is important to note that the pitch defect of amusical individuals does not seem to compromise music exclusively. The impairment extends to the discrimination of intonation patterns, when all linguistic cues are removed. This observation suggests that the pitch deficiency experienced by amusical individuals is not music-specific but is music-relevant. In effect, as mentioned previously, fine-grained discrimination of pitch is probably more relevant to music than to any other domain, including speech intonation. Music is probably the only domain in which fine-grained pitch discrimination is required for its appreciation. Accordingly, a degraded pitch perception system may compromise music perception but leave other domains, such as speech intonation in which meaningful pitch variations are coarse, relatively unaffected. Yet, the same pitch-tracking mechanism may subserve both domains.   (Ayotte et al., 2002, p. 250)[1]

In other words, it seems likely that the ability to discriminate intonation patterns in speech (prosody) uses the same pitch discrimination mechanism as is used for music, but that (1) the use of this mechanism by music is more refined than modern linguistic speech requires, because (2) in speech the intonation pattern may be coarse enough not to be very detrimentally affected

---

[1] They state in the abstract of their paper that 'Interestingly the disorder appears specific to the musical domain. Congenital amusical individuals process and recognise speech, including speech prosody, common environmental sounds and human voices, as well as control subjects [do]' (Ayotte et al., 2002, p. 238); this seems to be at odds with the main discussion and conclusions of the paper, as outlined here.

by the impairment of this ability, and/or linguistic elements of the speech signal may disambiguate the pitch signal.

Indeed, Patel et al. (2008) show that discrimination of speech intonation patterns is detrimentally affected in congenital amusics, to the extent that 30 per cent of congenital amusics studied (from British and French-Canadian backgrounds) had difficulty in discriminating a question from a statement—a discrimination that relies on correctly identifying pitch fall or rise. That the comprehension and production of the prosodic elements of linguistic speech often exist intact with the comprehension and production of the prosodic elements of music, and that when one of these functions is damaged, so too is the other (Mazzucchi et al., 1982; Peretz, 1993; Peretz et al., 1994; Patel et al., 1998; Patel et al., 2008), indicate that there are indeed shared common substrates between elements of speech and musical functions: namely, those concerned with prosodic intonation and melodic contour. The findings outlined by Brust (2003) and Peretz (2003) indicate that these functions are predominantly right-hemisphere localized. Zatorre (2003) reports that several experiments carried out in his own laboratory, both behavioural-lesion and functional imaging studies, have shown that secondary auditory cortex areas in the right hemisphere are more involved than left-hemisphere areas not just in the melody-processing tasks, but also in tasks involving the *imagination* of well-known melodies. Feedback in the production and perception of pitch is reliant on right-hemisphere superior temporal gyrus (STG) primary and secondary auditory cortex, but this task did not show activation of the left-hemisphere areas.

The degree of involvement of the left hemisphere in the processing of tones seems to be dependent upon the level of analysis required in the processing, with greater left-hemisphere involvement when greater analysis of the sound is required. Earlier research by Peretz and Morais (1980, 1983) showed that non-musicians (i.e. individuals not formally trained to a high level in Western musical performance) exhibited a right-ear (left-hemisphere) advantage when required to *analytically* process dichotically presented tones (where different information is presented simultaneously to both ears and the subject is required to selectively attend to one). Conversely, they exhibited a left-ear (right-hemisphere) advantage when required to *holistically* process the tones. Shanon (1982) and Schweiger and Maltzman (1985) also both found that the degree of tendency for right-ear/left-hemisphere dominance depended upon the complexity of the analytic (vs. holistic) task: greater analytical demands require greater left-hemisphere participation. When no conscious processing of the sound is required by the situation, it seems that there is little left-hemisphere involvement.

Reporting on PET scan data during dichotic tests, Bogen (1985) suggests that there may be a link between human voice processing and the processing of other sounds with a rich timbre. Melodies without much timbre or harmonic content (played on a wooden recorder) showed little if any left-ear/right hemisphere advantage; in contrast, an organ chord produced a distinct left-ear/right-hemisphere advantage. As noted, it is certainly the case that right-hemisphere damage can result in the inability to process timbral content in both speech and music (Brust, 2003). Bogen and Van Lancker and Canter (1982) observes that voice recognition is detrimentally affected by right-lesion damage, and proposes that the advantage for the right hemisphere in timbre processing might be related to voice recognition. That human voice recognition requires little or none of the analytical input of the left hemisphere is interesting, as it implies that the mechanisms for the recognition of voices are quite separate from those of linguistic analysis and comprehension. The mechanisms for voice recognition are almost certainly evolutionarily far older than those for linguistic processing, and it is possible that particular use is made of these more ancient mechanisms for processing timbre-rich musical sounds. Menon et al. (2002) have confirmed a functional asymmetry in timbre processing, although they found that both hemispheres were activated to a similar *extent*; activation areas of the temporal lobe 'were significantly posterior in the left hemisphere when compared with those of the right' (Turner and Ioannides, 2009, p. 164). The activation areas specific to timbre also appear to be different from those involved in pitch and melody processing, being deeper in the temporal lobe—this, suggest Turner and Ioannides (2009), may indicate involvement of emotive processing in response to timbral qualities.

As is the case with the control of affective (emotional) sound *production* (Chapter 6), the areas of the brain responsible for controlling processing of affective content in signals seem to be evolutionarily ancient. Karow et al. (2001) compared the abilities of cortical brain-damaged subjects with or without concomitant subcortical basal ganglia damage to process affective speech prosody, emotional facial expression, and linguistically based information. It is the cortical areas that have particularly developed over the course of human brain evolution, whilst subcortical areas retain structures that are evolutionarily ancient. Those experimental subjects with only cortical damage performed all tasks on which they were tested without significant difficulty, irrespective of the location of the lesion in the right or left hemisphere. However, amongst those with subcortical damage, there was a distinct lateralizational bias in their abilities: those with left-hemisphere subcortical damage had difficulty processing the linguistic information, whilst those with right-hemisphere subcortical damage had the greatest difficulty processing prosodic-emotional information and facial expression. As well as reinforcing the distinction between these elements of vocal content (linguistic-semantic and affective), the integration of neurologically ancient subcortical systems into the processing of emotive information of both vocalization and facial expression suggests (perhaps unsurprisingly) that the evolutionary roots for these abilities are ancient (see also Belin et al., 2004). Interestingly the findings also suggest that the processing of aspects of linguistic information relies on more ancient subcortical structures too, though in the left hemisphere rather than right. The relationship between tonal content and facial expression is explored in more detail in Chapter 8.

Neural pathologies also highlight distinctions between the processing required for linguistic, musical, and environmental sounds. In cases of auditory agnosia, the perception of speech, animal noises, music, and other environmental sounds may be selectively disrupted, depending upon the localization of the brain damage. In cases of generalized auditory agnosia (usually the result of bilateral damage), the perception of timbre and complex sounds, as well as rhythm and temporal information in the stimuli, is often severely disrupted (Marin and Perry, 1999). In cases where the brain lesion is in the right hemisphere only, preservation of normal speech perception can exist whilst the perception of non-verbal sounds (including music) is disrupted (Fujii et al., 1990). In contrast, there are also cases of unilateral left-hemisphere lesions in which the perception of non-speech sounds (including music) tends to be largely intact, whilst speech perception is disrupted (Marin and Perry, 1999).

Buchanan et al. (2000) used fMRI to observe brain activity during tasks involving discrimination of emotional tone and phonemic characteristics of heard spoken words. They found that in both the phoneme analysis and the emotional tone analysis tasks significant bilateral activity was observed in comparison to baseline activity, but that in comparison between the two tasks, there was hemispheric specialization. The task involving analysis of emotional characteristics resulted in significant activity in the right inferior frontal lobe, whereas the phonemic characteristic analysis task led to greater activation in the left inferior frontal lobe. Again this provides evidence of shared neurology for some elements of both linguistic and tonal-affective content analysis, with other elements of those functions being undertaken by left and right cerebral hemisphere specializations.

Some specifics of these specializations are suggested by earlier research by Benson (1985). Benson observed PET scans of the activity in the brains of subjects while they carried out various tasks relating to the production and perception of speech. His results show that the left hemisphere appears to be dominant with regard to semantic verbal meaning and syntactic sequencing and relationships, whilst the right hemisphere is dominant in prosodic melody and semantic visual images (Benson, 1985, table 3, p. 201). For the other aspects of gestural, prosodic, semantic, and syntactic language, both hemispheres participate.

The separation of linguistic meaning processing on the one hand from musical phrase structure and sentence structure on the other is also highlighted by Besson and Schön (2003). The processing of the appropriateness of words within semantic sentence structure and of notes or chords within musical phrase structure seems to generate the same neurological response (see also Steinbeis and Koelsch, 2008). However, the processing of the linguistic meaning of words generates a neurological response specific to that task (Besson and Schön, 2003). When dealing with what might be termed 'secondary level' information associated with sounds, learned associations such as meaning, imagery, and representations, it seems that some of the same structures are used in both speech and music processing: 'Secondary auditory regions (BA 22) are activated by hearing and understanding words (Falk, 2000) as well as by listening to scales (Sergent et al., 1992), by auditory imagery for sounds (Zatorre et al., 1996), and access to melodic representations (Platel et al., 1997). The supramarginal gyrus (BA 40) seems involved in understanding the symbolism of language (Falk, 2000) and the reading of musical scores (Sergent et al., 1992)' (Besson and Schön, 2003, p. 277).

So music and language share large neurological areas, but also exhibit differences in their neurological organization; in particular, music is more bilaterally organized and language functions are usually strongly left-lateralized (in right-handed males, particularly). Turner and Ioannides conclude that 'This is consistent with a view that the capacity to be affected by music, when compared to language, is more likely to be innate, and is supported by specialized brain areas. Language, by this argument, is a highly specialized subset of music cognition' (Turner and Ioannides, 2009, p. 174).

TONAL AND RHYTHMIC INFORMATION PROCESSING

Peretz et al. (1994) highlight the links between the systems processing musical and linguistic prosody on the one hand and their dissociation from systems processing musical and linguistic rhythm on the other: their patients with bilateral lesions of the superior temporal cortex exhibited simultaneously impaired perception of prosody in speech and perception of melody in music. In contrast, their perception of the rhythmic elements of speech and music was preserved. Patel et al. (1998) conclude from their study of two amusics that their findings 'support the view that the perception of speech intonation and melodic contour share certain cognitive and neural resources, as do the perception of rhythmic grouping in the linguistic and non-linguistic domain' (p. 136).

Zatorre (2003) also proposes that two different processing systems are involved in the perception of auditory sequences, one specialized for processing spectral (tonal frequency) elements of the signal, and the other more specialized for processing temporal information. The former predominates in the processing of tonal sequences, and the latter in the resolution of linguistic sequences, although both would be involved in processing both types of signal. If we possessed a single *linear* processing system which processed both types of information there would be a trade-off between the quality of information possible to extract from each type of signal, but having specialized *parallel* systems allows both types of information to be extracted at a high level of detail. This appears to be somewhat analogous to the functioning of rods and cones in the eye, the former specializing in contrast at the expense of colour, the latter the reverse (although in the case of temporal/ spectral sound processing we are discussing a part of the perceptual brain architecture rather than that of the sensory organ itself).

Research in Zatorre's lab has shown that corresponding regions of the auditory cortex in the left and right hemispheres appear to fulfil these parallel functions, with the left auditory cortex showing greater cerebral blood-flow response to temporal rather than spectral variation, and the equivalent area in the right hemisphere showing the opposite pattern (Zatorre, 2003).

Studies of the actual physiological structure of the brain also appear to support this proposed difference in function. The human brain has a larger volume of white matter underlying Heschl's gyrus in the left hemisphere than in the right, which appears to relate to the level of myelinization of the nerve fibres there. Myelinization allows faster transmission of information (heavily myelinated nerves fire faster), thus allowing greater temporal resolution of an input, but it results in less dense nerve bundles, which would produce lower spectral resolution. The less myelinated nerves underlying Heschl's gyrus in the *right* hemisphere are more densely packed, resulting in greater spectral processing ability, but slower firing and thus lower temporal resolution. This difference supports the hypothesis that the right-hemisphere primary auditory cortex has become specialized at processing the spectral elements of sound inputs, whilst the left-hemisphere primary auditory cortex has become specialized at extracting temporal information, in each case from both speech and musical signals. This allows both types of information to be extracted from the signal without one compromising the other (Zatorre, 2003). This reciprocal specialization of two symmetrical areas of the auditory cortex in the left and right hemispheres may be a human-specific development, as macaques' auditory cortical neurons are sensitive to *both* spectral and temporal information simultaneously.

Although there is specialization in the primary auditory cortex of the two hemispheres, both hemispheres are actually involved in processing of different aspects of rhythmic processing. Petsche et al. (1991) report a number of pathological cases illustrating the importance of the left frontal hemisphere in rhythm perception and production, though their own work using EEG scans illustrated that, in fact, many areas in both hemispheres are stimulated when processing rhythm, and that the two halves of the brain are functionally closely connected whilst doing so, more so than 'during any other sensory function' (Petsche et al., 1991, p. 326).

fMRI studies by Grahn and colleagues of links between rhythm and movement have shown that while listening to and processing rhythm, there is bilateral activation in the basal ganglia, the cerebellum, supplementary motor area, premotor area, and superior temporal gyrus—in other words both the auditory cortex and *motor areas* are activated in both hemispheres when processing rhythm (Grahn and Brett, 2007). The level of activation is affected by whether or not the rhythm attended to has a consistent beat—when it does there is greater response in the basal ganglia (motor areas) and auditory cortex (processing meaning) than when listening to a rhythm with no beat. These findings are consistent across both formally trained musicians and non-musically trained individuals, with the exception that trained musicians show greater communication between the auditory cortex and motor areas (Grahn and Rowe, 2009), as well as some activation in the left frontal short-term working memory area—the articulatory rehearsal area—during rhythm perception (Grahn and Brett, 2007).

Turner and Ioannides (2009) also report from their own recent MEG experiments, and that of Popescu et al. (2004), that both hemispheres are involved in rhythm tracking but that there is asymmetry in the level of activation, with the left hemisphere seeing more activation when regular rhythms are encountered; the involvement of the left hemisphere decreases as rhythmic stimuli diverge from a perfectly metrical tempo. They find that the involvement of right-hemisphere areas remains consistent whether tracking metrical or non-metrical rhythms: 'The activity in different brain areas reflects musical structure over different timescales; auditory and motor areas closely follow the low-level, high-frequency musical structure. In contrast, frontal areas contain a slower response, presumably playing a more integrative role' (Turner and Ioannides, 2009, p. 171). They suggest that the fact that music activates so many distant areas of the brain in a cooperative way, including auditory, motor, and integrative areas, is one reason why music has such a profound effect on us.

People spontaneously seek temporal regularity in the occurrence of sequences of events over a wide range of stimuli, not just music (Drake and Bertrand, 2003), and this relies on the same underlying neurology to extract the temporal regularity regardless of the nature of the auditory stimulus. People seem to be most sensitive to changes in regularity of stimuli with around 600 ms (0.6 seconds) between them (inter-onset interval, or IOI). This preferred IOI varies between individuals, with a range of about 300–800 ms, but the optimal zone clusters around 600 ms, over a wide range of stimuli (Drake and Bertrand, 2003). On the basis of existing research, and in the context of proposing further research to test these (and possibly more) apparent universals, Drake and Bertrand (2003, pp. 24–9) propose five candidates for universals in human temporal processing:

1: *Segmentation and grouping*: we tend to group into perceptual units events that have similar physical characteristics or that occur close in time;

2: *Predisposition towards regularity*: processing is better for regular than irregular sequences, and we tend to hear as regular sequences that are not really regular;

3: *Active search for regularity*: we spontaneously search for temporal regularities and organize events around this perceived regularity;

4: *Temporal zone of optimal processing*: we process information best if it arrives at an intermediate rate (discussed earlier);

5: *Predisposition towards simple duration ratios*: we tend to hear a time interval as twice as long or short as previous intervals.

Interestingly, the preferred IOI for metre processing corresponds closely with human gait duration, in which the duration of each stride for which the foot is on the ground in walking also clusters around 600 ms (Alexander and Jayes, 1980; Minetti and Alexander, 1997), which could betray a close relationship between the 'fine-tuning' of human rhythmic perception and control, and the development of bipedal locomotion. The relationships between rhythmic processing and motor control are explored in more depth in Chapter 9.

Whilst there is a distinction between rhythmic and *melodic* processing, there seems to be an association between rhythmic and *linguistic* function, consistent with these left- and right-hemispheric functional specializations. Research discussed by Samson and Ehrlé (2003) suggests that left temporal lobe function may both cover IOI processing in rhythmic (and musical) sequences and have an equivalent role in syllable processing. Alcock et al. (2000) studied a family (known as KE Family) with an inherited developmental speech and language disorder (related to a defect in the human version of the FoxP2 gene; see Chapter 6), who were tested on various tasks related to the production and perception of melody and rhythm. They found that whilst the affected family members were not deficient in the production or perception of pitch, be it in the form of melodies or single notes, they were deficient in the production and the perception of rhythm, both vocally and manually. Alcock et al. concluded that the oral and praxic defects of the family's condition could not be at the root of the impairment in timing ability, but that the reverse must be the case; i.e. rhythmic capacity forms an important component of oral/praxic ability. It seems that oral/praxic ability must have built upon existing rhythmic capacities—and thus that the capacity to perform planned sequences of complex muscular movements of rhythmic behaviour, both orally and manually, must pre-date oral praxic abilities.

Left- and right-hemisphere specialization for the processing of aspects of speech and musical sounds, respectively, appears to be innate, being present at birth, even before any linguistic capability is developed. Dichotic listening tests show that infants exhibit the same left-ear/right-hemisphere and right-ear/left-hemisphere advantages as adults do for processing music and speech sounds respectively, well before they actually understand linguistic content (Best et al., 1982; Bertoncini et al., 1989; Trehub, 2003)—i.e. the left-hemisphere dominance for speech sounds is, in this case, not a consequence of linguistic processing, but of patterning properties of the vocalization itself.

DOES THE BRAIN HAVE A NEUROLOGICAL MODULAR SPECIALIZATION DEDICATED UNIQUELY TO MUSIC?

The battle to demonstrate that music cognition has a biological, rather than purely cultural, foundation was once a rather uphill one. Obviously the production and processing of music involves neurological activity, but it was far from widely accepted that this activity was not just 'sequestered' by musical activity, as opposed to that neurology being specialized for musical activities. It tended to be thought that if musical activities 'simply' make use of neurological mechanisms which have other functions, then this would indicate that it was purely a sort of 'parasitic' cultural product (see Chapter 11 for more detailed discussion). Thanks to the work of numerous researchers in recent years, however, the view that musical capabilities are purely a cultural product is now held by few researchers into music cognition.

One significant question which remains, however, is whether, in order for music to be demonstrated to have a biological foundation, it need be demonstrated that there are biological mechanisms *exclusively* dedicated to it, such as a 'music module', as some researchers have suggested. For example, Peretz has tended to assert that the demonstration of a biological foundation for musical capabilities depends upon the identification of at least one music-specific 'module' or domain of cognitive processing—a functional neural network that is demonstrably dedicated exclusively to musical function. This has, despite some claims, proven difficult to demonstrate. But does it necessarily follow that any cognitive domain need be uniquely dedicated to music in order to demonstrate that music has a biological foundation? Peretz's (2003) rationale is that

> Brain specialization refers to the possibility that the human brain is equipped with neural networks that are dedicated to the processing of music. Finding support for the existence of such music-specific networks suggests that music may have biological roots. Conversely, the discovery that music may have systematic associations with other cognitive domains or variable organisation across individuals would support the view that music is a cultural product. (2003, p. 192)

In fact, it can be argued that these two proposals are not mutually exclusive: the first is that if there are *any* networks or domains used by music that are specific to music processing then this would indicate a biological basis, whereas the second is that if music uses *any* networks or domains that are also used for other cognitive processes then this would indicate that it is a cultural product. The possibility clearly exists that music draws upon some domains that are dedicated to it and some which are used for other cognitive processing too.

Finding a music-specific network would not necessarily reinforce the idea that music has biological roots; in fact, a network that is now music-specific may not always have been so; music may, in theory, continue to use a network originally responsible for, or shared with, a different but related function. Conversely, in theory, networks/domains that were once specific to music could subsequently have come to be shared with other behaviours. The discovery that music has systematic associations with other cognitive domains would not necessarily support the view that music is a cultural product.

Numerous structures will be involved in the processing of musical production and perception. Each of the individual structures themselves, as well as the *combination* of structures activated in music processing, may be either music-specific or non-music-specific. If music *only* uses domains dedicated to it then it would follow that it is likely that music is a specially evolved capacity building upon no other existing capacities and not functioning in the same way as any other capacities. However, the possibility that music 'has biological roots' is not ruled out by the converse: if music *only* uses domains which also have other cognitive functions, this would not rule out the possibility that the use of *that combination* of those domains was something specific and exclusive to music.

In other words, music may make use of systematic associations with other cognitive domains, but this would not necessarily indicate that it was a cultural product. First, those shared domains may be being used in combination with music-specialized domains. Second, the *combination* of cognitive domains used could be unique to music, whether or not the domains themselves are, and that combination could have arisen through biological selection. Third, there is the possibility that cognitive domains that are shared between music and other functions initially served musical or music-like functions and that they came to be used in other processes later—clearly this would not rule out a biological foundation for music either.

One difficulty in discussing what constitutes a module, and whether or not it need be uniquely dedicated to a particular category of task, is in setting the parameters of what constitutes an ability, mechanism, structure, network, or domain. These terms are not all interchangeable, either with each other or with 'module', but the limits (and possible overlaps) of their meanings are not consistently established. Another is in defining the category of task, for 'music' is clearly not a single task; anything we might categorize as a 'module' dedicated to it will have to be a mercurial combination of active processing mechanisms. The individual task-elements into which we might split music processing (e.g. pitch change perception, beat ascription) may not in any case correspond with any single one of the individual processing mechanisms in the brain.

So what does the neurological evidence considered bring to this issue? Peretz proposes that there are two principal 'anchorage points' of brain specialization for music: 'the encoding of pitch along musical scales and the ascribing of a regular beat to incoming events' (2003, p. 201). At present, however, it is not clear that either of these processing capabilities is specific to music, although music may use them in specific ways. On occasion it has seemed that a processing module uniquely dedicated to music has been identified, but findings such as those described in the preceding sections show that whilst these capabilities (encoding of pitch and ascription of a regular beat) are clearly absolutely fundamental to the appreciation of musical stimuli, it is far from clear that they are uniquely dedicated to music. It is very clear that very many of the functional brain structures used during music production and perception are also used during speech perception and production (see also Besson and Schön, 2003), and it is far from clear that there are any individual *structures* dedicated solely to music. It seems likely, however, that there are functional *combinations* of those structures that *are* unique to music production and perception. What we can look for is unique combinations of those brain structures that are being used for music, and in this sense these combinations would constitute 'music-dedicated modules'.

Peretz (2003) ultimately suggests that if we can identify a combination of musical abilities (or processing components) that are acquired by people in all cultures, a few of which are essential for the normal development of musical skills (by some presumably independent standard of 'normal development' and 'musical skills'), then we would have identified a music-specific network of abilities, or module.[2] I would contend that there would seem to be no reason for any of those individual networks/mechanisms used in musical skills to be uniquely dedicated to music, although the 'common core' should constitute a music-specific *combination* of processing networks/cognitive abilities; this potent combination, the fact that the human brain *can* integrate elements

---

[2] 'The music-specific neural networks should correspond to a common core of musical abilities that is acquired by all normally developing individuals of the same culture. This common core should also be universal, in forming the essence of the musical competence acquired by members of all cultures. This universal competence can hopefully be reduced to a few essential processing components that represent the germ of brain specialization for music. In this perspective there is no need for all musical abilities to have initial specialization. Brain specialization for a few mechanisms that are essential to the normal development of musical skills would suffice' (Peretz, 2003, p. 201).

such as pitch, affective information, temporal content, and so on, and process them as part of an entity ('music'), then, is the brain specialization that makes it 'music-ready'.

The contention, however, that this unique common core of musical abilities should be universal is an important one, and one that research within our own culture is addressing with respect to members of our own culture, but which requires more investigation across other cultures. The identification of the extent to which these capabilities are human universals is clearly an area to which ethnomusicology has the potential to make a critical contribution (Chapter 1 discusses some important shared elements), but as Peretz (2003) points out, ethnomusicologists, in recent years at least, in common with the majority of cultural anthropologists, have tended to avoid making generalizations across cultures (with a few very significant exceptions such as Bruno Nettl and John Blacking; see Chapter 2). This is clearly an area in which significant advances could be made in coming years.

CONCLUSIONS

In summary, whilst humans have developed the specialized ability to voluntarily control the duration, structure, and complexity of vocalizations, with precise control of the larynx and orofacial musculature, the process of vocalization nevertheless relies on activation of deep-rooted and evolutionarily ancient instinctive emotional motor control neurology used in all primate vocalizations. These systems are involved in human vocalizations of all types.

The *production* of both vocal melodies and speech contours expressing emotion and intention (speech prosody) draw upon related structures (concerned with affective-tonal vocal production), but semantic elements of linguistic speech draw upon different, specialized, neurological structures (related not to the physical act of carrying out the vocalization, but to the expression and comprehension of its meaning).

Similarly, the *processing* of tonal content in both speech and music seems to rely on the same structures as each other: it appears that the ability to discriminate intonation patterns in speech (prosody) uses the same pitch discrimination mechanism as is used for pitch processing in music, but that the use of this mechanism by music is very refined, more refined than modern linguistic speech requires. These mechanisms are located predominantly in the right-hemisphere temporo-parietal region, in the superior temporal gyrus and frontal areas, with neurons in the right auditory cortex being especially tuned to pitch perception. The analysis of emotional tone content in speech seems to rely on activation in the right inferior frontal lobe, as well as evolutionarily ancient subcortical structures in the right hemisphere which are also used for processing emotional content in facial expression.

There is some involvement of left-hemisphere structures when the processing of tonal input requires conscious analysis, but not when no conscious processing of the sound is required. This does not extend to timbre processing and voice recognition, which appear to rely only on right-hemisphere structures, deeper in the temporal lobe than pitch-processing mechanisms.

In terms of processing, whilst neurons in the right auditory cortex seem to be especially sensitive to spectral (tonal) information in auditory stimuli, those in the left auditory cortex seem to be especially sensitive to temporal information. Left-hemisphere areas are also implicated in the capacity to perform planned sequences of complex muscular movements of rhythmic behaviour. These are important functions of Broca's area and the areas around it in the left hemisphere, and these functions also form an important component of oral/praxic ability. The left hemisphere appears to be dominant with regard to semantic verbal meaning and syntactic sequencing and relationships; phoneme analysis relies on activation in the left inferior frontal lobe, and linguistic processing relies also on some input from subcortical structures in the left hemisphere.

So both music and language functions use both left- and right-hemisphere structures; certain sub-functions of music and language seem to be shared, whereas functional lateralization does seem to be the case for others (e.g. Borchgrevink, 1982; Schweiger, 1985; Marin and Perry, 1999; Brust, 2003). In particular, areas in the right hemisphere appear to be responsible for processing and production, in both melody and speech vocalization, of prosodic melody, pitch control, tonality of singing, timbre processing, and voice recognition. Left-hemisphere regions appear to be implicated in production and processing of semantic verbal meaning and syntactic sequences, as well as rhythmic production and perception, and some aspects of conscious auditory analysis. It is important to note that whilst some of the structures involved in specific aspects of auditory processing appear to be specifically lateralized to the left or right hemisphere, the overall process of sound perception involves activation of structures in both hemispheres, and in some cases specific tasks themselves also involve bilateral activation, albeit with some degree of bias towards greater activation in one hemisphere or the other.

The process of instigating vocalization draws upon deep-rooted structures involved in tonal-emotional expression, and the perception of emotional content in tonal information also involves structures that are used for extracting emotional information from other sensory signals. The specialized functions involved in linguistic verbal meaning have emerged later than these systems, apparently building upon some of the same structures in the left hemisphere that are required for the performance of planned sequences of complex muscular movements, including both vocalization (through laryngeal and orofacial muscular control) and rhythmic behaviours.

It is far from clear that any of the neurological structures that are used in processing the various aspects of musical stimuli are uniquely dedicated to that purpose, although it certainly appears that at least some of these structures have become finely tuned to the considerable processing demands of musical stimuli. Vocal tonal production and the processing of tonal information each use a combination of both evolutionarily ancient structures involved in primate emotional vocal signalling, and structures which (whilst also used for other forms of communication) have become finely tuned to the demands of musical activity. The structures that are used for music production and processing are also used in producing and processing aspects of other forms of communication, but this combination of neurological structures, and the interaction between them in producing and processing musical signals, represent a specialized use of those mechanisms.

Indeed, following their review of a large body of research, Marin and Perry proposed that 'The close correspondence between the networks of regions involved in singing and [linguistic] speaking suggests that [linguistic] speech may have evolved from *an already-complex system for the voluntary control of vocalisation*. Their divergences suggest that the later evolving aspects of these two uniquely human abilities are essentially hemispheric specialisations' (1999, p. 692—emphasis added). The extent to which music and language processing overlap and share neural resources in adults and children (Patel, 2003; Koelsch et al., 2003; Schön et al., 2004; Koelsch et al., 2005) lead Koelsch and Siebel to conclude that 'it appears that the human brain, at least at an early age, does not treat language and music as strictly separate domains, but rather treats language as a special case of music' (2005, p. 582). Following analysis of an extensive range of developmental and neurological studies, Brandt et al. (2012) elaborate this situation even more strongly. In their view, from a developmental perspective, language should be described 'as a special type of music in which referential discourse is bootstrapped onto a musical framework' (p. 1), because the ability to learn to speak language is predicated upon the ability to hear musically. In order to learn speech, infants rely on processing of timbre, pitch and tonal contour, dynamic stress and rhythm as a scaffold for the later acquisition of the linguistic specializations of sematic and syntactic content (Brandt et al., 2012); early in development human speech is heard in the same way as music, with many of the processing differences between the two emerging over the course of development.

The fact that the various elements of musical activities draw upon cognitive mechanisms that are also used in similar ways during other activities—or vice versa—does not undermine the importance of musical activity in an evolutionary perspective, its relevance to the development of human cognition, or its importance in human behaviour; on the contrary these overlaps can emphasize its fundamentally important relationship with other critical aspects of human cognition and behaviour. This in itself has great implications for the role of evolution in the shaping of musical capabilities and the role of musical capabilities in the evolution of other aspects of human behaviour. We can look more closely at some of the overlaps between aspects of musical processing and the processing of other sound information, including speech, in the following chapters; these relationships may provide some insight into how and why the functions emerged and developed.

8

Vocal Versatility and Complexity in an Evolutionary Context

> What is it that makes you write songs? In a way you want to stretch yourself into other people's hearts. You want to plant yourself there, or at least get a resonance, where other people become a bigger instrument than the one you're playing. To write a song that is remembered and taken to heart is a connection, a touching of bases. A thread that runs through all of us.
>
> Keith Richards (2010), pp. 277–8

INTRODUCTION

It is clear that not only do the physiological capacities for musical and linguistic production and perception have much in common, but also many of the neurological mechanisms for producing and perceiving elements of musical and speech stimuli are shared too, with specializations apparently having emerged from those shared foundations. The following chapters turn from looking at relationships between musical and speech behaviour principally in terms of the physiology (including neurophysiology) responsible, to looking primarily at relationships in terms of functionality, with a view to better understanding why such relationships might exist.

This chapter starts by looking briefly at early vocal behaviours in some other primates. Which of the elements that we see in human infants are also present in other primate infants? What roles do these early vocalizations play, and what might this tell us about how they emerged in our predecessors? We then go on to look at which elements of melodic and rhythmic perception are innate in humans; those elements that are possessed at birth must be a product of our biology and the consequence of evolutionary processes, rather than being culturally learned during life. Staying with infant behaviours, the next section of the chapter looks at how adults vocally communicate with prelinguistic infants, what effects those vocalizations have, and what they have in common with melodic vocalization. In interacting with prelinguistic infants, of course, any communicative properties of the vocalization must come from non-linguistic elements, such as tonal contour and rhythmicity. The similarities between tonal aspects of melodic and other communicative vocalizations have led to suggestions that they have a shared ancestry. The chapter goes on to examine those ideas, and then the roles that vocalizations play in other primate social communication. Finally, we look at rationales for how and why complex vocalization capabilities emerged in human ancestors, and how these may have related to social complexity.

EVIDENCE FOR AN INHERITED CAPACITY FOR THE PERCEPTION OF MELODY AND RHYTHM

The study of infant perceptual and vocalization abilities is useful in the context of researching the evolution of vocal complexity for several reasons. First, it allows for the identification of the extent to which these abilities are innate versus culturally learned. Second, it allows for the study of the ways in which vocal interaction can communicate information that is non-linguistic. Third, in combination these areas of investigation can lead to rationales for the evolution of those capacities for vocal production and perception that are, indeed, innate. Similarly, the study of vocalization capabilities in other primates can contribute to each of these areas.

Early vocal behaviours in primate infants

Primate infants display vocalization behaviours from an early age, and these may be valuable in informing us about which aspects of vocalization behaviours have an ancient foundation, and ways in which they develop and are useful. In those examples reported, primate infant vocalizations are dominated by social interpersonal concerns. Following his study of trill vocalizations in white-faced capuchin monkeys *Cebus capucinus*, Gros-Louis (2002) observed that they seem to facilitate social interactions, and such vocalizations were carried out to the greatest extent by immature individuals and adult females. Infants trilled most when approaching other individuals, and those that did so tended to interact with them in an affiliative way afterwards; Gros-Louis proposes that infants' trilling action may have an immediate effect on the listener's behaviour towards them, particularly with regard to such socially important activities as receiving grooming, touching, and inspecting food.

Similar findings were made by Elowson et al. (1998b), observing infant pygmy marmosets; response from a caregiving adult was far more likely to be given to an infant when vocalizing than when it was not. Elowson et al. (1998a) also made particular study of the 'babbling' behaviour (a form of early vocal behaviour, which is prelinguistic in human infants) of non-human primate infants; they identified considerable parallels with human infant babbling behaviour, on a number of counts. Elowson et al. (1998a) review and outline seven key features of human infant babbling:

Babbling is universal and frequent irrespective of the infant's cultural background;

Babbling is rhythmic and repetitive;

Babbling begins between six and ten months, peaking at seven months of age;

Babbling comprises a subset of the phonetic sounds found in adult speech;

Babbling has well-formed units with the consonant-vowel structure of adult speech;

Babbling lacks apparent meaning with respect to how syllables are used by adults in language;

Babbling is instrumental in moulding the caregiver–child bond in the infant's first year.   (p. 33, box 2)

In their long-term study of Amazonian pygmy marmosets *Cebuella pygmaea*, Elowson et al. (1998a) observed that '(1) both humans and pygmy marmosets have evolved a family-based social unit of cooperating individuals; (2) both humans and marmosets have an open, plastic system of communication where subtle changes of vocal production occur over the life-time of the individual; and (3) both go through a period of babbling as part of their vocal development' (p. 36).

In particular, 'the long, complex strings of vocalisations produced by the infant monkeys' (Elowson et al., 1998a, p. 32) have much in common with the babbling of human infants outlined earlier, such as 'universality, repetition, use of a subset of the adult vocal repertoire, recognisably adult-like vocal structure and lack of a clear vocal referent' (p. 31).

Future research will, hopefully, show whether this type of vocal development is exclusive to these particular primates, or whether equivalent babbling behaviours are exhibited by other primate infants. It is conspicuous, however, that although they are repetitive, rhythmicity appears to be lacking from the vocalizations of the monkeys. What is especially interesting about this case is that the pygmy marmoset and other species in the *Callitrichidae* family are unique amongst non-human primates in that they live in clearly-defined groups of extended family based on a sophisticated social system. These groups include 'in addition to infant twins, the two parents, older juvenile and subadult siblings, and often unrelated individuals... the analogy to human families is striking' (Elowson et al., 1998a, p. 32). In these groups, all individuals contribute to the care of the infants, in the form of carrying them, grooming them, and close physical contact (huddling together at night and as protection against predators). It is possible that it is this form of social organization that leads to the early development and importance of such socially orientated vocal behaviour.

For Elowson et al. (1998a) the most interesting element of their findings is that it indicates that similar processes underlie human and (at least some) other primate vocal learning, and they accordingly propose that pygmy marmoset babbling has a relevance to understanding the evolutionary processes of our vocal development. In the context of the current research the most interesting implication is a phylogenetic, or evolutionary, one. If this is a form of infant vocal behaviour and learning that is phylogenetically ancient and genuinely shared with other non-human primates, then it is interesting in that right alone, indicating that we could have the expectation that such behaviours were carried out by intervening ancestral hominin species. If, on the other hand, it is a type of behaviour that has emerged by convergence, it is interesting as it could inform about the circumstances in which such behaviours are selectively useful to the individual and the group. In either case, it indicates that lower primates have the cognitive capacities to develop and use vocalizations in such a way, that such vocalizations have an important social function at an early stage, and that there is no issue with foundations of such behavioural capabilities having an ancient provenance. There is no reason for us to believe that any of our hominin ancestors could not also have exhibited such behaviours. Whether there would have been selective advantages to doing so is discussed further later in this chapter (and in the wider context of musical behaviours, in Chapter 11). If so, differences from species to species would presumably be in terms of the vocal range, control, and capabilities, which would ultimately bear upon the complexity of the utterances possible, and normally used. Other social vocalizations in higher primates are discussed below.

Infant-directed speech, music, and vocalization

If a sense of rhythm and a sense of melody exist in neonates (newborn children) and infants, this would indicate that perception of these phenomena (inherent in music) is not simply a product of cultural influences, but instead has some basis in innate hereditary factors. A wide range of research has now demonstrated that this is exactly the case: infants are born with abilities fundamental to musical processing, including the perception of frequency, timing, and timbre, and, in fact, these abilities appear to be more finely developed than is necessary for the purposes of musical perception alone (Trehub, 2003).

Further, prelinguistic (and specifically, pre-lexical) infants have been shown to understand some content from rhythm and tone alone, which provides some confirmation of the ability of non-verbal (rhythmic and melodic) vocal utterances to communicate important information in their own right, independent of linguistic content. In tests by Fernald (1989a, 1992a, 1993; Fernald et al. 1989) 5-month-olds heard utterances of approval and of prohibition in four languages, and displayed significantly different affective (emotional) response to the two types of utterances, regardless of language, and regardless of whether they had heard that language before. The test was also carried out with nonsense-syllables, with the same result, and in all instances intensity of utterance was controlled for, so that the only variations were in rhythm and tone.

This inherited capacity to perceive melody and rhythm and to gain affective information from them was not necessarily selected for by music; however, it does suggest that the ability to comprehend non-linguistic emotional utterances was selectively important. Study of the nature of infant-directed vocalization can shed some light on why this might have been the case.

As a form of vocalization which exaggerates the prosody of normal speech (Trainor et al. 2000), infant-directed (ID) speech (or 'motherese', or 'parentese') has received an increasing amount of attention in recent years, in the context of language evolution, the development of linguistic capacities, and the evolution of music (e.g. Fernald, 1992a, 1992b, 1993; Werker et al. 1994; Papousek, 1996a and b; Lewkowicz, 1998; Dissanayake, 2000; Mang, 2000; Trainor et al., 2000; Falk, 2004b).

ID speech is used most intensively toward children between the ages of 3 and 5 months, but continues to be used up to around the age of 3 years (Falk, 2004b; Stern et al., 1983), and is characterized by particular frequency ranges and contours. It has a higher frequency (F_0), and thus pitch, than adult-directed (AD) speech; pitch contour is exaggerated in utterances which have varied pitch; it has a larger pitch range, slower tempo, and is more rhythmic than typical AD speech (e.g. Fernald, 1992b; Trainor et al., 2000). In particular, when soothing an infant, infant-directed vocalizations have a low pitch with a falling pitch contour, when engaging attention and eliciting a response a rising pitch contour is usually used, and when attempting to maintain the attention of the infant a bell-shaped contour is usual (Fernald, 1992b). These properties have a high correlation across cultures, and this use of exaggerated vocal range and stereotyped patterns seems to be a universal trait of human parental behaviour, having been observed in cultures in Europe, Africa, and Asia, in tonal and non-tonal languages (Greiser and Kuhl, 1988; Fernald, 1992b, 1993; Werker et al., 1994; Kitamura et al., 2002). For example, English and Mandarin mothers use very similar ID speech to speak to their babies (Greiser and Kuhl, 1988), and English and Cantonese infants aged 4½ to 9 months both showed preference for ID over AD speech, irrespective of whether it was coming from English or Cantonese speakers (Werker, et al., 1994); these findings attribute a particular significance to the properties of ID speech, irrespective of cultural origin.

In addition to these attention-gaining or -maintaining vocalizations, there is also great consistency across cultures in the form of ID vocalizations communicating approval, disapproval/prohibition, or calming/soothing—essentially, emotive vocalizations communicating the affective state of the speaker, or attempting to elicit such a state in the receiver. Across English, French, German, and Italian, these are highly consistent in form (Fernald, 1992b). Infant-directed vocalizations of approval have a high mean F_0, a wide F_0 range, and a rise–fall contour. Prohibition and disapproval vocalizations, in contrast, have a low mean F_0, narrow F_0 range, and are short with an abrupt onset. Comforting vocalizations also have a low F_0 and narrow F_0 range, but they are longer, have a softer onset, and are described by Fernald as having a 'legato' quality.

At a rudimentary level, one would expect precisely this difference between prohibitive and calming vocalizations, in terms of onset and duration. The former takes the form of a sudden, arousing, shocking sound, eliciting a state of sensory arousal which must be similar to that provoked by any loud and unexpected noise from the environment, and when coming from the infant's own parent is perhaps particularly potent as a consequence of coming from a source normally associated with comfort and security. The latter, calming vocalizations, act in exactly the opposite way: gentle onset would result in little autonomic arousal, and a narrow frequency range over a longer duration would maintain this low state of positive arousal with no unexpected stimuli. As an aside it can be observed that the very common 'ssshhh' sound has precisely these characteristics and intended effect; this is perhaps the most extreme version of such a stimulus uttered in these circumstances, and minimizes the stimulus information by avoiding engaging the larynx at all (as well as featuring reduced bodily movement and a neutral facial expression). A continual sensory input of a consistent form is comforting precisely because it contains nothing unexpected, and the fact that it is coming from a source already associated with such calm and security must add to this effect.

As noted, accompanying onset and duration in importance are cues from contour and average pitch. These properties of exaggerated contour and raised frequency seem to fulfil important roles in addressing the initial limits of an infant's perceptual and cognitive abilities. Infants are more sensitive to sounds at higher pitch ranges, and continuity in pitch contour is an important cue in attending to a single speech source; exaggerating these contours makes it easier for an infant (who cannot rely on linguistic content for this purpose) to attend selectively to the sound source (Fernald, 1992b). The signal is thus 'high in perceptual salience and relatively easy to process' (Fernald, 1992b, p. 419).

The overall F_0 of the utterance is clearly also important in communicating affect (emotion), with high frequency consistently being associated with positive sentiment, such as expressing approval and stimulating and maintaining attention. This is consistent with the findings of Scherer (1985, 1986) that increases in mean F_0 and F_0 range are typical of vocal expressions of enjoyment and happiness. In contrast, F_0 decreases, especially with a harsh voice, are typical of expressions of irritation and anger. This appears to be consistent not only across human cultures, but in various other species too. Morton (1977) found that across a diversity of animals high tonal vocal sounds were associated with appeasement, submission, friendliness, or fear whereas low sounds were associated with threat. This has led Trainor et al. (2000) to suggest that another factor in the higher frequency of ID vocalizations might be that vocalizations directed towards an infant naturally underplay aggression.

Whilst various reasons might be suggested for this association between pitch and level of aggression, there is a significant factor influencing the frequency produced by the action of the vocal chords, which does not rely on any external environmental explanations: facial expression.

Orofacial musculature has a fundamental role in shaping the upper vocal tract and modulating the output of the vocal chords, in all speech, whether infant-directed or adult-directed (see Chapter 5). It is also responsible for the facial expression of emotion. It is thus perhaps unsurprising (though highly significant) that there is a correlation between given facial expressions and characteristics of vocalizations made at the same time. Facial expressions of disgust, happiness, sadness and anger are not learned and culturally determined, and the capacity to produce and recognize them seems to be innate and universal (e.g. Ekman and Friesen, 1971; Ekman, 1980; Carlson, 1994). In humans, Tartter (1980) showed that vocalizations made whilst smiling were of a higher mean frequency than those made with a neutral expression due to the effect of smiling on altering the shape of the upper vocal tract; smiling increases second formant frequency (Tartter and Braun, 1994). Tartter and Braun also showed that vocalizations made whilst frowning had lower formant frequency and longer vowel duration than speech with a neutral expression; listeners were able to discriminate speech made whilst frowning, smiling, and with neutral expression from each other with no visual input. Furthermore, they were able to do so in both vocalized and whispered speech, suggesting that the same effects of facial expression on aspects of affective content of vocalization would be applicable to vocalizations even in the absence of great vocal chord versatility (which accords well with Morton's (1977) findings regarding vocalizations in other animals). Given the universality of certain fundamental facial expressions and the correspondence between these and characteristics of vocalizations, we can also expect characteristics of particular emotional vocalizations to be universal and innate too.

This correlation between facial expression and vocal quality apparently has an ancient provenance, being shared by our nearest evolutionary relatives. Chimpanzees also frequently couple particular vocalizations with particular emotional facial expressions, the vocalizations being moderated by alterations of the size and shape of the mouth, and thus the resonating upper vocal tract (Falk, 2004b). Amongst bonobos, utterances always occur along with facial expressions, gestures, and tactile communication (Falk, 2004b; Bermejo and Omedes, 1999), which is true for human ID speech too (Falk, 2004b). Humans also use facial affect (emotional expression) and vocal affect to inform judgement about the affective content of each other; they seem to be interdependent systems in both production and perception (DeGelder and Vroomen, 2000), with the former having had a significant impact on the development of the nature of affective communication in the latter. Indeed, Belin et al. (2004) suggest that the human voice is an auditory equivalent of the face, in terms of both recognition of identity and extraction of emotion; the extraction of paralinguistic information from human vocalization (such as emotional content and identity information) may use an equivalent neurocognitive processing path to that used to extract such information from faces (a proposal that is supported by the findings of Karow et al., 2001; see chapter 7).

As Trainor and Schmidt (2003) observe, one peculiarity of the relationship between music and emotion is that some emotions, such as happiness and sadness, appear to be easy to express through music, whereas others, such as anger, are much more difficult to express through this medium. With regard to the suggested relationship between music, vocalization, and the facial expression of emotion, it is an intriguing possibility that this might be the case because music best expresses those emotions whose expression in vocalization is most influenced by facial expression. Were this the case then it could be predicted that the auditory characteristics of the vocal expression of anger are not mediated by facial expression of anger to the same extent that the auditory characteristics of the vocal expression of happiness and sadness are mediated by facial expression. This hypothesis is one that future research could certainly test.

It is very important to reiterate at this stage that infant-directed vocalizations are not something completely different from other vocal behaviour, they are a variation of it. Whilst ID speech is easily distinguishable from AD speech, especially non-emotional AD speech, the prosodic contours of ID speech are an exaggerated version of those that occur in normal speech, and are not something independent and unique (Trainor et al., 2000). Trainor and colleagues demonstrated that ID speech essentially reflects free expression of emotion, in comparison with typical AD speech, in which emotional content is more inhibited (although still present): 'When AD speech does express emotion, the same acoustic features are used as in ID speech' (Trainor et al., 2000, p. 188). They found that the emotions 'love-comfort', 'fear', and 'surprise' were equally distinguishable across ID and AD samples, with few differences between the properties of ID and AD samples (although a significant difference exists between the emotions that each is principally used to express). Observations such as 'through the melodies of the mother's voice, infants could gain early access to her feelings and intentions' (Fernald, 1992b, p. 423) are equally applicable to adult speech prosody, for adult listeners. The difference is that ID speech is tailored to the emotional and perceptual needs of infants. The fact that the prosodic features of vocalizations (in particular pitch, pitch contour, and tempo) are exaggerated in ID speech allows infants to attune to them better (see earlier), but they are not fulfilling a role fundamentally different from the role of prosody in AD speech.

Falk (2004b) proposes that since, in contrast to chimpanzee mothers, 'human mothers continually produce affectively positive vocalisations to their infants' (p. 498), this behaviour derived from an initial evolution of prosodic and instructional vocalizations in early hominin mothers. Falk argues that as hominin babies were born increasingly early (due to increases in cranial size), and thus increasingly helpless, they would have been unable to cling to their mothers. They would thus have relied on their mothers to carry them, which would have left the mother unable to use their hands to carry out the manual tasks in which they would normally have been engaged. Hominin mothers would thus have often been required to temporarily put the infant down whilst engaging in their manual tasks. For the infant that has been put down and separated from its mother, emitting prosodic emotional cries would be a good way of attracting the attention of the mother, and soliciting care. From the point of view of the mother, uttering emotive vocalizations to reassure the infant that the mother is near and solicit its attention would also be beneficial, certainly in stemming its cries, the utterances being 'disembodied extensions of mothers' cradling arms' (Falk, 2004b; p. 501).

From this scenario, Falk argues that this situation of 'parking the infant' led to the development of the characteristics of ID speech, and that this form of vocalization then formed the foundation for the development of later, complex linguistic vocalizations; 'Over time, words would have emerged in hominins from the prelinguistic melody (Fernald, 1994: 65) and become conventionalised' (Falk, 2004b, p. 500).

Whilst Falk's scenario of the necessity of setting infants down seems quite likely (that is, assuming that early *Homo* did not use technology such as a skin sling or papoose to assist with carrying infants—an assumption that is not necessarily justified), and emotive vocalizations would almost certainly have become increasingly important in this situation (as they are amongst other primates when infants are unintentionally separated from their mothers), there is no particular reason to believe that these should have formed the *basis* for affective vocalizations. Likewise it seems quite reasonable that protolanguage and then more complex language emerged from affective vocalizations (an idea explored in much greater detail later), but there is no particular reason to believe that it was from this specific situation. Almost all of the situational examples that Falk uses might apply equally well to interaction between *any* individuals within a hominin group. For example, the idea that distance between emotionally conjoined individuals would elicit emotive vocalization might apply equally to any removal of physical proximity in a species which is used to relying on grooming activity to build and reinforce affective relationships. The foundations for such vocalizations are used across all members of a group, not just between mothers and infants, in higher primates and modern humans; this is discussed in more detail in the following chapters.

In fact, the premise that human mother–infant interaction is unusual in that it is characterized by almost continual affective vocal interplay may be overstating the case of mother–infant interaction relative to that of other interpersonal interaction. It is a challenge to identify a situation where *any* two emotionally conjoined humans would sit in complete silence without any verbal interaction (with the possible exceptions of reading, watching television, or in a theatre, all of which are very recent developments in evolutionary terms; even in these situations it is relatively unusual to sit in silence, and it could be argued that the characters in the book, programme, or play are providing vicarious interaction anyway). It might be counter-argued that this is because our adult vocalization behaviours are derived from these mother–infant interactions, but considering that these themselves appear to be derived from vocal behaviours which span all members of higher primate groups, this seems a less likely explanation than that emotive vocalizations of all forms are derived from interpersonal vocalizations between all members of a group. It seems more likely that infant-directed vocalizations are a specialized perpetuation of a sophisticated form of non-linguistic interpersonal interaction which was used earlier between all individuals.

This said, the observation that vocalizations can constitute a surrogate for physical contact represents a very significant property of vocalization. The evidence that Falk cites, and the evidence investigated here, concurs with her penultimate paragraph, which is not specific to ID vocalizations:

> It is reasonable to speculate that by the time individuals across social groups began to originate and conventionally share simple instructive utterances, protolanguage was in the process of emerging from the prelinguistic melody. Whatever its precise nature, however, protolanguage and the other languages that eventually evolved would, forever after, retain some of that melody. Thus, rather than being totally separate from language (Burling, 1993), tone of voice represents a signature from its very origin that, as transpired for the cosmic microwave background signature left over from the big bang, should be recognised and investigated. (Falk, 2004b, p. 503)

So what significance does all this investigation of proto-language and prosody have for musical evolution? Terms such as melody, tempo, and pitch are conspicuously prevalent in descriptions of prosodic affective vocalizations and ID speech, and the evidence analysed in Chapter 7 illustrated that prosody and melody are produced and processed by closely related or shared neural structures. Tellingly, ID *singing* also appears to be a human universal, with shared characteristics (such as gentle dynamics, gradual pitch contours, and repetitive motifs) across cultures (Trehub et al., 1993); lullabies and playsongs also convey emotional rather than linguistic meaning, and ID speech and singing share many of the same features and characteristics, in terms of pitch and tempo (Fernald, 1989b; Papousek et al., 1991; Trainor et al., 1997; Trehub et al., 1993). Infants listen for significantly longer to infant-directed singing than to normal singing, and even engage more with ID singing than with ID speech (Trehub, 2003).

The abilities of infants to perceive frequency, timing, and timbre are present at birth (Trehub, 2003). Dissanayake (2000), like Papousek (1996a), has proposed that the musical perceptual abilities of infants were selected for because they are used in the non-verbal elements of vocal communication between mother and infant. Whilst ID speech does have lexical (linguistic) content, which plays a role in the linguistic development of the child (Burnham et al., 2002), this seems to be secondary in role and motivation to the prosodic elements. As discussed earlier, these utterances rely on tone, inflection, and intonation, and have a wide vocal range; their principal role is to communicate information about well-being, emotional status, needs, approval, and intention, as is evident from their cross-cultural, panlinguistic similarity. Dissanayake hypothesizes that this mother–infant behaviour emerged as a consequence of hominin infants being born progressively earlier as brain size increased, and thus more helpless, increasing the need for emotional conjoinment at this early stage of life (a rationale later echoed by Falk in her scenario). This would, she suggests, have become particularly important at the time of 'archaic *Homo sapiens*' (those hominins directly ancestral to both modern humans and Neanderthals, often collectively called *Homo heidelbergensis*), which exhibited a large increase in brain size. She proposes that these vocalizations then came to be adopted increasingly into daily life and ritual, eventually becoming the basis for musical behaviour, because they 'were found by evolving human groups to be emotionally affecting and functionally effective... to emotionally conjoin and enculturate the participants' (Dissanayake, 2000, p. 401).

The benefits of such social-emotive interaction between individuals (emotional conjoinment and enculturation) are well argued and supported by much of the evidence considered here, and the connection between musical behaviours and affective vocalization is an important one (as illustrated by the evidence considered earlier in this chapter, in Chapter 7, and further explored in Chapter 9). However, it is a little difficult to envisage hominins beginning to use a behaviour in novel situations that was originally evolved as a 'baby-talk', because they realized it 'promoted affiliation and congruence in adult social life' (Dissanayake, 2000, p. 401). As discussed earlier, there are precedents for the importance of affective-communicative elements of such vocalizations in adult vocalizations, as well as in primates without language (see later); it is important that we remember that when we observe that traits are innate in infants, we are actually observing that they are innate in *humans*, and it is not necessarily the case that these innate abilities were selected for because they only directly benefit infants.

Additionally, the fact that the vocal tract of *Homo heidelbergensis* seems to have been effectively modern (see Chapter 5) would suggest that selective pressure for its use for wide-ranged tonal expression had come rather earlier than Dissanayake proposes. This form of communication might gradually have evolved into music as Dissanayake suggests, or at least provided shared foundations, but it could also have been the basis for communication amongst all of a population (discussed later, cf. Aiello and Dunbar, 1993; Scherer, 1991; Brown, 2000). It may be that the use of this mode of emotive utterance for communication between (prelinguistic) infants and adults today is a vestige of this earliest form of social communication; after all, it is potentially still selectively useful at least until the child gains full linguistic abilities, and is also still inherent in tonal elements of *any* emotional vocal communication (Trainor et al., 2000; see also Wharton, 2009). Thus an alternative, and perhaps more parsimonious, explanation to Dissanayake's suggestion is that social-emotive vocalization was a form of communication that came to be used *throughout* the social group at a much earlier time, without preference, both adult–adult and infant–adult, but is now perpetuated, in this predominantly non-lexical form, in adult–infant interactions, and in the prosodic content of adult speech.

Trainor and Schmidt (2003) suggest that musical stimuli tap directly into evolutionarily ancient neurological circuits related to approach and withdrawal behaviours, and that they do so as a consequence of the use of musical traits in infant-directed vocalizations (ID singing and ID 'musical' speech), used by adults towards infants to communicate positive and prohibitory emotional information essential for survival. Such vocalizations would perhaps be especially important in human infants, who are helpless and dependent for such an extended period. They propose that this sensitivity to these emotion-eliciting cues in vocalization 'goes underground' with the development of linguistic capabilities, but is perpetuated into adulthood, and remains stimulated by musical cues. (They go on to suggest that it may even continue to hold a survival value in adulthood as a medium in which to 'practise' the experience of emotion without risk—an idea closely allied to explanations proposed by Cross, 2001; see Chapter 11.)

Whilst many of the emotional responses to musical stimuli are apparently the result of deep-rooted and evolutionarily ancient neurological responses, Trainor and Schmidt's (2003) rationale has a certain circularity to it, in that it begs the question as to why these 'musical' traits of vocalization should elicit such useful responses in infancy at all. Musical traits of vocalization would be no use for communicating emotionally with infants if they did not *already*, at that age, elicit emotional responses. The ability to perceive and respond to these particular traits is clearly innate, activates deeply rooted and ancient mechanisms, and must be evolved from foundations in the responses to vocalizations present in our primate ancestors. It is difficult to separate the perpetuation into adulthood of a behavioural or cognitive trait that is useful in infancy, from the innate presence in infancy of a behavioural or cognitive trait that is useful in adulthood. Indeed, in some cases such a distinction might not be necessary or desirable at all: the cognitive or behavioural trait could be useful in both infancy and adulthood for related reasons.

A particularly interesting observation regarding ID speech is its level of apparent 'redundancy', or repetition. In the words of Fernald (1992b), 'mothers repeat over 50% of their phrases when interacting with 2-month-old infants. Prosodic repetition is common too (Fernald and Simon, 1984), often with slight melodic variations, which keep these repetitive runs interesting as well as highly predictable for the infant (Stern, 1977)' (p. 419). Of course, because of this, the repetition is actually not 'redundant' in the sense of unnecessary, but instead is important in fulfilling the role of the vocalization, namely, the communication of affect, modulation of arousal, and eliciting of attention and affective response. These properties of ID speech, repetition with variation, along with highly variable pitch contours and high rhythmicity, are characteristic features of the melodic content and structure of music too. It is perhaps unsurprising that music frequently has these same effects on the listener/participant.

It is from these parallels that much insight can be gained regarding the origins and foundations of musical behaviour. What it is important to note is that these are parallels not just with ID speech but with the prosodic affective elements of all emotional vocalization, whether accompanied by linguistic structure or not. Interestingly, Trainor et al. observe that 'Because the emotions music conveys to, and induces in, listeners constitute the meaning of music (Meyer, 1956), speech that adopts musical features might also be expected to be good at communicating emotional information' (2000, p. 194). Setting aside all the other relationships between meaning and emotion in music for the moment (discussed in Chapter 10), in considering that at least some linguistic speech *adopts* musical features, Trainor et al. evidently consider speech as post-dating the tonal-emotional elements of music. The consensus is indeed that these tonal elements of speech pre-date the linguistic, syntactic elements. However, given the precedents for the importance of tone and affect in many other primate communicative vocalizations, perhaps Trainor et al.'s observation should more accurately read that speech *shares* musical features, rather than *adopts*. Whilst the vocal interaction between parent and infant is often described as 'musical' (e.g. Fernald, 1992a; Papaeliou and Trevarthen, 1998; Trainor et al., 2000) Lavy (2001) points out that what the interplay between mother and baby is actually very much like is positively valenced, emotionally-charged vocal utterance; it is referred to as musical because *many musical sounds are very much like the vocal utterances that humans have an innate ability to produce and perceive.* Moreover, parameters along which music operates are the same as those which encode emotional cues in acoustic signals and are perceived and quite probably responded to as such. (p. 51, emphasis in original)

It is certainly the case that this form of interaction has a direct and physiological effect upon emotional state and arousal. ID singing and speech appear to be effective at modulating levels of arousal in infants, resulting in decreased cortisol levels in infants' saliva after a 20-minute period, and even after 25 minutes in the case of ID singing (Shenfield et al., 2002, cited in Trehub, 2003). Trehub (2003) suggests that the ability for infant-directed vocalizations to be used for modulating arousal parallels the deliberate use of music for the regulation of emotional state in adolescents and adults. Singing to infants seems to enhance maternal (parental/adult) well-being as well as that of the infant, and apparently plays a role in strengthening emotional bonds between the adult and infant, as does musical activity between adults in other contexts (Trehub, 2003; see also Chapter 10).

A shared heritage between this type of vocalization and singing-type behaviours is further suggested by the observations of Mang (2000). Mang's study of the vocalizations of preschool children over a 42-month period concluded that in young children who were at the initial stages of gaining vocal control, it was often very difficult for judges to make distinctions between song and speech vocalizations. As the children started to make their own clear distinctions between singing and speaking they would sometimes *purposefully* alternate between singing and speaking vocalization 'to communicate in novel forms of contextual intermediate vocalisations'. The immediate implication of this is expressed by Mang (2000): 'the "fuzzy" boundary between preschool children's song and speech brings out the question as to what constitutes a conception of singing and speaking both acoustically and contextually' (p. 116).

PROTO-MUSIC/LANGUAGE: RATIONALES FOR A SHARED ANCESTRY

The significant overlaps between the form and function of melodic and non-linguistic communicative vocalizations discussed in the preceding sections have led to suggestions that they share an evolutionary, functional, heritage. Some authors have argued that language must be the precursor of music, and that music makes use of mechanisms originally selected for by language or other forms of vocal communication (e.g. Pinker, 1997; Calvin, 1996; Sperber, 1996) and others have argued the reverse (e.g. Darwin, 1871; Livingstone, 1973; Vaneechoutte and Skoyles, 1998). The authors in the former category generally assert that musical abilities could not have been selected for in their own right. This idea is discussed in more depth in Chapter 11. The suggestions of Falk (2004b), Dissanayake (2000), Papousek (1996a) and Trainor and Schmidt (2003), regarding musical and other vocalization capacities having emerged from the selective benefits associated with adult–infant communication, have already been discussed. Others have suggested that they share a common origin in a form of 'musilanguage' (Rousseau, 1781; Scherer, 1991; Brown, 2000; Morley, 2000, 2002a, 2003; Mithen, 2005). Underlying such suggestions should be a consideration of the nature, purposes, and advantages associated with such proto-vocalization abilities.

The musilanguage idea was suggested by Jean-Jacques Rousseau in his paper 'Essai sur l'origine des langues' (1781), in which he suggested that speech and song were not originally distinct from each other, and that early languages were melodic and poetic rather than being practical or prosaic. Rousseau believed that specialized language and music emerged out of a common ancestor, with the earliest language fulfilling the role of emotional expression, being 'sung' rather than spoken, and facilitating the organization of human societies (Rousseau, 1781; Besson and Schön, 2003). Apart from the important shared neurological and physiological mechanisms for the production and perception of linguistic vocalizations and melody (discussed in Chapters 5–7), and the parallels between them in infant and infant-directed vocalizations (discussed earlier), language and music also share several structural and functional features which reinforce this hypothesis of a shared heritage. From the perspective of emotive (affective) communication, the possibility that the earliest forms of language and music might have had a common foundation has also been suggested by Scherer (1991):

> one might reasonably speculate that both proto-speech and proto-music might have used affect vocalisations as building blocks.... It may well be, then, that the externalization of affect or emotion via vocalisation is at the very basis of music and speech. (p. 147)

Affect (emotion) is also externalized via other media, such as posture, gesture, and facial expression, and the extent to which these should be considered a single related system, with vocalization a specialized subset, is explored in Chapter 9. A common ancestral 'musilanguage' has also been proposed by Brown (2000), and elaborated in much greater depth than by previous authors. He suggests that music and language are 'reciprocal specializations of a dual-natured referential emotive communicative precursor, whereby music emphasizes sound as emotive meaning and language emphasizes sound as referential meaning' (Brown, 2000, p. 271). Brown points out that the major parallels between linguistic speech and melody are combinatorial syntax and intonational phrasing. The latter is based on scales of discrete pitch levels in both melody and speech (although in relation to a fixed pitch in Western music). In both music and speech, phrase is the basic unit of structure and function, and both speech phrases and musical phrases are 'melodorhythmic' structures, with expressiveness of the phrasing being very important.

According to Brown, the melodic and rhythmic elements of melodic and linguistic vocalization are derived from:

1) acoustic properties of the fundamental units of music and speech: pitch sets, intensity and duration in the case of the former, and phonemes and phonology in the latter,
2) sequential arrangement of these properties in a given phrase,
3) expressive phrasing mechanisms that modulate the basic acoustic properties of a phrase.

These acoustic properties constitute a 'phonological level' of meaning to melody and speech, an 'acoustic mode' involving emotive meaning and interpretation. There is additionally, of course, a level of higher-order interpretation, or 'meaning level', involving referential meaning. Both music and language have this 'meaning level' (or 'vehicle mode' as Brown calls it), but it is at this level that they diverge, with their meaning being determined by the culturally and contextually specific elements of music and language.

Brown thus suggests that the common features of music and language evolved before the distinct features; namely that lexical tone, combinatorial syntax, and expressive intonation were shared ancestral features of music and language. More specifically, he suggests that initially, ancestral to both music and language, there was a 'simple system involving a repertoire of unitary lexical-tonal elements' which evolved into a 'less simple system based on combinatorial arrangements of these lexical-tonal (and rhythmic) elements' (p. 290). This system of lexical-tonal units being strung together into sequences results in the creation of 'simple, unordered phrases having higher-order meanings' (p. 285), where the phrases as a whole have a global emotive and/or pragmatic meaning, due to the overall contour of the phrase. Examples in modern speech are surprise and question intonations. Such phrases have a rhythmic as well as a melodic structure (due to the temporal arrangement of the constituent units); because of the relationship between the individual units, they communicate on a compound basis as well as the global basis described earlier.

It is surprising that Brown considers that the unitary elements of such utterances would have been *lexical*-tonal, and does not mention *emotive*-tonal utterances, when he originally highlights the emotive-communicative elements of music and language as being a major parallel. He describes lexical-tone as referring to the use of pitch in speech to convey semantic (lexical) meaning, but makes no mention of the use of pitch and tone to communicate emotional state and reactions. For these utterances to have a principally *pragmatic*, rather than semantic role (see Wharton, 2009), initialy at least, would seem more consistent with other aspects of the scenario. As discussed in some detail in Chapter 9, a single discrete vocalization can, through tonal contour, express a great diversity of information regarding emotive state and reaction, without being part of a larger sequence, and without having a specific referent other than being an expression of personal state. This does not undermine Brown's suggestion of the emergence of the importance of lexical-tonal sequences, or of the importance of tonal contour in the phonological meaning ('acoustic mode') of such utterances as a whole, but it seems most likely that the use of lexical-tonal units would only have grown out of emotive-tonal units. In these, tonal contour is already important, and there is no requirement for an additional explanation as to how the importance of tonal contour of a phrase could develop out of discrete units which were not themselves tonally contoured. Tonal contour is already important, at a discrete utterance level, and the combination of tonally communicative utterances could lead to pitch contour being expressive at a global-phrase level too.

An alternative order to this process, which is perhaps more parsimonious than Brown's (2000), has been proposed by Wray (1998, 2000). Rather than arguing for the initial combination of short tonal utterances, she argues that it is more likely that holistic tonal phrases would have formed the foundation for more complex communication, and that these could then have subsequently become segmented into elements with individual meaning; indeed, holistic tonal utterances still form important components of vocal communication today, alongside lexical language. This model is also more consistent with a derivation of human vocal behaviours from precedent forms such as those that we see in other primates today, than are models emphasizing an initial occurrence of discrete, lexical, utterances. The fossil evidence considered in Chapter 5 suggested, though, that the ability to produce short utterances controlled for pitch and intensity emerged before the ability to produce extended versions of such utterances. This might support Brown's proposed order or could, alternatively, suggest an increasing length of holistic utterances before their later segmentation (cf. Wray 1998, 2000).

The emergence of the use of vocal contour *is* something that Brown discusses, however. Indeed, the importance of the production and perception of pitch in vocalizations is central to the musilanguage model. Brown considers that since most primate vocalization systems rely heavily on unpitched grunts and pants, a pitched vocalization system (shifting from one pitch to another) must have emerged at some stage in the development of human music and language. In fact, there is little need to argue for the emergence of an entirely new system of pitch control in humans (or human ancestors). Brown's description underplays the use of pitch in primate vocalizations: his emphasis on 'unpitched grunts and pants' implies a lack of laryngeal engagement in vocalization in chimpanzees and monkeys, which is not the case. Whilst grunts and pants are important vocalizations in higher primates, it is not the case that they (higher primates) do not use pitch moderated by the larynx (the various forms and roles of these vocalizations are discussed later). What *is* the case is that these are of limited range, and do not tend to shift in pitch much over the course of a single vocalization (with the exception of the pitch-glides of gibbon 'song'). What must have emerged in the course of the evolution of pitched-contoured vocalizations is the integration into vocalization of an *increased range and control* of pitch contour (Chapters 5 and 7), but this need not have emerged as a new system of vocalization; it could instead have been building upon the limited pitch control used for emotive-tonal-social vocalization already used by primates (as discussed later).

These proposals are in fact more in keeping with the cross-modal 'sentic modulation' (Clynes, 1977) which Brown takes as an underlying mechanism of emotional expression in humans and other animals. This is the idea that expression of emotion across the media of tone of voice, facial expression, dance step, and musical phrase (for example) can all be seen as equivalent along three spectra: tempo modulation (slow–fast), amplitude modulation (soft–loud) and register selection (low–high pitch). Brown interprets this as 'a general modulatory system involved in conveying and perceiving the *intensity* of emotive expression across a continuous scale' (2000, p. 287, emphasis in original). This appears to be invariant across modalities of expression in humans (discussed in greater depth in Chapter 9) and, as mentioned previously, seems to function similarly in emotive expression in other animals (Morton, 1977, 1994). It is particularly interesting in the light of the aforementioned observations of the strong correlation between vocal affect and orofacial expression (e.g. Tartter, 1980) and the functioning of the Intrinsic Motor Formation (Trevarthen, 1999) (Chapter 9). This idea of cross-domain expression and action closely parallels Donald's (1991, 2001) conception of 'mimetic' cognition, in which a single controller is responsible for expression across facial, vocomotor, manual, and corporeal media. Mimesis describes a system of abilities which manifest themselves in expression and control of emotion through public gesture and action, and allows the sharing of attention and knowledge 'by means of gesture, body language and mime, any of which can communicate an intention quite effectively, without words or grammar' (p. 263). It thus forms the foundations for complex social interaction and structure, and the enactment and re-enactment of events and actions. These elements of communication remain extremely important even alongside words and grammar (Wharton, 2009); the implications of the development of such cognitive and communicative abilities are considerable, and are discussed further in subsequent chapters.

Regarding the divergence of music and language from a musilanguage ancestor, Brown (2000) suggests that the common ancestral form of *language* was most likely to be tonal, as is the case with Chinese and, in fact, the majority of the world's languages (2000, p. 281), rather than non-tonal (like intonation languages such as English). As far as Brown himself is concerned,

> The [musilanguage] model's principal contribution to the study of language evolution is to provide a new chronology for the development of language's structural features: language evolved out of a sophisticated referential emotive system; phonological syntax preceded propositional syntax; tone languages preceded intonation languages; speech could have evolved early, due to its exploitation of lexical tone instead of enlarged segmental inventories; lexical tone, combinatorial syntax, and expressive intonation were ancestral features of language that were shared with music; broad semantic meaning preceded precise semantic meaning; and language's acoustic modality preceded its representational modality. (p. 294)

In terms of music evolution, its contribution is to highlight the roots of melodic vocalization, and melody in general, in this interpersonal affective communication. Brown seems to consider that entrainment to rhythm as a feature of musical behaviour is something that was integrated into it afterwards, following the divergence of music and language; evidence considered in Chapter 9 indicates a more fundamental relationship between vocalization and rhythmic muscular control.

Despite these critiques, Brown's musilanguage model is in general convincingly argued and, although there is little overlap between the evidence he considers and that dealt with in this book, the model accords extremely well with the diverse data examined here. It does not, however, include any scenario for the original use of such a musilanguage, its relationship to skeletal or cognitive development, or any *rationale* for its development into the 'reciprocal specialisations' of music and language (although he does describe the possible mechanism in depth).

Attempts to situate these changes in an evolutionary context, and address those issues, have been made since (e.g., Morley, 2000, 2002a, 2003; Mithen, 2005; Fitch, 2000b, 2009a). Mithen (2005) emphasizes in particular the holistic, mimetic, manipulative, multi-modal, and musical aspects of this non-linguistic ancestral form of communication, referring to it by the appropriately onomatopoeic acronym 'hmmmm', in a neat and accessible scenario. He takes a multidisciplinary approach to set out a potential timeline for the development of such communication, including relating it to models of linguistic evolution that emphasize the significance of holistic utterances over discrete lexical meanings at an early stage (e.g. Wray, 1998, 2000; see also 2006). Like the previous authors mentioned, he ultimately suggests that lexical and grammatical linguistic behaviour emerged late from this earlier vocal-melodic form of communication, and that our modern, specialized, use of music is also derived from this origin. Although he does not discuss in detail the direct Palaeolithic evidence for musical activities, he does extrapolate further from other types of evidence in order to describe scenarios in which versions of this musilanguage might have been used at particular points in human evolution.

SOCIAL VOCALIZATION IN PRIMATES

In the words of Fernald, 'human symbolic communication builds on our primate legacy, a foundation of affective communication established in the preverbal period' (1992b, p. 423). As noted earlier in the chapter, vocalizations play an important role in social interaction and bonding activity amongst a variety of non-human primates. As Seyfarth and Cheney observe, 'In the wild, nonhuman primate vocalizations signal the presence of different predators, provide information about the group's location and movement, facilitate friendly interactions, and lead to reconciliation between individuals who have recently exchanged aggression' (1997, p. 249). The former case has recently received much attention (and debate) as a possible form of symbolic referential behaviour and as a precursor to syntax (e.g. Seyfarth and Cheyney, 1992) but it is perhaps to the latter two cases that we should look for evidence of the roots of complex vocalization in inter-individual interaction.

As already noted, amongst bonobos, utterances always occur along with facial expressions, gestures, and tactile communication (Falk, 2004b; Bermejo and Omedes, 1999), and chimpanzee mothers also vocalize softly whilst examining their infants (Falk, 2004b; Nicolson, 1977). Interestingly, Seyfarth and Cheyney (1997) conclude from their analyses of the circumstances under which such socially facilitative vocalizations occur that there is no indication that those individuals interacting have any knowledge of their conspecifics' mental state, knowledge, beliefs, or desires. Instead they suggest that in the case of reconciliation calls, recipients have learned from experience that individuals producing reconciliation-type calls usually follow them with friendly rather than aggressive behaviour. Whether or not they are correct in this belief, the significance of such behaviour should not be underestimated—it is not merely 'stimulus-response' Pavlovian conditioning. Whether or not the individuals involved have knowledge of complexities of their conspecific's mental state, they clearly have a set of expectations about how the interaction is going to proceed following such an utterance. That such utterances come to be a predictive precursor to friendly interaction establishes them firmly in a repertoire of inter-individual behaviour. That they are then followed by activities further consolidating such interpersonal relations suggests a group or repertoire of related behaviours based on expectations about the other's behaviour that cannot easily be viewed simply as a sequence of stimulus-response reactions.

Gelada baboons also use a series of vocalizations for 'social grooming' purposes, to establish vocal relationships and social bonds with conspecifics (Richman, 2000). Interestingly, these are similar in form and vocal detail to human vocal formulae; in both humans and geladas friendly vocalizing is produced in units of an average length of nine to ten syllables, about five syllables per second, three or four strong beats per unit, with an intonation contour, and the end of the vocal unit being signalled by tonal change (Richman, 2000). The findings of Elowson et al. (1998a and b) and Gros-Louis (2002) regarding primate infant vocalizations also illustrate an early and important social function to vocal behaviour; such vocalizations can be very important in eliciting care responses from mothers to infants in various primate species, in instigating affiliative activities with others in the group, and they operate in a variety of very specific contexts. It seems that vocalizations can form a very important component of repertoires of social inter-individual behaviour in both infant and adult primates.

EVOLUTIONARY RATIONALES FOR COMPLEXITY OF VOCALIZATION: PROTO-MUSIC, PROTO-LANGUAGE, AND SOCIAL VOCALIZATION

Aiello and Dunbar (1993) note that there is a proportional relationship amongst the higher primates and humans between degree of encephalization (cortical development) and group size: as group size increases, so does neocortical development. This strong correlation between group size and neocortex size exists also in carnivores and cetaceans (whales and dolphins) (Kudo and Dunbar, 2001). Links between neurological development and group size are well attested in studies of primate neurology. The size of the neocortex and the striatum increase with the mean number of females in a group; the amygdala and the number of neurons in the parvocellular lamina also increase with overall group size (Mondragon-Ceballos, 2002). Furthermore, lesions of the prefrontal cortex in primates lead to extreme social apathy; it is clear that cortical development is implicated in social function (Mondragon-Ceballos, 2002). Treating the relationship between group size and neocortical development as a direct one, Aiello and Dunbar (1993) calculated the probable group sizes of the various Palaeolithic hominins on the basis of their brain size.

Amongst primates such as chimpanzees, grooming is a very important function in maintaining an individual's social network, alliances, and coalitions within a group; such a network is important in providing support in power, mate, or food resource contests. However, beyond a certain group size, it would become impossible to maintain a social network effectively through manual grooming alone, as the time taken would be too great (Dunbar, 1998). Evidence considered already has illustrated the social importance of vocalization in many primates, in particular in instigating relaxed proximity and grooming. The importance of sensitivity to emotional content of vocalizations has also been discussed. Dunbar (1998) suggests that physical grooming began to be supplemented with utterances that increased its efficiency, and that as group size increased there would be a selective pressure for the 'grooming' utterances to become progressively more efficient and expressive. This could then feed back allowing group size to increase further, and so on. The integration of affective vocalization, which already forms an important part of interspecific interaction and affiliation in primates, into grooming activity would not be a great step. Note that it is not asserted that neocortical development is the determining factor in group size, but that it places a constraint on group size. The major influences on group size are ecologically imposed local costs and benefits, but it is hypothesized that the cognitive mechanisms that allow individuals to live together in coherent stable groups have evolved to support the group size imposed (Kudo and Dunbar, 2001).

Whilst the evidence supporting a correlation between group size, social complexity, and encephalization is strong, the relationship between encephalization and group size may not be a linear one as Aiello and Dunbar (1993) suggest. In this case it would be impossible to accurately predict the group size of earlier hominins from the degree of encephalization they exhibit, and thus impossible to assert at what point physical grooming would have become impractical without the presence of 'vocal grooming'. However, more recent research by Kudu and Dunbar (2001) has further reinforced the connection between grooming and neocortical development in primates. In examining a wide cross-section of primates (including modern humans) they found that coalition size/grooming clique size and neocortical development are strongly correlated, and that coalition size/grooming clique size is proportional to overall group size. It seems that there is consistently a strong relationship between cortical development and social group size across different primates, and that social relationships are an important stimulant to cortical development.

Hopefully, future fossil and archaeological evidence will add further data regarding the group size of hominins. Several sites containing the remains of numerous individuals exist; for example, *Australopithecus africanus* at Sterkfontein and *Paranthropus robustus* at Swartkrans (both South Africa), *Australopithecus afarensis* from Hadar (Ethiopia), and *Homo heidelbergensis* from the 'pit of bones' (Sima de los Huesos) at Atapuerca, Spain. The latter contains the remains of more than thirty-two individuals, which were initially interpreted as being from part of the same community (Arsuaga, 1997), and thus as having the potential to provide clues as to the group size of *Homo heidelbergensis*. However, it has proven difficult to demonstrate whether the remains accumulated suddenly (through some catastrophic event), or gradually; if they accumulated over some time (either naturally or via the agency of other hominins), it is impossible to extrapolate group size from them (Andrews and Fernandez-Jalvo, 1997; Bermúdez de Castro and Martinón-Torres, 2004). For the moment our best indicators of group size, beyond the degree of encephalization, remain the archaeological evidence for intensity of landscape and resource exploitation, and analogy with other species.

In any case, evidence that the use of language to establish and maintain social networks far outweighs its use for 'higher' discussion provides important support for the idea that social use was one of the fundamental foundations of language—for example, Dunbar et al. (1997) and Emler (1992) both found through analysis of natural conversations that 60–70 per cent of all conversation time in humans today is devoted to social information exchange. The evidence discussed in Chapter 9 and elsewhere in this chapter also reinforces the proposed early importance of the associations between vocal behaviour and interpersonal social-emotional expression (versus lexical or signalling content).

It is also interesting to note that the development of complexity in social vocalizations as an adjunct to grooming behaviour provides an intriguing explanation for the systematic interrelationship between vocalization and manual movement—specifically, between fine laryngeal, orofacial, and manual muscular control. If complex vocalizations developed as an adjunct to grooming behaviour, then one would expect the action of the hands (in particular, precision movements) and vocalization to be intimately related. As vocalizations increasingly substituted for manual grooming, the movements of the hands might become more and more structured in their accompaniment of the vocalizations; out of this structure they could form the foundations of syntactic structure as hypothesized by Armstrong et al. (1994) (see Chapter 9), which then came to be incorporated into vocalizations too.

If it is indeed the case that social-affective content was initially the most important component of vocal communication, then users of this system would have derived considerable selective benefit from the development of increasing vocal agility and control. As the variety of sounds they could make increased, so too would the variety of sound sequences (and thus expressiveness); with increased control ambiguities would be minimized and efficiency increased. With greater expressiveness and efficiency, social grooming would become more effective and less time-consuming, and the individual's social network could become larger and more cohesive. An individual able to maintain a large cohesive social network that provides support in power, mate, or food resource contests would have considerable selective advantages over its peers.

As Vaneechoutte and Skoyles (1998) observe, modern language use does not *need* to use all of the vocal tract's range (although it often *does* use it all). In fact, with syntax and grammar to disambiguate meaning, one can speak on a monotone without much fear of the syntactic meaning being misunderstood. The utterance would have no pragmatic content, no emotive content through tone, so would lack this dimension of its semantics, but even this could be substituted for to a large degree by careful choice of words. On the other hand, the use of the voice as an instrument, singing, *does* make use of the entire range of the vocal tract. Vaneechoutte and Skoyles (1998) conclude that this suggests that the vocal tract evolved to the extent that it has in order to support (animal-like) song rather than speech. In some animals, song is used to declare territorial information, or readiness to mate. Amongst other animals (specifically, tropical songbirds, whales, porpoises, wolves, and gibbons), song has evolved as a means of establishing pair and group bonding. Vaneechoutte and Skoyles (1998) assert that this is an older requirement than linguistic speech and hypothesize that the ability to sing evolved in humans for the same reasons that it did in other animals. Consequently, they suggest that language now uses mechanisms selected for by song, concluding that 'the ability to sing provided the physical and neural respiratory control that is now used by speech' (Vaneechoutte and Skoyles, 1998, p. 1).

That the establishing of pair and group bonding is an older requirement than full linguistic speech is evident from many animal behaviours, but this is not a function of vocalization that has been lost since the development of full linguistic capacities; as discussed earlier, language also fulfils important functions with regard to aiding social behaviour, and this social bonding property is a very important common feature of linguistic and musical behaviour.

As discussed in Chapter 2, ethnographic evidence from a wide diversity of hunter-gatherer and other 'traditional' cultures parallels this finding for music. Many of the most important roles of music amongst Plains Native Americans (Nettl, 1992; McAllester, 1996), African Pygmies (Kisliuk, 1991; Turino, 1992; Ichikawa, 1999), Yupik and Inuit Eskimos (Nettiez, 1983; Johnston, 1989), and Australian Aborigines (Breen, 1994; Myers, 1999) are social, interactive, and integrative, and the performers themselves often see these as the most important consequences of the activity. Blacking (1995) also observes, from his studies of the Venda peoples, that performance of a musical pattern 'may announce social situations, recall certain feelings and even reinforce social values' (Blacking, 1995, p. 39). Blacking himself believed that social bonding was not the limit of the function of music, but that there is actually a relationship between music and social structure, with particular types and elements of music being used by different groups within an integrated culture.

There is an alternative explanation to Vaneechoutte and Skoyles's suggestion that the vocal tract evolved to support animal-like song. This is that the reason that language can today function without full tonal use of the vocal chords is that lexical complexity and syntax have removed ambiguity from our vocal utterances, allowing meaning to be expressed at an additional linguistic level. This would not have been the case in a proto-language without modern lexical complexity and syntax (similar to that proposed, for example, by Sperber (1996), Aiello and Dunbar (1993), Wray (1998) or Brown (2000), and discussed earlier), in which meaning was tonally generated, and would also not be applicable to melody. In these full use of the vocal tract would maximize communicative potential. It seems more likely, then, that the vocal tract developed to the extent that it did, not to support song before language (as Vaneechoutte and Skoyles (1998) suggest), but instead to support tonal prosodic affective utterances that were a precursor to both language and melodic music (as Brown, 2000, suggests), *and were useful as social-emotional communication.* In this case, it is only since complex lexicon and syntax developed that speech has been able to be used without recourse to the entire human vocal range.

Complex *vocalization* could exist long before complex *language* in the form of an increasingly complex communicative system ('proto-language') based on tone-dominated social affective utterances. Aiello and Dunbar (1993; e.g. also Dunbar, 1998) have proposed that early vocal utterances were initially used to facilitate and then replace social grooming; the evidence of the preceding sections suggests that such utterances would have been prosodic tonal emotive vocalizations, and these would indeed seem most likely to be effective in facilitating social grooming. Interestingly, the findings of Watt and Ash (1998), that music has an action on the mind similar to the action of interacting with a person, and that people attribute human-like qualities to music, also complement the idea that melodic content in music could have developed from the use of socially based vocal interaction. This social hypothesis, along with the observations and findings relating to infant-directed and emotional vocalization, suggests that human use of non-verbal song has far more in common with emotional, social, and emotive-social vocalization than with the song of birds, which is territorially or reproductively orientated rather than expressive and socially bonding (Slater, 2000).

CONCLUSIONS

Human infants are born with abilities fundamental to musical processing, including the perception of frequency, timing, and timbre, and are able to extract different emotional content from vocalizations, on the basis of tone and rhythm alone. In the emotional state that they express, some vocal sounds and frequency changes are fundamental, invariant across cultures, and even species. Part of the reason for this is that facial expression has a fundamental influence on vocal quality, as orofacial musculature helps determine properties of vocalization such as frequency and vowel duration. Given the universality and innateness of certain fundamental facial expressions and the correspondence between these and characteristics of vocalizations, we can also expect characteristics of particular emotional vocalizations to be universal and innate too. This correlation between facial expression and vocal quality also apparently has an ancient provenance, being shared by our nearest primate relatives, and similar correlations between vocal sound and emotional expression are also exhibited by several other species. We use facial affect and vocal affect to inform about the content of each other, interdependently within both production and perception.

The universality of the vocal sounds and frequency changes that express particular emotions is especially evident in infant-directed (ID) speech, where the exaggeration of these elements of the vocalization is a characteristic feature. ID vocalizations can tell us a great deal about the nature and role of the prosodic elements of speech and their relationship to musical melodic behaviour, as many of the properties of ID speech are shared with music. There are numerous parallels in terms of variable pitch contour, high rhythmicity, repetitive motifs, and the communication of affect, modulation of arousal, and eliciting of attention and affective response. It should be noted that the characteristic features of ID speech are also characteristics of the tonal (non-linguistic) elements of adult-directed (AD) speech, and they apparently share the same foundations and roles in emotional expression. Vocalizations produced by preschool children themselves are often difficult to classify as either linguistic or musical.

It seems that the best explanation for the collective phenomena described here is that the shared prosodic pitch and tempo-related properties of emotional speech (ID and AD) and music are not borrowed from one to the other, in either direction, but are, and always have been, a shared fundamental component of both. The music-like characteristics of ID vocalizations act upon cognitive-perceptual mechanisms that respond emotionally to emotional cues, and the characteristics of musical stimuli act upon the same mechanisms, not because these responses of the perceptual mechanisms are *perpetuated* into adulthood, but because their *function* is to respond in this way to these cues in *all* vocalizations, ID and AD. This emotional response to emotional cues is the foundation of empathy and successful interpersonal interaction, and musical stimuli act upon the mechanisms responsible as a reified form of the cues inherent in human emotional interaction. Response to such cues is as essential in adulthood as in childhood and infancy—but the use of those cues towards infants is more pronounced as a consequence of the need to develop and nurture those all-important interactive skills. Both music and planned use of vocalization (especially ID but also AD) make use of an innate set of emotional responses to particular properties of vocalizations—properties of vocalizations that were extended from the communicative vocalization activities of our primate and later hominin ancestors.

Already important in forming and maintaining social relationships, such vocalizations may have increased in importance initially due to the need to increase the efficiency of physical grooming activities, and may have subsequently come to be used in broader spheres of behaviour, though nevertheless maintaining a core role in the formation and maintenance of social relationships—as, indeed, full linguistic speech still does today.

Linguistic speech and melody share common features in the form of intonational phrasing and combinatorial syntax, and share a 'phonological level' of meaning, an 'acoustic mode' involving emotive meaning and interpretation (Brown, 2000)—what might be termed 'intonational semantics'. Whilst music often does use the full range of sounds producible by the vocal tract, full language does not need to use the whole range to communicate effectively, as linguistic structures provide an additional source of semantic content and disambiguate meaning. This would not have been the case for a pre-syntactic, pre-lexical proto-language.

Tonal-contoured units expressive of affective state could partition into more discrete, smaller units (Wray, 1998) or become combined into progressively larger globally contoured units (Brown, 2000), or both. In either case, what must have emerged in the course of the evolution of pitched-contoured vocalizations is an *increased range and increased control* of pitch contour, allowing greater vocal versatility, expressiveness, and thus efficiency, in proto-linguistic vocal affective communication. This need not have emerged as a new system of vocalization initially, however; instead, it probably built upon the type of limited pitch control already used for emotive-tonal-social vocalization amongst higher primates. The selective advantages associated with the possession of such capabilities, such as the formation of optimal cooperative, mating, and parent–infant relationships would have resulted in the continued refinement of such capabilities through the evolution of the vocal tract and control over it, through the lowering of the larynx and increased innervation of the associated laryngeal and upper vocal tract musculature (see Chapter 5).

In these circumstances, the socially important, emotionally communicative elements of such vocalization would have remained the dominant element initially, with iconic and then abstract (symbolic) lexical associations subsequently increasing in importance. It is the latter—iconic and abstract lexical content—which relies on symbolic and analogical capacity and probably would have been a late-emerging element of communication, perhaps not prior to *Homo sapiens*; the former—emotionally communicative elements—do not rely on symbolic capacity, and, as illustrated, would have been selectively important at a much earlier time. Even with the emergence of full lexical and syntactic language, the social-affective communicative foundations of prosodic contour remain a fundamental element of vocal communication—the same elements of vocalization that form a foundation of melodic musical behaviours.

9

Vocal Control and Corporeal Control—Vocalization, Gesture, Rhythm, Movement, and Emotion

> Vrindavan [pleasure abode of Krishna] is permeated with the moods of divine sweetness and intimacy. It is an eternal, blissful reality where every movement is dance and every word a song.
>
> Bhudhara Das (2006), p. 30

INTRODUCTION

Whilst the preceding chapters have been principally concerned with vocalization ability as an obvious prerequisite of melodic behaviour (and communication), wider bodily movement is also a fundamental component of musical behaviour and, indeed, of vocal behaviour and expression of emotion. Gesture and vocalization appear to be in some respects, if not entirely, contiguous and interdependent systems, with neurological associations apparently existing between the fine control of orofacial and laryngeal musculature, and that of manual gesture. Investigation into why these systems appear to be related in this way should give insight into how they are able to function in the ways that they do and how this relates to interpersonal interaction. Further investigation of the relationship between gestural control and vocalization may help to shed light on the apparently strong association between rhythmic and melodic behaviours in music. This chapter looks first at general relationships between vocal content and manual gesture, goes on to consider how gesture and vocalization seem to be related in prelinguistic infants, as a further indicator of innate, biological elements to those relationships, and then examines proposals for how gestural communication may have had a role in the development of complex vocal communication. Finally it goes on to examine how bodily movement, rhythm, and emotion are related, and the role of entrainment—the ability to synchronize with a pulse—in human musical experience.

VOCAL CONTENT AND MANUAL GESTURE

The involvement, often involuntarily, of manual gestures in vocal expression will be familiar to most people. This has prompted research into the interrelationship of these functions (e.g. McNeill, 1992; Iverson and Goldin-Meadow, 1998; McNeill, 2000), and hypotheses that manual gesture and vocal control are functionally linked and share important neurological and/or evolutionary foundations. Some authors have suggested that manual gesture formed the foundation for syntactic elements of language (e.g. Hewes, 1973, 1992; Corballis, 1992; Armstrong et al., 1994; Stokoe, 2000), whilst others have carried out experimental observation of interrelationships between manual and vocal control (e.g. Petitto and Marentette, 1991; Locke et al., 1995; Feyereisen, 1997; Locke, 2000; Mayberry and Jaques, 2000; Nishitani et al., 2005).

Research by Mayberry and Jaques (2000) regarding gesture in normal speech and stuttered speech casts considerable light on the interrelationship between vocalization and gesture. They found that when disfluency occurs in stuttered speech it is accompanied by fewer gestures than accompany fluent speech—in fact, at the time of stuttering, gesture ceases entirely. This is in contrast to disfluency in normal speech, in which case at a pause gesturing tends to increase along with vocalizations like 'um' and 'ur'. These vocalizations actually maintain the elements of prosody in normal speech disfluencies, and gesture, likewise, is unaffected. In stuttering, on the other hand, the whole of speech prosody and rhythm ceases, as does gesture.

So, in normal speech disfluency there is no disruption of the *motor* and *prosodic* elements of speech—these are maintained in 'ums' and 'urs', non-lexical vocalizations—the disfluency originates in a lexically induced pause as one seeks the correct word or phrase, which does not affect gesture. This is not the case in stuttered disfluency. In the words of Mayberry and Jaques: 'gesture and speech are an integrated system in language production. When speech stumbles and stops as a result of stuttering, the hand always waits for speech so that the meanings being expressed by the hand in gesture always coincide with the meanings being expressed by the mouth in speech, even when the gesture must wait for a very long time. Gesture and speech are clearly not autonomous' (2000, p. 208).

To exclude the possibility that this correspondence of gesture and vocalization cessation in stutterers was caused by a more general shutdown in the motor system, Mayberry and Jaques carried out an experiment in which stutterers were required to carry out a button-pushing task whilst describing a cartoon. That they proved able to carry out the motor task of button-pushing with their arm (which clearly uses the same motor systems as gesturing) uninterrupted throughout bouts of stuttering indicates that motor shutdown is not at the root of the usual co-occurrence of gesture cessation with stuttering. In fact, stuttering individuals are able to carry out all manner of manual non-gesture movements whilst stuttering, but do not execute gestures that are speech-related. Clearly it is possible for vocalization and manual movement to operate independently of each other, but when the hands and arms are not otherwise occupied, they are preferentially engaged with vocalization, and when this occurs it seems to be at a deep and early stage of the process.

Mayberry and Jaques observe that 'the fact that the temporal concordance between gesture and speech execution is always maintained throughout stuttered and fluent speech suggests that the complex neuromotor patterns of gesture and speech are coordinated and integrated prior to their production in extemporaneous expression' (2000, p. 209). In fact, the concordance between speech and gesture does not appear to be instigated by the lexical components of the speech, but by cyclical motor control. Franz et al. (1992) found that when subjects were required to produce repetitive movements with the finger, forearm, and jaw, and vocal repetition of a syllable, there was significant correlation within subjects of the cycle duration of each task; i.e. each subject reached a 'default' timing cycle for the repetitive muscular action, irrespective of which task was being performed. Franz et al. conclude that common timing processes are involved not only in movements of the limbs, but also in speech and non-speech movements of oral structures, and suggest this indicates a governing cognitive 'muscular timing module' responsible for instigating all rhythmic cyclical muscular activity. This is relevant to the discussion of entrainment, later in the chapter, and also accords well with the findings of Alcock et al. (2000) (described in Chapter 7) that the capacity to perform rhythms, both manually and verbally, forms an important foundation of oral/praxic ability. Franz et al. (1992) also found that simultaneously produced finger, arm, and oral movements concur whilst carrying out repetitive non-linguistic motor repetition, which Mayberry and Jaques (2000) suggest could be the mechanism by which gesture-speech co-expression occurs. Whilst the concordance of gesture and speech control appears to be instigated on a motor-control basis rather than a lexical basis, the content of an utterance clearly has some influence on the nature of the gesture. As Mayberry and Jaques observe, the harmonized complex motor patterns of the gesture and speech system must ultimately, subsequently, be integrated with the output of the conceptual linguistic systems.

The question still remains of which cycles in vocal control are being integrated with the gestural cycles. Mayberry and Jaques suggest from their research, and from that of McClave (1994) and Nobe (1996), that it is the prosodic patterns of speech that contain the oscillatory cycles of muscular control. McClave found that it is with the nuclei of tone groups that gesture co-occurs, rather than with syllable stress as was commonly thought, and Mayberry and Jaques's (2000) observations of the onset of gesture and vocalization in normal speakers and stutterers accord with those of McClave.

The implication of all this research is that the cross-modal coordination of gesture and speech does not require a central representation or linguistic input initially; such integration occurs whether utterances are linguistic or not. It seems that there is a cyclical rhythmic motor coordinator which instigates such muscular sequences irrespective of the musculature that is used, and that the complex patterns of gesture (finger, hand, arm, shoulder, and joint musculature) and vocalization (orofacial, laryngeal, and respiratory musculature) are coordinated (Mayberry and Jaques, 2000). The cycles of vocalization that are integrated with the gestural cycles are prosodic, tonal ones; in the case of speech, as opposed to non-linguistic vocalization, linguistic meaning and narrative sentence structure are integrated into the gesture-speech system subsequently, before their physical manifestation.

GESTURE AND VOCALIZATION IN INFANTS

The vocal and gestural motor behaviour of infants and children confirms a deep association of these functions. From the earliest stage, infants' babbling and gesture seem to be interrelated: Masataka (2000) found that in (Japanese) babies at the babbling stage their utterances were more syllable-like when they were able to move their arms than when they were not (Mayberry and Jaques, 2000) (see Chapter 8 for a description of babbling behaviour). Locke et al. (1995) observed a strong association between the onset of babbling behaviour in infants and their exhibiting of lateralization of motor control. They found that in children who had not yet begun to babble, and in those who had been babbling for some time, there was no evident dextral bias in their use of a rattle; however, in children who had recently begun to babble, their use of a rattle was much greater if placed in their right hand than in their left. Locke et al. propose that this is as a consequence of a greater involvement of the left hemisphere (responsible for right-hand control) in the production of repetitive vocal-motor activity, such as is the basis for babbling (an association backed up by the neurological evidence discussed in Chapters 6 and 7 regarding the functions of Broca's area and the areas around it). They suggest that development of the skill and coordination of the vocal sequences results in the increased development of the coordination of the manual gestural activity.

The gestures made by babbling infants are not iconic; they are rhythmic and emotionally determined movements (Trevarthen, 1999; Falk, 2004b) that accompany the vocalizations, but do not add meaning or symbolism to them. As language begins to develop, the gestures become more intentional and instrumental, however (Messinger and Fogel, 1998). From the point when children first combine gesture with a meaningful word in the same utterance (around the age of 14–17 months), gesture-alone combinations begin to decline and synchronous gesture/speech combinations begin to increase (Butcher and Goldin-Meadow, 2000). Butcher and Goldin-Meadow (2000) also observed that after this integration of gesture and lexical speech had occurred, a novel type of combination emerged in which speech was combined with gesture conveying related but different information, in particular, combinations of object and action. It seems the integration of gesture and speech may provide the foundation for the combination of object and action, with either gestural object and verbal action, or gestural action and verbal object.

Furthermore, the first children to produce these gesture-word combinations were also the first to subsequently produce two-word combinations. In all the children, the former occurred before the latter: 'all of the children we have observed thus far were able to concatenate elements of a proposition across gesture and speech at a time when they were unable to accomplish this feat in speech alone' (Butcher and Goldin-Meadow, 2000, p. 254).

Shifting the emphasis on this observation, perhaps it is the case that gestural-vocal integration must be in place before the ability to integrate two words, and more specifically, noun and verb or adjective, can emerge. At least, it seems that the integration of two concepts (e.g. object and action) occurs across two media before it occurs within a single medium, and it is possible that the former is an important stepping-stone towards the latter.

This evidence suggests that the interrelationship between manual gesture and vocalization, both production and perception, is a deep one. That the two functions not only seem to co-occur in development but can also interfere with each other further reiterates this relationship. Feyereisen (1997) found that (adult) subjects asked to carry out a verbal task had a longer vocal onset time when they were asked to concurrently carry out a manual gestural task than when they were not carrying out a manual task at the same time, supporting the idea that there is a competition for shared resources between gestural and speech-production systems (Feyereisen, 1997). The tendency to minimize orofacial and oral movement by pursing the lips or protruding the tongue when carrying out a precise manual task is well known anecdotally, and suggests a benefit to minimizing resource competition between those muscular systems. Having said that, Feyereisen (1997) interestingly found that *no* delay to *manual* initiation time was observed when subjects were concurrently carrying out a verbal task, perhaps suggesting that manual tasks have preferential access to this neurological resource, which vocalization has come to use latterly, in evolutionary terms.

GESTURE, VOCALIZATION, AND MEANING

Considering the size of the body of research investigating language origins, and of that regarding interdependence of gesture and vocalizations, it is surprising that the twain have not more often met. There have been some such connections made, however, leading some authors to suggest that manual gesture formed the foundation for syntactic elements of language (e.g. Hewes, 1973/1992; Corballis, 1992; Armstrong et al., 1994; Stokoe, 2000). It is not the purpose of this book to explore the origins of fully developed language per se, so this issue will not occupy as much space as it undoubtedly could; the origins of language have been, and continue to be, extensively explored and debated elsewhere across a number of disciplines. However, as illustrated earlier, connections between vocalization, manual and corporeal movement are important in any form of vocal behaviour, and probably have a long evolutionary history, pre-dating modern linguistic behaviour.

Clearly, manual, digital, and corporeal movement can communicate important information, either as an adjunct to vocal language, or on their own. As Hewes (1994) observes, 'Gestures of the hands, fingers and other upper body parts, even when they are not elements in a developed sign language, can indicate locations and directions of movement, the shapes and sizes of things, and the effects of actions on other things in ways that vocal sounds not part of a vocal language cannot match' (p. 361).

Armstrong et al. (1994) observe that (some) gestures can be construed as either words or sentences, and point out that a single gesture can represent an entire sentence's meaning. An example they give is of the American Sign Language (ASL) gesture for 'seize': 'the upper arm ... rotates at the shoulder to bring the forearm and hand across in front of the body until the moving hand closes around the upright forefinger of the other hand' (p. 355). This is an iconic gesture, actually representing the activity being communicated, but Armstrong et al. caution against missing the cognitive importance of such communication: the single gesture can have a subject, a verb, and a direct object (or agent, action, and patient, in semantic terms) and to produce or interpret such a gesture requires cognitive foundations that can process these relationships. They propose that such 'word/sentences, in the form of visible gestures, could have provided the behavioral building-blocks associated with neuronal group structures for constructing syntax incrementally, both behaviourally and neurologically' (p. 356).

It is interesting to note too that the capacity of gestures to act as sentence/words is a property that is also shared by some vocalizations. Another example of a gesture representing a whole sentence is given by Armstrong et al. (1994); in ASL, when a signer responds to another by bending their elbow and touching their forehead with their fingertips, this means 'I know that' (not just simply the verb 'know' as many ASL dictionaries imply) (Armstrong et al., 1994). This can also be true of non-linguistic vocal utterances, however. Such utterances can also function as an 'embryo word/sentence' in this sense, and perhaps could have done so during hominin evolution. For example, a particular tonal contour of 'mmm' (usually involuntarily accompanied by certain body language, though not essentially so to its communication) would communicate 'I know that' as opposed to 'I agree with that', 'I understand that', 'I'm not sure about that', 'I grudgingly accept that', 'what did you say?'/'I didn't hear that', or 'I like that'. These meanings are communicated via tonal contour, intensity, and duration, rather than any other property of the utterance. The number of such meanings potentially attributable to an utterance would increase with greater vocal tract versatility and control. In expressing a state of mind, or reaction, such an utterance expresses what can, using lexical language, only be communicated in a sentence using several words.

That such different expressions of 'mmm' are also accompanied by quite specific body language, involuntarily on the part of the speaker, betrays their affective origin in a system in which vocalization and corporeal expression, or to put it another way, vocal and corporeal gesture, are intimately linked. As Blount (1994) puts it, 'Vocalised phonetic "gestures" are accompanied by a priming of the motor system, producing coordinated movements of other parts of the body. Perhaps the capacity to align the components is what evolved as underlying the capacity for gesture, vocal or otherwise' (p. 359) (cf. Clynes's (1977) 'sentic modulation', discussed in Chapter 8, and Donald's (1991) 'mimesis', discussed in Chapter 11). This accords well with the evidence discussed earlier regarding the interaction of gesture and vocalization; it would also be at least equally applicable to body movements accompanying musical vocalization, and vocalization accompanying dance. The fact that seeing body language is not *essential* to understanding an utterance (although potentially helpful in disambiguating it) is perhaps an indication of the degree of sensitivity to vocal contour, and of control over it, that we have since developed.

In reality our spoken language is a combination of lexical and prosodic modes of communication, and the *pragmatic* content of an utterance can be far more significant than its semantic content. The importance of pragmatic elements in communication has been recently emphasized in detail by Wharton (2009), who discusses in particular the significance of tone of voice, facial expression, and bodily gesture in adding layers of meaning to linguistic communication, a very welcome addition to analyses of human communication that so often focus principally on the purely linguistic.

The tonal contour of an emphatic 'I *know* that!' seems very similar to the tonal contour of an '*mmm*!' communicating the same thing, and the same can be said of a sympathetic or a placatory expression of the same. The lexical form of this type of utterance is most useful in disambiguating it in a sensorally deprived context, such as over a distance. In contrast, it seems that the more intimate the circumstance, the more inclined we are to use the non-linguistic, essential, tonal, efficient 'mmm' in preference to the linguistic equivalent. This may be evident, for example, when tasting (or smelling) fine food, when one might utter '*mmm*!', in preference or precedence to *saying* 'I *like* that!'. The same can be said of the familiar vocal response to a firework display. Such vocalizations are visceral, and can be more immediate, more expressive, and consequently a more valuable currency of communication; furthermore, they often cross linguistic and cultural boundaries.

Of course, expressing personal state is only one of the many uses to which *lexical* language is put. It may be noted that all the various meanings attributed to these utterances of 'mmm', express relations (in particular, affective reaction) between the first person and an object or concept, 'I —— that', and this is a limitation of that form of utterance. Interestingly, whilst certain non-linguistic vocalizations seem to be able to communicate relations between self and object/other, they cannot do the same for other and object/self. It is very difficult to express any equivalent statement of 'you like that', 'you know that', 'you agree with that', etc. in a single tonal utterance in the same way as described for first-person statements of the same. The closest we can get is a tonal contour of 'mmm' which expresses 'you like that?' or 'you understand that?'. These are not the equivalent of statements of the relation between other and object/self, but are invitations to the other to utter their own such statement about their own state.

As long as the subject is the self, the verb is a personal state, and the object is present, monosyllabic contoured tonal utterances can communicate very important information between individuals; they can also solicit the communication of such information from another, with all the social value that such a solicitation holds for both individuals. But whilst these non-linguistic vocal expressions can act as statements about reactions/relationships between the self and other things or persons, they are limited to that purpose. However, it is quite reasonable to suppose that such utterances formed the first *vocalized* form of *concepts* relating the self to object/other, and that this ability to vocalize such concepts was subsequently built upon by more complex (combined) vocal and gestural sequences. This does not undermine Armstrong et al.'s (1994) illustration of the importance of syntactic properties of gestural communication, but it shows that, within a limited repertoire, simple nonverbal vocalizations can communicate similar relations between subject and object, as long as the subject is the self. Hopefully it is also a reminder of quite how expressive such simple non-linguistic vocalizations can be—they can constitute statements about personal state and thought, and can constitute enquiries about others' state and thought, both of which are hugely important social skills.

A fundamental difference between these vocalizations and the gestures of which Armstrong et al. speak, however, is that such utterances do not have separate components that are being interrelated; for Armstrong et al. (1994) the important feature of gestural signs is that they have a potential for decomposition into meaningful agent/action subunits. As such it seems quite plausible that they could have bridged the gap between the vocalizations described earlier, and multi-part vocalizations with syntactic structure; indeed, this seems to be what occurs ontogenically (Butcher and Goldin-Meadow, 2000), as already discussed.

Armstrong et al. (1994) are at pains to reiterate that their proposal is not for a *gestural origin of language*, but that gesture may have played an important part in the development of the cognition necessary for modern language. They recognize the importance of social vocalization from the outset in hominin evolution, and see no reason for positing a phylogenetic discontinuity between primate vocalization and speech. Armstrong et al. (1994) also propose that language had multiple causes and multiple purposes. What they ultimately suggest is that there was an evolutionary stage at which visible gestures took the lead 'with respect to flexibility of output and, critically, the elaboration of syntax'. Subsequently, 'the cognitive apparatus necessary to support these developments would then have been available in spoken language as well'. Underlying this model is a belief that vocal 'gesture' and manual 'gesture' are essentially part of a continuum of finely-controlled body movements 'broadcasting information to the (social) environment' (Armstrong et al., p. 358). It seems likely that vocal and manual fine muscular control were always contiguous, but that (in linguistic speech) a vocal dominance emerged. What Armstrong et al. propose is that there was an intervening period where the balance shifted towards manual gesture, facilitating the development of syntactic structures.

This is in contrast to Hewes (1973/1992), who argues that 'a preexisting gestural language system would have provided an easier pathway to vocal language than direct outgrowth of the "emotional" use of vocalisations characteristic of non-human primates' (p. 65). This latter rationale makes an artificial distinction between gestural and vocal communication, and does not acknowledge the joint importance of both syntactic and prosodic elements to vocal language. What seems more likely from the evidence already considered here and in the preceding chapters is that both were always used together, with corporeal gesture and vocal prosodic gesture both having their roots in corporeal emotional motor systems, but that gestural complexity, and the integration of iconic gesture with vocalization, could have facilitated the development of syntactic elements of language, coupled with tonally-expressed affective content.

Further, that the capacity to perform rhythms, both manually and verbally, forms an important foundation of oral/praxic ability (Alcock et al., 2000), and that the motor sequences instigated in vocal and manual gesture are coordinated by a cyclical (i.e. rhythmic) motor controller (Franz et al., 1992; Mayberry and Jaques, 2000), highlight not only shared neurological foundations between the ability to execute *linguistic* and manual *gestural* sequences, but also the link between *vocal* and manual *rhythmic* capability. Whilst the involvement of a shared cyclical motor coordinator in the instigation of both manual and vocal gesture may have a very ancient provenance, its continued involvement in the ongoing prolonged vocalization—the continued coordination between the rhythmicity of the vocalization and accompanying gesture—may be something that is peculiar to human vocal expression. According to Geissman (2000), although all other primate vocalizations have tonal affective content, human vocalizations are apparently unique in their rhythmic content—even the twenty-six other species that 'sing' do not make deliberate use of rhythmic content—perhaps because of the way that vocal gesture and corporeal gesture are used by us contiguously in ongoing *extended*, planned sequences.

RHYTHM, CORPOREAL MOVEMENT, AND EMOTION

The relationships between rhythmicity, gesture, and affective expression extend beyond vocalization and manual gesture. It is clear that whilst rhythmic information and melodic information are apparently processed in neurologically specialized areas of the brain which are somewhat independent of each other (as discussed in Chapter 7), there is also fundamental integration of the systems involved with rhythmic physiological movement, vocalization, and emotional state. Music and dance make systematic use of this interdependence (Trevarthen, 1999). As discussed in Chapter 1, and as Mitchell and Gallaher (2001) observe, music and dance are historically interdependent developments, with many common features: they are both temporally organized, and described in terms of rhythm, tempo, beat, pace, movement, choppiness, or fluidity, for example. In fact, one rarely exists without some form of the other. Each of these properties can be observed and experienced across numerous modalities, in vision, audition, and kinaesthesis (the proprioceptive feedback associated with voluntary body-movement).

Particular pieces of music can elicit consistent emotional responses in listeners, and movements that often occur in response to music, which are kinaesthetically experienced, can also be a part of emotional experience. Body posture and emotional state are strongly interrelated: our posture and movements can express a great deal about our emotional state, intentionally and unintentionally, and others' body-posture and movements thus provide important cues as to their emotional state. As well as being able to observe such cues, we can empathically experience something of their emotional state in mirroring them with our own bodies. This relationship is explored further in Chapter 10.

Musicality and rhythmic movement, Trevarthen (1999) points out, involve deliberate control and sequencing of this system (see also Malloch and Trevarthen, 2009b). This can result in a self-directed feedback from movement into emotional state and, importantly, feedback and interaction between individuals, in terms of synchrony of movement and of emotional state.

Trevarthen (1999) elaborates on a functional system existing as the brain stem, basal ganglia, and the limbic structures of the Emotional Motor System, which he calls the Intrinsic Motor Formation (IMF). This is responsible for integrating functions of attention, learning, and physiological expression of emotion, including the synchronization and coordination of sequences of movements; it seems to be the case that there is a close interrelationship between the emotional-controlling elements of the limbic system and the areas responsible for the coordination of motor sequences and posture (see also Panksepp and Trevarthen, 2009). In Trevarthen's own words, the 'movement-creating reticular networks and nuclei are intricately combined with the neurochemical systems of emotion. The same activating neurones that select movements and control their energy and smoothness also cause changes in the emotions felt and the intensity and "colour" of consciousness' (Trevarthen, 1999, p. 161). This system is active automatically in all types of interpersonal interaction, but is made deliberate use of in musical activities.

Findings relating to facial expression and proprioceptive freedback may have important implications here. As discussed in Chapter 8, facial expression and emotional content of vocalization are closely related, with particular orofacial musculature configurations resulting in particular vocal qualities (and the production of certain vocal tones and contours being dependent upon the instigation of certain facial expressions). Furthermore, core emotional facial expressions seem to be human universals (e.g. Ekman and Friesen, 1971; Ekman, 1980). What is especially interesting is that feedback from facial expression actually *affects* our emotional state. Levenson, Ekman, and Friesen (1990) found that asking subjects to carry out various facial movements to create the expressions of fear, anger, surprise, disgust, happiness, and sadness caused distinct changes to the activity of the autonomic nervous system, such as changes in heart rate, galvanic skin response (skin conductivity due to perspiration), and temperature. The subjects were not informed as to the facial expression being replicated, they were simply told to carry out a sequence of muscular movements which resulted in the formation of a given expression, for example, to raise their eyebrows, then pull them together, whilst tightening the skin under their eyes and stretching their lips horizontally (producing an expression of fear). This particular expression resulted in increased heart rate and reduced skin temperature, whilst a facial expression of anger increased both heart rate and skin temperature, and a happy expression decreased heart rate whilst leaving skin temperature unaffected (Carlson, 1994).

Furthermore, humans have an innate tendency to imitate the facial expression of others with whom they are interacting. Even at the age of 36 hours, infants can imitate facial expressions that they see (Field et al., 1982); clearly this is an innate, and not a socially conditioned, tendency, and results in some 'contagion' of the emotion being expressed. Kraut and Johnston (1979) similarly found that people have a greater tendency to smile in the presence of others; we respond to a prevalent mood and this in turn influences our own emotional state. Wild, Erb, and Bartels (2001) found that even when exposed to images of another's facial expression (showing sadness, disgust, happiness, anger, surprise, fear, or pleasure) for as little as 500 milliseconds, subjects reported corresponding changes in emotional state. Carlson (1994) suggests that 'imitation provides one of the channels by which organisms communicate their emotions. For example, if we see someone looking sad, we tend to assume a sad expression ourselves. The feedback from our own expression helps put us in the other person's place and makes us more likely to respond with solace or assistance' (p. 351). Being sensitive to the emotional state of others is an extremely important skill in that it helps us to respond in the most appropriate way to them (Schmidt and Cohn, 2001; Sloboda and Juslin, 2001). This can assist in the avoidance of harm at the hands of an angry individual, allow us to anticipate and prevent harm to a third party (Sloboda and Juslin, 2001), as well as being fundamentally important in more positive situations related to the formation of social bonds, networks, and procreation, such as providing comfort when appropriate and pursuing and maintaining interest from the opposite sex.

The implications of all this for the present discussion are threefold:

- First, the production of particular vocal tones normally relies upon adopting certain facial expressions, which correlate with the emotion being expressed;
- Second, the production of a particular facial expression whilst producing a particular vocalization will result in some degree of feedback which actually affects emotional state;
- Third, there is a natural inclination to mimic such expressions and to feel such associated emotions; i.e. such physiological-emotional feedback may occur not only during production, but also during *perception* of such a stimulus.

We are thus equipped with a set of mechanisms that allow us to automatically experience something of the emotional state of others when we hear sounds that they produce. The non-verbal corporeal communication of affective state is not limited to facial expression, but incorporates the whole body and posture; the work of authors such as Trevarthen indicates that these findings of feedback and 'contagion' of emotion, and their implications, apply equally to whole-body expressions of affective state.

This may go a considerable way towards explaining physical and emotional response to rhythmic and dance stimuli, both proprioceptively and visually. Krumhansl and Schenck (1997) found that subjects who listened to a piece of music and different subjects who observed a dance to that piece of music, showed great concordance both within and between the two conditions as to their interpretation of the timing of tension and emotion in each, and the emotion that they experienced at given points. That the dancer interpreted the music in such a way that crescendos in the music corresponded with high movements, high notes with leaps, and staccato and legato sequences with matching physical movements perhaps renders that congruence in interpretation between the two less surprising, given the literality, or *iconicity*, of the interpretation; it nevertheless remains interesting that such physical movements are the (apparently) obvious way of physically representing such auditory phenomena, and that interpretation of them as such occurs. The main implication of the findings relevant to the current discussion is that both media equally represent tension, release, and particular emotions, underlining the cross-modality of such affective expression and interpretation. The affective content is apparently interpreted equivalently in visual, auditory, and kinaesthetic media.

As with facial expression, this is not restricted to perception, but is also a two-way process. Mitchell and Gallaher (2001) review a considerable body of research which illustrates that 'music prompts kinesthetic (motor) responses in both children and adults that often match some aspect of the music' (Mitchell and Gallaher, 2001, p. 66), and that 'kinaesthetically experienced movements that sometimes occur in response to music can also be a normal part of emotions' (pp. 66/7). This is a phenomenon with which most people are familiar already, from their own experience (which does not diminish the value of such research, but adds to it). That this coordination of emotional communication and of rhythmic behaviour is innate is evidenced by studies of infant–parent interactions (Chapter 8) and infants' responses to rhythmic stimuli (Zentner and Eorola, 2010, discussed later).

This innate interrelationship between rhythmic and vocal action, and emotional experience, has potentially powerful effects when systematically stimulated in coordinated bodily rhythmic experience. McNeill (1995) reports how the use of rhythm alone may be significant in inducing group cohesiveness. He observes that when people move together in dance (or military drill) this elicits a physiological arousal in the participants, and a great sense of emotional commonality and solidarity. He calls this 'muscular bonding', and proposes that such effects would have important selective benefits to any hominin group that practised them. On the basis that all surviving cultures of humans make use of music and dance of some form, he suggests that those groups that had practices of moving together rhythmically developed stronger communities and became better at cooperating together in other spheres, particularly hunting.

Direct evidence for the group selective benefit of this would be difficult to identify, but the principle of 'muscular bonding' itself has a strong foundation; it is deliberately used in military circumstances the world over, and ethnographic evidence of the action of dance on evoking group solidarity is very widespread (e.g. Nettl, 1992; Locke, 1996; Johnston, 1989; Blacking, 1995; see Chapter 2). Further, the very close association of the areas of the brain responsible for complex vocalizations and complex sequences of muscular movements, which seem to have evolved in tandem (see Chapters 5–7), may go some way to explaining the association of rhythm, dance, and tonal vocalization in almost all cultures.

Merker (1999) has suggested an alternative, or additional, selective scenario. He suggests that coordinated synchronous rhythmic and melodic group behaviour could be derived from synchronous chorusing activities carried out to attract female hominins from neighbouring territories. This is on the premise that, like chimpanzees, Lower- and Middle-Palaeolithic populations may have been exogamously organized (females leave their natal population group and join another where they settle to raise young). In this situation, males can make distance-calls to attract prospective mates from neighbouring groups; Merker points out that if this activity was carried out by several individuals at once, synchronizing their calls (as is the case with several insect species such as crickets and, visually, glow-worms), then the area reached by the sound from the call increases proportionally. Having attracted females from neighbouring groups in this way, it would then be in the interest of individual members of the calling group to have an individual distinct calling activity to attract females to themselves rather than to the rival males in their group.

Distance-calls amongst higher primates in the wild are almost invariably accompanied by simultaneous body movements such as branch-waving and beating of the ground. Merker proposes that such behaviours, in a group-synchronized form, were the root of musical behaviours in hominins. This is an interesting idea which in many respects makes sense; however, it does remain the case that there is no known precedent for this type of *cooperative* behaviour amongst primates—it is principally known amongst insects. This is not to say that there is any reason to believe that it *could not* have developed amongst primates, simply that there is no precedent yet known to support the idea. Furthermore, in Merker's scenario, there is the issue of two conflicting manifestations of the same behaviour, firstly cooperative, then the same behaviour being used competitively. Initially there would have to be a group selective advantage to carrying out the calling behaviour in synchrony, followed by an individual selective advantage in creative variation of the vocal behaviour to distinguish oneself from one's rivals, according to Merker's proposal; i.e. the selective pressure for synchronous chorusing would by necessity impose an equal selective pressure for *asynchronous* calling, if this were indeed the behaviour used to woo prospective mates having attracted them to the territory. Further research and evidence from primatology may prove useful in testing the plausibility of such a process.

The fossil record cannot directly tell us when organized rhythmic behaviours emerged; in contrast to vocalization, which has specifically dedicated and specialized physiological features, rhythm production does not have any dedicated physiological correlate of which there will be fossil evidence. In terms of the basic physical ability to deliberately make percussive sounds, we know already from the evidence of stone tool manufacture, that very early humans had the coordination and dexterity necessary to strike one item with another quite precisely (e.g. Wynn, 1993; Johanson and Edgar, 2006; Roux and Bril, 2005). This would result, intentionally or unintentionally, in percussive sound (recent research by Cross, 1999b, n.d., and Blake, 2011, has investigated the potential use of lithics for intentional sound-making). Even without coordination to knock two objects together, percussive sequences could potentially be created by foot stamping, an activity obviously physiologically possible long before the advent of the genus *Homo*. But musical rhythmic behaviours (MRB) (to use Bispham's, 2006, term) constitute more than a prolonged sequence of percussive sounds. Whilst the ability to produce such a sequence would be a necessary prerequisite to exhibiting musical rhythmic behaviours, it is not enough in itself to constitute musical rhythmic behaviours. These incorporate rhythmic patterns with a consistent periodicity, or metre, and furthermore, as McNeill's (1995) and Merker's (2000) scenarios both emphasize, coordination of rhythmic behaviour between individuals. Neither flint knapping nor foot stamping capability indicates the ability to deliberately produce extended sequences of muscular movements coordinated with a consistent metre. As was discussed in Chapters 6 and 7, neurological structures in the left temporal lobe in and around Broca's area are involved in the execution of complex sequences of muscular movements, both manual and vocal-orofacial, and these have rhythmic qualities. It may be that the ability to deliberately carry out prolonged sequences of muscular movements with rhythmic qualities developed in tandem with the ability to carry out longer and more precisely controlled vocalizations, but to understand the development of the type of rhythmic capabilities that we associate with musical behaviours we will need to look more closely at the nature of their metrical properties and how they are used in an interactive way.

ENTRAINMENT

Having the ability to perceive and produce melodic and rhythmic stimuli is not enough to allow participation in musical activities. A fundamental component of our ability to undertake the musical and dance activities that we do is our ability to coordinate the timing of our own sound production and movement with either an internally generated pulse or with the percept of a pulse within an externally generated stimulus—a process known as entrainment.[1] This is no small task: not only does a pulse have to be identified as a stimulus (often within the product of someone else's bodily action, or generated internally), but motor action must be executed with the same pulse, and must be constantly modified to account for any variation in both the stimulus and the motor action. It is our ability to entrain that makes possible the achievement of a commonality of action, experience, and purpose which characterizes musical, dancing, and marching behaviours and which makes possible the types of effects (and benefits) described, for example, by McNeill (1995).

There are many naturally occurring (endogenous) rhythms within the human body, at various timescales, including heartbeat, blinking, respiration, locomotion, circadian cycles and hormonal (including menstrual) cycles (Clayton et al., 2005). Most of these are also, obviously, features of the bodies of other animals too. In fact, internally generated rhythms may be central features of all life (Clayton et al., 2005), and there is even a journal dedicated specifically to their study, *Biological Rhythms Research*. Some biological rhythms correlate directly with arousal levels. 'Indeed, all movements of animals, even those such as jellyfish or worms, without jointed body members, have a cyclic time control of rhythm (Llinas, 2001). For all animal movement, increases in physiological arousal are accompanied by increases in frequency, providing music-makers with a uniquely natural means of modulating arousal in their listeners (Molinari et al., 2003), encouraging a response in the form of rhythmical movements or dance (Cross and Morley, 2009)' (Turner and Ioannides, 2009, p. 165). This latter effect in humans is only really possible because of our entrainment with the external stimulus; it can also be added to this observation that whilst increasing or decreasing rhythmic frequency has concomitant effects on arousal level, the maintenance of a tempo at a fixed level has the potential to have a profound regulatory effect on arousal—inducing a degree of continuity in arousal level which would be difficult to replicate under normal ecological conditions.

---

[1] 'the concept of entrainment... describes the interaction and consequent synchronization of two or more rhythmic processes or oscillators' (Clayton et al., 2005, p. 4). Clayton et al. (2005) carry out a very useful and comprehensive overview of the various forms of entrainment and biological rhythms, the history of their investigation and relevance for musicological study.

Entrainment can occur between different physiological processes, such as motor movement and respiration and, as noted previously, between the timing of elements of speech and manual/upper body gesture (Franz et al., 1992; Mayberry and Jaques, 2000). What seems to be peculiar to humans, however, is the ability to voluntarily identify an internally generated rhythm which may be independent of these other physiological rhythmic cycles, and to coordinate our own other muscular movements with that internal rhythm. This internal rhythm may apparently be spontaneously generated with no reference to an outside stimulus, or may itself be generated to coordinate with an identified rhythm in an external stimulus.

Bispham (2006) has carried out a valuable assessment of the capabilities which appear to underlie human musical rhythmic behaviours (MRB), the extent to which these are analogous to or homologous with rhythmic behaviours in other animals, and the extent to which they appear to be dedicated to musical function. Both newborn human infants (who are prelinguistic) and non-human primates (who are non-linguistic) are able to distinguish between different rhythmic patterns within vocalization (specifically, between languages in different rhythm classes), indicating that this ability to extract rhythmic patterns from an auditory stimulus is not predicated on language ability and did not evolve for that purpose (Hauser and McDermott, 2003).

Zentner and Eorola (2010) have shown that infants (5–24 months) engage spontaneously in rhythmic movement in response to rhythmic stimuli (music, rhythmical patterns with a regular beat, and isochronous drumbeats). It is the beat within the stimulus, rather than other aspects of it, which elicits the rhythmic behaviour. They did not find that speech (infant-directed or adult-directed) elicited consistent rhythmic movements in the infants, with the exception of 6-month-old infants, who did produce rhythmic movements in response to infant-directed speech. This coincides with the period when infants are generally most stimulated by and engaged with infant-directed speech.

Although infants do not entrain their movements with rhythmic stimuli, in the sense of each gesture synchronizing with beats in the music, faster movements were related to faster stimuli, and there was also correlation between acceleratory passages in the stimuli and acceleratory sequences of movements in the infants (Zentner and Eorola, 2010). The degree to which synchronization occurs seems to be a product of the clarity of the rhythmic stimulus (Zentner and Eorola, 2010), although it may be the case that children do not possess the necessary degree of motor control to precisely coordinate gesture and stimulus beat until about 4 years of age (Provasi and Bobin-Begue, 2003; Kirschner and Tomasello, 2009; Zentner and Eorola, 2010). Zentner and Eorola (2010) also found that accuracy of synchronizing with rhythmic stimuli had a positive affective effect—i.e. there was a correlation between the degree of synchronization with rhythmic stimuli and the duration of smiling in the infants.

Interestingly, there was no age-related effect to any of these observations regarding engagement with rhythmic stimuli, with no difference in performance between different age groups of infants between 5 and 24 months of age. The tendency to engage with a rhythmic stimulus by producing rhythmic gestures, and for this to produce a positive emotional effect, would appear to be present in early infancy and to remain unchanged at least up until 2 years of age. The ability to synchronize movements with rhythmic auditory stimuli (i.e. to entrain) would appear to require greater motor coordination than is possessed by very young infants, although they are sensitive to—and physically respond to—the tempo, and changes in it, from a very early age; the ability to precisely entrain develops spontaneously subsequently (Kirschner and Tomasello, 2009; Patel et al., 2009).

Whilst gorillas (Schaller, 1963) and chimpanzees (Goodall, 1986) both undertake rhythmic manual drumming in signalling behaviours, using physiological mechanisms obviously very similar to those used by humans in drumming, there is as yet no evidence that these actions are coordinated with reference to an internally generated pulse (as opposed to their rhythmicity being derived purely from biomechanical function or learned motor sequences). Nor is there evidence that they are coordinated with reference to external rhythmical stimuli, including the actions of another individual—specifically 'coordination between the actions—overt or perceptual—of two or more individuals based on a mutually manifest and regularly structured temporal framework' (Bispham, 2006, p. 127). There is thus currently no evidence in other primates of the form of entrainment involving interaction with others that is such an integral part of musical behaviours (Bispham, 2006). Whether this represents a genuine incapacity to do so or a lack of ecological impetus remains an open question.

Whilst beat perception and synchronization (BPS) behaviour seems to have emerged in humans since our last common ancestor with chimpanzees, the capability of entraining to an external rhythmic stimulus is not unique to humans when the wider animal kingdom is considered (Schachner et al., 2009; Patel et al., 2009; Fitch, 2009b). Interestingly, the animals that also exhibit entrainment are also those that exhibit vocal mimicry behaviours. Genuine BPS has now been observed in a sulphur-crested cockatoo (*Cacatua galerita eleonora*) (Patel et al., 2009; Schachner et al., 2009) and a grey parrot (*Psittacus erithacus*) (Schachner et al., 2009). Schachner et al. (2009) also carried out a systematic search of videos posted on YouTube claiming to show animals who were involved in 'dancing' behaviour; they found that although more than 50 per cent of the animals featured were non-vocal mimics (e.g., cats, dogs, chimpanzees, dolphins), all of the examples which featured *genuine* evidence of entrainment were of species that exhibit vocal mimicking behaviour, all birds (including macaws, parrots, parakeets, cockatoos) with the exception of one Asian elephant.

Clearly this is not a case of shared evolutionary heritage—the separation between these species and humans is far too great and the same capabilities are not exhibited by intervening species—but a possible case of convergent evolution, with an analogous solution emerging to an analogous problem. It could be that entrainment capability emerges as a by-product of the capability to carry out mimicry of vocal sounds, and that it emerged in humans as a consequence of the increasing importance of vocal learning in human evolution; alternatively, it could be a by-product of the emergence of mimicking ability at a more general level. Further research demonstrating whether entrainment capability is exhibited by all species that carry out vocal mimicry, and conversely by none who do not exhibit vocal mimicry, will help to test these associations. The preliminary survey of Schachner et al. (2009) points in this direction, and it is also consistent with the strong association in humans between vocal and rhythmic gestural coordination discussed earlier. As Patel et al. (2009) put it, 'More generally, future comparative work can help determine what neural abilities are necessary foundations for BPS. In this regard, it is important to note that vocal learning may be a necessary but not sufficient foundation. For example, BPS may require the neural circuitry for open-ended vocal learning (i.e., the ability to imitate novel sound patterns throughout life) as well the ability to imitate nonverbal movements, two evolutionarily rare traits shared by parrots and humans' (Patel et al., 2009, p. 829). It may be that the human aptitude for vocal mimicry and learning builds upon, and constitutes a specialized use of, a wider ability for mimicry of gesture and corporeal expression.

It has also been suggested that our ability to entrain forms part of our ability to extract temporal information from stimuli, which in turn also allows us to form expectations about future events (Jones and Boltz, 1989; Bispham, 2006). It appears to be utilized in turn-taking in interaction between individuals, in a looser, often subconscious way (Clayton et al., 2005), as well as allowing the 'tighter', precise temporal coordination of actions that is used in musical behaviours.

Processes of entrainment have been characterized as being either symmetrical or asymmetrical (Clayton et al., 2005). Asymmetrical entrainment refers to the rhythm in entity B entraining to the rhythm occurring in entity A with no two-way interaction—i.e. the rhythm in entity A is not affected by the rhythm in entity B. Symmetrical entrainment refers to the rhythms in entities A and B adjusting to some degree to each other in order to achieve entrainment. Rather than being viewed as absolutes these should perhaps be viewed as the extremes of a continuum of degree of reciprocity in the process of entrainment. The degree of reciprocity in 'symmetrical' entrainment may vary, from being absolutely symmetrical, with both entities constantly adjusting to variations in the other, to being virtually asymmetrical, with one entity doing most, but not all, of the adjusting to the other.

In participating in musical activities a variety of different degrees of symmetry (reciprocity) in the entrainment can occur. At one extreme, one may give oneself over entirely to the stimulus, 'get lost in the music', become totally subservient to the rhythm, as a listener or dancer, with no attempt (or potential) to influence the stimulus; at the other extreme one may engage in an entirely cooperative entrainment, where the distinction between performer and listener becomes meaningless. In the former case there is the appeal of submersion of individual choice, action, and control, 'losing oneself', in a (usually) safe, bounded, and largely predictable context. In the latter case there is the appeal and value of reciprocal cooperation, synergy with another individual, or a group. In fact many musical experiences will not fall at the extremes of this continuum of symmetry of entrainment, but will fall somewhere in between, a powerfully appealing combination of reciprocal cooperation and 'losing oneself'. They may also feature a combination of both extremes—for example, when two individuals cooperatively (symmetrically) entrain their movements with each other whilst both entraining subserviently (asymmetrically) to music being played; i.e., dancing together. The rhythmic concordance, physical cooperation and proximity, and more general achievement of commonality of experience required mean that it is perhaps not a coincidence that music, dancing together, and sexual relations are frequently associated—they share much in common.

The role of entrainment in music should be understood in the context of the whole-body experience of musical stimuli, and this discussion relates directly to the previous discussions of embodied motor control, corporeal expression of emotion, and Trevarthen's Intrinsic Motor Formation and Pulse. Entrainment is a fundamental aspect of not just our production of music but also our perception of it. It is evident that perception of music is essential to its production, but it is increasingly evident that the reverse is also true, in that it seems that the experience of listening to music also results in a motor activation in the listener (Grahn and Brett, 2007; Grahn and Rowe, 2009). Relationships between the perception of vocal gesture, body language and emotion have already been discussed (see also Chapter 10). There is increasing evidence that in perceiving musical stimuli, activation occurs of neural systems that are also involved in the production of these stimuli.

Keil (1995) has referred to this phenomenon as 'kinaesthetic listening', in which listeners to music who are experienced performers experience in their muscles a sensation of what it would be like to play the music (Clayton et al., 2005). Sacks (2008) also describes various anecdotal cases suggestive of a motor component to the mental experience of music: one of Sacks's correspondents describes that after a day of having a musical 'brainworm' (a piece of a tune going around repetitively in the head) his throat felt as uncomfortable as it would had he sung all day. Another found that systematically relaxing the speech apparatus (the activity of which he associated with 'auditory thinking') resulted in dispelling the 'brainworm' tune. Another correspondent—an instrumentalist—described that when she had a 'brainworm' she could not help but to think about it spatially and kinaesthetically, and would feel the fingerings of the piece in her hand and arm (although she did not physically act them out) (Sacks, 2008).

Such cases further 'emphasize the significance of entrainment in listening to music, and... suggest that the study of entrainment in listeners is as important as that in performers' (Clayton et al., 2005, p. 8). Whilst these examples involve Western individuals who are experienced performers, they have been corroborated in wider ethnomusicological contexts too, and as Clayton et al. (2005) point out, ethnomusicologists are very aware that the distinction between listener and performer is not always an easy one to make—'participant' is often a more representative moniker than either.

There are several sources of evidence that auditory processing of rhythmic input is importantly implicated in corporeal muscular control, and that the systems are linked in some respects. Thaut et al. (1997) demonstrated that the use of a metronomic stimulus has a remarkable effect upon the rehabilitation of gait control in stroke patients. It seems that the external rhythmic stimulus fulfils an important cueing role in gait control, in terms of stride rate, length, and lateral (left, right) rhythm, and particularly when internal mechanisms for generating such cues are no longer functional due to the pathology. It seems that it is not merely the case that the patients are moving in time with the metronomic cue, but that this is actively aiding in the stimulation of motor control; when the rhythmic stimulus is removed, motor control also degenerates. Thaut et al. suggest that 'auditory rhythmic timekeepers may enhance more regular motor unit recruitment patterns... [this effect] underscores the existence of physiological mechanisms between the auditory and motor systems. The ability of auditory rhythm to effectively entrain motor patterns and also influence nontemporal parameters such as stride length, may help to assign rhythmic auditory stimuli a larger role in motor control than previously assumed' (Thaut et al., 1997, p. 211).

Thaut (2005) proposes that there is a direct neural connection between auditory rhythm perception and motor tissue activation. He states that 'The basic neural network underlying isochronous pulse synchronization consists mainly of composite motor and auditory areas with no clearly designated functionally separate area for synchronization' (Thaut, 2005, p. 58); i.e. the process of synchronization between an auditory rhythmic input and motor action occurs not via a dedicated synchronization area, but through overlap between the auditory and motor areas of the brain involved in both the perception and production of the rhythm. Thaut proposes that 'the neuronal activation patterns that precisely code the perception of rhythm in the auditory system spread into adjacent motor areas and activate the firing patterns of motor tissue' (Thaut, 2005, p. 58). Such associations between locomotor control and rhythm lend weight to suggestions (e.g. Trevarthen, 1999) that the shift to bipedal locomotion may have been an important factor in the development of rhythmic control and entrainment in humans (cf. also walking gait duration in humans: Alexander and Jayes, 1980; Minetti and Alexander, 1997, see chapter 7).

Successful entrainment and coordination may also turn out to be reliant upon 'mirror neuron', or action-observation, systems, systems of neurons activated in perceiving a motor action that are closely related to those that would be activated if actually carrying out those activities (e.g. Rizzolatti and Craighero, 2004; Buccino et al., 2001). These were first observed in monkeys observing grasping actions (Gallese et al., 1996; Rizzolatti et al., 1996), but have subsequently been implicated in many other complex motor activities (Buccino et al., 2001), including having a proposed role in the emergence of linguistic speech (Rizzolatti and Arbib, 1998; Rizzolatti and Craighero, 2004). Mirror neuron/action-observation networks appear to play a critical role in both the understanding of the actions of others and in learning through imitation, both fundamental prerequisites for human social organization and culture (Rizzolatti and Craighero, 2004). Both are important aspects of musical learning and experience, and it may also be that the several previous observations of 'sympathetic' experience which forms part of music perception (including gesture, posture, and entrainment) may be attributable to (or, at least, retrospectively relabelled as) mirror neuron activity.

As Turner and Ioannides (2009) observe, 'pertinent for understanding the brain mechanisms of musical communication and art is the increasing evidence that mirror systems in the brain are conveying information between participants about dynamic emotional states, qualities of purpose expressed in movement, and the goals of actions (Adolphs, 2003; Gallese, 2001)' (Turner and Ioannides, 2009, p. 165). For example, the areas of the right superior temporal cortex that experience increased blood flow during processing of melodies (Zatorre et al., 1994) form part of the mirror system of the brain, for sympathetic representations of others' activities over all sensory modalities (Decety and Chaminade, 2003; Jeannerod, 2004; cited in Turner and Ioannides, 2009). Similarly, Sakai et al.'s (1999) fMRI study of neural activation in short-term memory tasks related to rhythmic sound sequences showed that areas that are important for 'mirroring' movement intentions were activated (Turner and Ioannides, 2009).

In their discussion of emotion and music, Panksepp and Trevarthen (2009) relate mirror systems, interpersonal understanding, and emotion directly in an evolutionary context, and are worth quoting in full here:

> In the course of the evolutionary process that established social communication of intentions and feelings between animal *selves*, polymodal areas of the higher regions of the forebrain come to act in resonance with the intentions and feelings

of other subjects, constituting what are called 'mirror' representations.... Being emotionally controlled, these *other*-within-*self* intersubjective representations establish sympathetic resonances, and intersubjective contagions, probably by intrinsic affective systems situated much lower than the neocortex (Watt and Pincus, 2004), making complementary adjustments to the intelligence and feelings expressed in gestures of other bodies sensed by sight, sound and touch through neocortical processes that are epigenetically programmed by experience. This cerebral machinery of emotional *self-other* awareness (Thompson, 2001; Reddy, 2003) is the ancestor of much more than a 'language acquisition device' (Rizzolatti and Arbib, 1998); it is the motivator of sociocultural existence and its moral foundations, and of each individual's urge from infancy to learn cultural skills, including those of language (Trevarthen, 2004; Bråten and Trevarthen, 2007). (Panksepp and Trevarthen, 2009, p. 115)

Musical experience makes systematic use of this cerebral machinery.

The majority of the foregoing discussion of entrainment and musical experience has focused on the interactive aspects of the processing, the foundations in interpreting—and achieving parity in—the emotional states and intentions of others.

It might be quite reasonably observed that, in addition to the widespread communal experience of music, many people gain a great deal of enjoyment from performing music alone—the solitary shepherd with their pipe, and the popular activity of singing in the shower spring to mind. It may be suggested that this is the case for precisely the same reasons—these situations gain much of their value from drawing upon exactly the same 'interactive' mechanisms. Music is experienced in many of the same ways and using many of the same mechanisms as are employed in interacting with others, but in these cases, entraining with one's own internally-generated cyclical metre. Interpersonal interaction is in this sense inherent within entrainment. Musical activity, including rhythm, is inherently interactive, and provides a vicarious social interaction when alone.

CONCLUSIONS

Gesture and speech are interdependent. Both are affected simultaneously in stutterers, and gestural and vocal behaviours are interrelated from the earliest babbling in infants. They can operate independently, but when the upper limbs are otherwise unoccupied they are sequestered into speech-related gesture. Furthermore, it seems the perception, as well as production, of vocalization can be linked with gesture, and this is true from birth.

Common timing processes are involved not only in movements of the limbs, but also in speech and non-speech movements of oral structures, suggesting that there is a cognitive rhythmic motor coordinator that instigates such muscular sequences irrespective of the musculature that is used, and that the complex patterns of muscular gesture (in fingers, hands, arms, shoulders, and joints) and in vocalization (orofacial, laryngeal, and respiratory musculature) are coordinated. The concordance between gesture and speech is instigated early in the vocalization process, by cyclical motor control, with gestural movements being associated with the nuclei of tone groups—prosodic rhythm—rather than the lexical elements of speech. This is also evident in the gestures accompanying infant vocalizations. Greater lateralization of gesture occurs after babbling begins, implying left-hemisphere development at this time in the integration of the two actions. These earliest gestures are emotive and rhythmic rather than iconic (which accords with the finding that gesture corresponds with prosodic rhythm rather than lexical content) and only start to be used iconically and in combination with words when lexical behaviour has started to develop. In the case of speech (as opposed to other, non-linguistic, vocalizations), linguistic meaning and narrative sentence structure are integrated into the gesture-speech system *after* the integration of gesture with prosody, but before their physical manifestation.

Whilst extremely important in the type of information they express, nonverbal vocal utterances, in meaning, are limited to acting as expressions of personal state and relationships between the self and other. Such utterances can express personal state (well-being, approval, disapproval, disgust, etc.) and in this sense contain a subject-verb-object relationship, in that the subject is the self, the verb is a personal state, and the object is present as a focus of shared attention. They can also solicit such information from another individual. Both the expression and solicitation of such information are obviously important and advantageous skills for mediating social relationships. In terms of the emergence of language from such vocalizations, however, it seems likely that such utterances formed the first *vocalized* form of *conceptualizations* relating the self to object/other, and that this ability to vocalize such concepts was subsequently built upon by more complex vocal and gestural sequences. In fact, it is at that stage that the syntactic potential of the coupling of vocal and *iconic* gestural communication may have become fundamentally important, bridging the gap between non-lexical vocalizations of affect, and multipart vocalizations with syntactic structure and iconic associations.

Production of complex vocalization relies on priming of the whole motor system. Particular non-lexical vocal utterances (and non-linguistic content of speech) are accompanied by quite specific involuntary body-language; they share an affective origin in a system in which vocalization and corporeal expression, or to put it another way, *vocal and corporeal gesture*, are intimately linked. Note that this is at least equally applicable to body movements accompanying musical vocalization, and vocalization accompanying dance. There are not only shared neurological foundations between the ability to execute vocal and manual gestural sequences, but also a link between vocal and manual *rhythmic* capability. The capacity to perform rhythms, both manually and verbally, forms an important component of oral/praxic ability—detriments to one result in detriment to the other. This integration occurs whether utterances are linguistic or not.

Although the production and perception of rhythm and melody involves some neurologically specialized and distinct areas of the brain which are in some respects independent of each other (as illustrated in Chapter 7), there is also clearly important integration of these systems, with rhythmic muscular movements being coordinated with prosodic elements of vocalization in their production. Specifically, they are interdependent in the planning and execution of sequences of muscular movement associated with instigation of vocalization, rhythmic physiological movement, and expression of emotional state in these media. The production and perception of tonal *content* in vocalizations do not appear to require any input from rhythm-controlling systems in the left hemisphere, but the planning and execution of the muscular sequences themselves do.

Affective content can apparently be interpreted equivalently in visual, auditory, and kinaesthetic media, each of which can represent tension, release, and particular emotions, underlining the cross-modality of such affective expression and interpretation. Vocal quality is directly influenced by facial expression, and the production of particular facial expressions and particular body postures actually causes us to experience some emotional response as a consequence. Furthermore, we tend to do this to some extent automatically when *witnessing* facial expressions and body language in others, and whether or not we fully physically manifest those expressions or postures ourselves, it is likely that action-observation-network neural firing occurs which replicates some of the brain response to adopting such a posture ourselves. There seems to be a close interrelationship between the emotional-controlling elements of the limbic system and the areas responsible for the coordination of motor sequences and posture—the same systems that select and control movements also cause changes in the emotion-controlling elements of the limbic system. This can result in a self-directed feedback from movement into emotional state and, importantly, feedback and interaction between individuals, in terms of synchrony of movement and of emotional state. In other words, this physiological-emotional feedback may occur not only during production, but also during *perception* of such a stimulus. This means that in producing, and even to an extent in perceiving, a particular sound we generate some emotional response in ourselves due to the kinaesthetic feedback from the physiology and neurology required to produce that sound. It also means that we should expect there to be some consistent correlation across all humans between our emotional response to particular sounds, and the facial expression required to produce them.

Note that feedback can also occur to some extent between the modes of experience. As is well known, our posture and movements can express a great deal about our emotional state, intentionally and unintentionally, and others' body-posture and movements thus provide important cues as to their emotional state. The interrelationships already discussed mean that as well as being able to observe such cues, we can empathically experience something of others' emotional state in mirroring them with our own bodies. Musicality and rhythmic movement involve deliberate control and sequencing of this system, requiring us to adopt particular expressions and poses in the creation of the stimulus, to carry out particular vocal, facial, and corporeal gestures, and furthermore they encourage the adoption of equivalent forms of these between individuals, which leads to a sharing of emotional state. Such reactions occur automatically whether one is fully participating in musical activity over all modalities, or only one (e.g. listening).

The ability to synchronize movement with an internally or externally generated pulse is a critical component of musical participation. Changes in frequency have direct effects on arousal level, and consistent frequency can effectively moderate level of arousal too. Although they appear not to have the level of coordination to match beats exactly, infants engage in rhythmic movement in response to rhythmic stimuli, their movement rate correlating with rates in the stimuli, and their level of synchronization positively correlates with positive emotional response. The ability to genuinely engage in entrainment may be directly related to the development of the abilities for sophisticated mimicry of gesture and corporeal expression, as well as having implications for abilities in turn-taking (critical in social interaction) and holding expectations about future events on the basis of patterns of events.

The process of entraining can be cooperative or subservient, or some combination of the two, and has the potential to allow both 'losing oneself' in the stimulus, and/or a profound sense of physical cooperation, and synchronization of arousal. Many musical experiences feature a powerful combination of both these effects, as individuals cooperatively (symmetrically) entrain their movements with each other whilst both entraining subserviently (asymmetrically) to music being played. The value of this experience is related also to the physicality of bodily gesture—the physical expression of emotional state—as well as direct overlaps in the mechanisms for the perception and production of musical stimuli. Entrainment in the experience of music makes systematic use of our systems for understanding the emotional states and intentions of others through physical gesture, and is an inherent part of all musical experience.

The idea that rhythm systems and melody systems have *come to be used together* over the course of the development of musical capabilities in humans would seem to be inaccurate, artificially separating these functions. Whilst they clearly do rely on some specialized processing mechanisms, there are fundamental overlaps between them in that vocal control, rhythmic muscular movement, bodily gesture, and emotional expression all rely on integrated systems which are activated in both production and perception of musical stimuli. Musical experience relies upon systematic use of a gestural system, including vocal tonal gesture relying on rhythmic cyclical muscular control, which exists to allow the expression of emotional state and the understanding of emotional state in others.

10

Emotion and Communication in Music

> Since nought so stockish, hard and full of rage,
> But music for the time doth change his nature.
> The man that hath no music in himself
> Nor is not moved with concord of sweet sounds
> Is fit for treasons, stratagems and spoils;
> The motions of his spirit are dull as night
> And his affections as dark as Erebus:
> Let no such man be trusted.
>
> Lorenzo, in Shakespeare,
> *The Merchant of Venice* (V. i. 81–8)[1]

INTRODUCTION

That music and emotion are strongly associated is without doubt, and the nature of this relationship has been the focus of debate and research since at least Classical times—and probably rather longer. The interrelationships described in the preceding chapters between tonal contour, rhythmic control, and non-linguistic social and emotional vocalizations, in humans and in other primates, lead to emotional and social associations in our experience of music. The preceding chapters have highlighted that critical aspects of musical experience—its production and perception—are rooted in systems of gesture (corporeal, including vocal) that are involved in expression, interpretation, and regulation of emotional state. But these systems are used for far more than straightforward communication of emotion—they are critical in the formation of interpersonal bonds and the establishment of empathy and interpersonal skills, in other primates and, in a developed way, in *Homo sapiens* infants and adults. Music is clearly far more than 'the language of emotion'. The production and perception of musical stimuli make use of social interactive capacities

---

[1] Also quoted by Turner and Ioannides, 2009.

but, although rooted in these capacities, much more can be said about the way that music has an effect on emotional state; the ways that these stimuli interact with our sensory mechanisms and our modern cognition, and the contexts in which we experience them, mean that our experience of music, and reactions to it, have become far more complex than merely 'pressing the buttons' of human interaction.

This chapter looks in more depth at how these associations manifest themselves in the relationships between music and emotion, and elaborates on some of the associations noted in earlier chapters. It starts by examining ways in which types of emotional content in musical stimuli have been understood and classified, how social content and context influence the emotional experience, and discusses more of the neurological and physiological mechanisms involved in emotional experience in music.

In the past, theories about the interrelationship between emotion and music often seem to have been characterized by impositions of mutual exclusivity on various explanations, when in fact there is a number of likely causes for the eliciting of an emotional response to a given piece of music, which can operate together at different levels within the listener. In recent years the study of the relationship between music and emotion has expanded considerably, encompassing the disciplines of psychology, neurology, and philosophy, as well as musicology, including a dedicated volume giving good coverage of the approaches and conclusions from each of these disciplines (Juslin and Sloboda, 2001) as well as numerous book chapters and scientific papers.

INTRINSIC AND EXTRINSIC EMOTIONAL CONTENT OF MUSIC

Discussing emotion and music is potentially confusing, as the relationship can be considered along a number of dimensions which are interrelated; separating those dimensions for the purposes of discussion is difficult and even potentially undesirable (see Lavy, 2001). In terms of the actual *processing* of the musical input, Lavy (2001) points out that the various rigid distinctions made between different categories of emotional content associated with music are somewhat artificial, bearing little resemblance to our understanding of emotion in other domains. Nevertheless, in discussing the *derivation* of such content they are useful. Whilst considering the relationship between emotion and music along these dimensions, the following sections will also illustrate the extent to which these classifications are interdependent.

The dimensions are as follows: music may have *intrinsic* or *extrinsic* emotional content (Juslin and Sloboda, 2001), which may be *iconic, indexical,* or *symbolic* (Dowling and Harwood, 1986; following Peirceian semiotic categories[2]), and may *represent* and/or *elicit* emotion (arousal theory vs. expression theory, Davies, 2001). The expression of emotion and eliciting of an emotional response by music can occur simultaneously and may, but need not, correspond—for example, whilst a given piece of music might be perceived by two listeners to be *representing* a particular emotion, the emotion(s) *elicited* in each listener need not be the same, either as that being expressed or as that elicited in the other listener. This is because emotional responses are elicited in the listener for a variety of other reasons too.

Sloboda and Juslin (2001) and Sloboda (2001) categorize the roots of emotional effects of music in terms of *intrinsic affect* and *extrinsic affect*. The intrinsic emotional properties of the music are those which elicit in the listener an emotional response as a direct consequence of structural properties of the music itself (i.e. with no outside referent). These are the *symbolic* properties of music, its syntax and style (Dowling and Harwood, 1986). Into this category fall characteristics such as the fulfilment or violation of expectations (Meyer, 1956), intensity, and perceived tension and release. *Extrinsic* affect is the emotional response elicited as a consequence of association of properties of the music with previous events or experiences (its *indexical* properties) or through the music's resemblance to other phenomena (its *iconic* properties).

An influential theory proposed by Meyer (1956) argued that music expresses emotion—or induces emotional response—through the creation and resolution of expectations, in much the same way that emotional responses are elicited by other stimuli and events. Expectations about the progress of a melody may be due to learning conventions through personal experience (through exposure to particular forms of music), or may be innate expectations grounded in gestalt laws of cognition, such as the expectation that movement in a given direction will continue in that direction (applicable to an ascending scale, for example). It seems that some of these expectation-violations cannot be overridden by subsequent familiarity with a given piece of music, being

---

[2] Peirce's semiotics of signs—symbols, indices, and icons: Peirce divided 'signs' (forms of representation) into three types, according to the way in which they communicated information—the relationship with their referent:

Icon: A sign which refers to the object that it denotes merely by virtue of characters of its own—relationship by *similarity* to the referent (e.g. a sculpture of a horse, a painting of a fire)
Index: A sign which refers to the object that it denotes by virtue of being really affected by that object—*causal* relationship with the referent (e.g. a horse's hoofprint, smoke from a fire)
Symbol: A sign which is constituted a sign merely or mainly by the fact that it is used and understood as such—*conventional* relationship with the referent (e.g. the words horse, fire) (summarized in White, 1978).

Dowling and Harwood (1986) consider music as a type of sign, and consider its abilities to communicate information in the light of these Peirceian categories.

'hard-wired', such that irrespective of the number of times the piece is heard it may still 'surprise' those gestalt perceptual mechanisms, and thus stimulate arousal in the listener. Steinbeis and Koelsch (2008) have shown that neurological areas used in semantic processing in language are also used in the processing of music, suggesting that meaning can indeed be extracted from structural features of music, when the listener is familiar with the basic structural properties of (that type of) music—that the fulfilment and violation of expectations about the structural progression of the music indeed provide meaning independently of other sounds or events connected with the stimulus.

Whilst this provides some explanation for why a reaction occurs, it does not account for why specific feelings are elicited—joy, elation, or weepiness, for example—in *response* to the subversion of expectations. In explanation, Sloboda and Juslin (2001) suggest that intrinsic emotional properties are responsible largely for the intensity, or *valence*, of the affect (i.e. a particular level of emotional arousal), whilst extrinsic emotional properties of the music are largely responsible for determining the emotional *content* of the music.

As already noted, these extrinsic emotional properties of music are its *indexical* and *iconic* properties. Iconic properties may just represent emotional content, or may elicit emotional response as well. The indexical properties are likely to elicit rather than represent emotion, as they refer to personal and cultural associations. These are called extrinsic emotional properties of the music because such emotional content exists by virtue of its reference to events or phenomena outside the music itself.

Confusing things somewhat, Sloboda and Juslin (2001) classify iconic properties of music as extrinsic, whilst Sloboda (2001) classifies them as intrinsic. Sloboda and Juslin's position seems more appropriate, in that iconic properties of music definitively have a referent other than the structure of the music itself; i.e. they simulate something else to some extent, be it a thunderstorm, large animal, or properties of human emotional expression (examples of these are given later in the chapter), and thus can elicit in the listener a response appropriate to such a stimulus. Sloboda (2001) claims that 'a necessary consequence of iconic recognition is a cognition such as "this is happy music"' (unnumbered page), and whilst this may lead to the music making the listener feel happy, it need not do so. But what if the response to the iconic stimulus is essentially instinctive and does not involve cognizing of the stimulus as referring to something else? This is perhaps where the difficulty in classification arises—if this property of the music elicits an emotional reaction without any cognition on the part of the listener as to its referent, should it be considered to be an inherent property of the music (*intrinsic*), or should it still be taken to be iconic of something else (*extrinsic*)? In this situation the iconic stimulus actually has more in common with symbolic, intrinsic, characteristics of the music. This may especially be true in the case of properties of music processed by mechanisms involved in gestural communication, as these properties may be an inherent foundation of music, as well as seeming to have human emotional expression as a referent; they thus have the possibility of being both *iconic* and *symbolic*. This idea is discussed further later in the chapter.

It is certainly the case that music can induce emotional response without 'higher' cognitive processes being involved: Peretz and Gagnon (1999) describe a patient IR whose neurological damage resulted in a loss of ability to recognize or discriminate melodies, but whose ability to make judgements about their emotional content was unaffected; this appears to be paralleled by an inability to recognize faces, but a preserved ability to extract emotional information from facial expression (Peretz, 2001). In each case it is not necessary to recall concrete information about the emotional stimulus in order to extract emotional information from it—in fact, the two processes appear to be independent but, interestingly, the process of extraction of emotional information from two different media (facial expression and auditory information) seems to be related (see discussion in Chapters 8 and 9).

ECOLOGICAL CONTEXT, SOCIAL CONTEXT, AND THE HUMAN IN MUSIC

As an example of the ways in which music may elicit or express emotions, consider a listener sitting in a movie theatre. This may seem a curious example to choose in a discussion seeking to elucidate evolutionary connections between emotion and music, but it is a good illustration of the number of levels at which emotion and music can be associated. The music accompanying a particular scene in the film could have the following emotional associations:

1: as index, as a consequence of association with past circumstances in which that music has been heard,

2: as symbol, in tensions and release in structures of the music itself, or

3: as icon, being loud and fast, and thus suggesting energetic emotion such as excitement (Sloboda and Juslin, 2001), or resembling a thunderstorm, for example.

Note that in terms of its iconicity it can represent a physiological state, or ecological environmental stimuli. As will be seen in subsequent sections, physiological state and emotion are fundamentally interrelated, and at a deeper level than mere iconic resemblance. However, at a more literal level, the iconic representation of ecological stimuli also has evolutionary implications, as it may resemble the types of sounds that would be threatening in an ecological environment; an example is outlined by Cross (2003a):

> For example, a passage which incorporates a fairly rapid crescendo from *pp* to *fff*, combined with an upward expansion of tessitura has the same properties as a sound produced by something large in the real world that is either approaching us (hence getting louder and with an increasing amount of higher-frequency energy), or that is staying in the same location relative to us but is increasing in the energy it is expending. Hence we might experience arousal not through any breaching of expectation but simply because of the 'thrill' that is likely to accompany any real-world encounter in which a large sound-producing entity approaches us, or when we are in close proximity to a large sound-producing entity that is becoming rapidly more energetic. (Cross, 2003a, unnumbered page)

This could, in theory, apply equally to ecological sounds that are pleasing rather than threatening. This example illustrates well the earlier point that the listener need not actually cognize a resemblance between the music and an ecological stimulus in order for it to elicit the appropriate emotional response; i.e. the listener need not think 'this sounds like a large sound-producing entity approaching me, how exciting' in order for the music to elicit the appropriate thrill reaction. They may, in fact, make such a rationalization *after* their reaction to the stimulus.

In fact, the association between tempo and level of arousal is absolutely fundamental, with cyclic time control of rhythm and a direct relationship between arousal and frequency of movement being present in all animals (Turner and Ioannides, 2009; Llinas, 2001; Molinari et al., 2003).

As noted in Chapter 9, it can be added to this observation that whilst increasing or decreasing rhythmic frequency has concomitant effects on arousal level, the maintenance of a tempo at a fixed level has the potential to have a profound regulatory effect on level of arousal—inducing a level of continuity in arousal level which would be difficult to replicate under normal ecological conditions.

To return to the movie theatre, there is also another layer of *contextual* emotional association with the music: it is directly associated with a scene in the film, the emotional content of which is connected with the simultaneous experience of both the scene and the music. It is also associated with the context of the viewer themselves, who might have their hand entwined with that of their romantic partner in the dark at the back of the theatre. This is neither an intrinsic or extrinsic property of the music itself, being determined by the context in which the music is experienced (although such associations may subsequently constitute indexical extrinsic properties of the music when it is experienced again and they are reawakened). These contextual associations may feed into each other, forming part of the initial experience of emotion elicited as the various incoming sensory stimuli are associated, and are likely to affect the extent to which emotional response is elicited in the listener. In evolutionary terms, we have until recent history never been presented with circumstances where our personal context and state is dissociated from the visual and auditory stimuli with which we are faced. Situations where these inputs *are* engineered to be dissociated (such as watching theatre, when one's personal situation is divorced from that which is the principal sensory input) are a relatively recent historical invention. Consequently we naturally assimilate sensory stimuli simultaneously with our mood and level of arousal, each sensory input either influencing or being strongly associated with each other and with the affective state consequently experienced. Such contextual associations may feed back into future experiences of the same piece of music too, becoming part of its *indexical* communication of emotion (as described earlier).

The human in music

There is also the possibility of being aware of emotional expression on the part of the composer or performer of the music (*expression theory*), which may or may not elicit the same emotion in the listener. Expression theory views the expression of emotion not as an act of the music itself, which is not sentient, but to occur as a consequence of either an expression of emotion by the composer, performer, or some *perceived* persona that is represented in the music itself (Davies, 2001). Whilst an emotion can be elicited in a listener in a variety of ways, as discussed, when it is a response to an emotion being *expressed* in the music, it may be useful to view it in terms of sympathy and empathy. We can be sympathetic to another's emotional state without experiencing it ourselves. However, if we empathize with another's emotional state, we experience (at least something of) the emotion that they are experiencing themselves. The same might be said of music. We can be quite aware of the emotion being expressed, and in many circumstances not experience that emotion ourselves; in this case, we are in sympathy with the music. When we find ourselves experiencing the emotion being expressed, then this is an empathic response to the music.

Sacks (2008) observes that 'One does not need to know anything about Dido and Aeneas to be moved by her lament for him; anyone who has ever lost someone knows what Dido is expressing [in "Dido's Lament", from Purcell's *Dido and Aeneas*]. And there is, finally, a deep and mysterious paradox here, for while such music makes one experience pain and grief more intensely, it brings solace and consolation at the same time' (Sacks, 2008, p. 330). It should be noted that this is described in the context of a Western listener experiencing Western music, with their experience and expectations of forms of musical expression, and the adherence to those expectations in the piece itself, shaping the emotional response to a greater or lesser extent—it is an open question as to whether a listener previously exposed only to other musical traditions would be moved in the same way. It does seem, however, that a composer can, within the conventions and constraints of their musical tradition, create a piece of music which elicits in the listener a particular response appropriate to the narrative context of the piece, without the listener being consciously aware of specifics of that narrative. This response seems to be akin to genuine empathy—empathy with the music as a genuine expressive 'organism'. And herein lies the resolution to Sacks's 'paradox'—empathy is generated by a sense of shared experience, and from that sense comes the solace and consolation.

The idea that music is naturally interpreted by listeners as having human-like properties has been supported by experiments carried out by Watt and Ash (1998). They concluded that music has an action on the mind similar to the action of interacting with a person. Their experiments involved subjects hearing a selection of pieces of music, and attributing a variety of properties to them, selected from a list. Some of the adjectives were people-traits (e.g. male/female), some people-states (e.g. gentle/violent), some described movement (e.g. leaden/weightless), and some were adjectives that are rarely applied to people (e.g. sweet/sour). A summary of the research cannot do justice to the experimental controls which Watt and Ash imposed on their procedure and analysis, but they found that between subjects there was strong statistically significant agreement with regard to person-like attributions given to the different pieces of music, in contrast to the non-person-like attributions. They conclude, from this and other analyses, that music has an action on the individual similar to the actions of a person, and that when music is perceived, it is assigned attributes that would normally be assigned to a person, including trait-qualities, such as age and gender, and state-qualities such as emotions.

Although apparently unaware of Watt and Ash's research, Davies (2001) observes that 'Registering music's expressiveness is more like encountering a person who feels the emotion and shows it than like reading a description of the emotion or than examining the word sad' (p. 30). Having made this observation, it is slightly odd that Davies considers dubious Kivy's (1989) suggestion that expressive instrumental music recalls the tones and intonations of emotional content in speech. Whilst acknowledging that instrumental music can sometimes imitate singing styles, and that singing styles sometimes recall ordinary [vocal] occasions of emotional expressiveness, Davies states that 'For the general run of cases, though, music does not sound very like the noises made by people gripped with emotion' (p. 31).

Whilst this is probably true, a different assertion is being made, however. First, emotive prosodic contour of speech is not necessarily the same thing as the noises made by people *gripped with emotion*. Second, to recall something is not necessarily the same as to consciously resemble something—the sound could be processed in the same way without the listener being cognizant of a similarity between the musical stimulus and human vocal emotional expressive sounds. In contrast, the key is that it is at least partly being processed by the same cognitive mechanisms (as discussed in Chapters 7–9), and thus elicits, in the listener, similar responses; as already discussed, one need not cognize a resemblance in order for it to stimulate the same response. Lavy (2001) succinctly observes that 'Humans have a remarkable ability to communicate and detect emotion in the contours and timbres of vocal utterances; this ability is not suddenly lost during a musical listening experience' (p. v).

Interestingly, Davies (2001) goes on to say that 'if music resembles an emotion, it does so by sharing the dynamic character displayed either in the emotion's phenomenological profile, as Addis (1999) maintains, or in the public behaviours through which the emotion is standardly exhibited' (p. 31). This has been called *contour theory* (Davies, 2001), and holds that our interpretation of music as expressive of emotion is due to the resemblance of its contours and forms to physical expressions of emotional state. This falls closer to an arousal rather than expression model, and parallels previous suggestions of associations between qualities of music and gestural and other physical expressions of emotion (Clynes, 1977; Scherer, 1991; Chapter 9). Cross (2003a) elaborates Davies's position:

> Music, it is claimed, operates at the emotional level by some system of resemblances. Listening to 'sad' music simply leads to an emotional or affective state in the listener (by virtue of the some correspondence between features of the music such as a descent and features in the real world that would be interpretable as expressing an emotion such as downturned corners of the mouth and a slackness of posture that makes it appear that the body is somehow more earth-bound than usual). (Cross, 2003a, unnumbered page)

Such physiological manifestations of emotional state are an extremely important stimulus in interpersonal interaction; sensitivity to them and interpretation of them are essential social skills. However, to explain emotional response to music as Davies (2001) does, as occurring solely in terms of *iconic* resemblances between features of the music and physical expressions of emotion, is under-representing the complexity of the response to this aspect of the music. It unnecessarily separates physical and vocal expression of emotion. Amongst these physiological manifestations of emotional state is tonal-vocal expression of the emotion, which has, after all, a physiological basis, in laryngeal and orofacial muscular control; as discussed in Chapter 9, the physical and auditory expression and comprehension of emotional state are intimately integrated. Resemblances, in the iconic sense that Davies uses the term, are insufficient to account for the response to tonal qualities of a melody, which frequently occurs at a more fundamental level with less conscious cognizing required than making direct associations between features. A downturned mouth, slackness of posture, lower voice, and reduced pitch contour in vocalization, for example, are all part of a gamut of physiological responses associated with sadness and/or depression. They are not understood in terms of their *resemblance* to each other—they have features in common because they are physiological reactions—forms of gesture—derived from a common cause; they are expressive facets of that common cause.

Lavy (2001) summarizes the findings of Kappas et al. (1991), for example, that 'boredom is characterised by low fundamental frequency and a low intensity [of vocalization]; irritation leads to an increase in fundamental frequency and intensity, coupled with downward intonation contours; sadness and dejection lower the fundamental frequency significantly; by contrast, fear evokes a vast increase in fundamental frequency' (p. 41); these vocal characteristics all have clear corollaries in the other, physical, aspects of expression of the same emotions. The tonal-vocal qualities manifested (such as timbre, frequency, and tempo) and the physiological properties manifested (such as posture) are intimately interrelated, and dictated by levels of autonomic nervous system (ANS) arousal (Wagner, 1989; Lavy, 2001). The properties of music which seem to express sadness (for example) do not merely *resemble* these expressive physiological and vocal expressions of emotional state, but are part of the same system and are processed by some of the same mechanisms as the tonal-vocal elements (see Chapters 7 and 9). In this sense, this property of music is more *intrinsic-symbolic* than extrinsic-iconic.

Emotion of all kinds is expressed via numerous media at once; although varying in intensity, the vocal-tonal and physiological manifestations of emotional state rarely occur in isolation from each other in natural situations, and we interpret one as indicative of the presence of others. For example, if we see someone standing in the corner of a room, with slumped shoulders and downturned mouth, we have very strong expectations about the pitch and contour of their voice when they speak to us; likewise, if we hear a voice on the radio at a relatively low pitch and with little variation in pitch contour, we have strong expectations, and probably a mental image, regarding what their facial expression and posture would be. The same applies for other emotional states too, of course, such as elation, excitement, and anger. We do not need to resort to discussing *resemblances* between the dynamic character of public physiological expression and musical contour; they are part of the same system. This is consistent with Clynes's (1977) theory of sentic modulation and Donald's model of mimetic cognitive systems (discussed in Chapters 8 and 11). This also overcomes the problem of the cross-cultural applicability of contour theory, because it relies on a universal human mechanism of emotional processing rather than resemblance between emotional expression (universal) and musical features (culturally specific).

This explanation does not imply that we treat the *performer* (or composer, or other imaginary entity inherent in the music) as expressing to us this state in him/herself (as suggested by *expression theory*—although this is not to rule this out as a *different* possible cause of attribution of emotional reaction to music). Instead, it suggests that this auditory cue is interpreted *in the same way* as physiological and corresponding auditory expressions of emotional state. Thus we do not need to resort to positing the creation by the listener of a 'persona' within the music, to whom the cause of the emotional content being perceived is attributed (as is a suggestion of the *expression theory*).

Whatever the philosophical or semantic concerns, the implication of much of the preceding evidence remains that we process emotion in musical sounds, whether vocal or instrumental, in the same way as in vocalizations (and other expressions) of affective state (indeed, the majority of music in many traditional cultures is vocal rather than instrumental anyway; see, for example, Chapter 2). The acoustic signals used in the production and perception of emotion in instrumental music appear to be the same as those used in vocal utterances, such as high intensity and tempo and harsh timbre for anger, low intensity and tempo and slow vibrato for sadness, whilst variation in timing and intensity typifies fear (Gabrielsson and Juslin, 1996; Lavy, 2001).

Autism, Asperger and Williams syndromes

That the processing and production of music involve important social-emotional capabilities is further evidenced by studies of performance of music by individuals with autism, and Williams syndrome. A well-known feature of autism (exhibited also to a milder extent in Asperger syndrome) is an inability to emotionally relate to or communicate socially with other human beings; typical autistics appear to have no empathic ability (Davison and Neale, 1994), and an aversion to social interaction (Huron, 2001). People with Asperger syndrome often experience a 'flattening' of emotional response to stimuli in general, and this includes music. Music may be strongly appreciated in technical terms by Asperger sufferers, but not elicit an emotional reaction. This may correlate with relatively less than normal development of parts of the brain connected with experiencing 'deep' emotional responses, in particular the amygdala (Sacks, 2008). Sloboda et al. (1985) report on the case of a young autistic man, Noel Patterson. Whilst able to reproduce musical phrases and sequences of notes with a high degree of accuracy after they had been played to him, he reproduced them with no expression, playing the sequences of notes mechanically. Similarly, Sacks (1996) describes Temple

Grandin's[3] consideration that music is 'pretty', but her lack of emotional response to it (Huron, 2001).

A complete contrast to the autistics' experience is that of individuals with Williams syndrome; Williams sufferers are typified by a gregarious sociality and highly developed verbal capacities, despite retardation of development of some other aspects of cognition. They also exhibit a great tendency to undertake musical activities, especially communally (Levitin and Bellugi, 1997; Huron, 2001). Williams syndrome is caused by specific genetic defects impairing some forebrain development, and much of the resultant set of abilities and inhibitions appears to be a consequence of visual-related skills being principally affected whilst those dependent more on hearing are spared. Normal-level spatial-cognitive skills are prevented from emerging, and Williams syndrome individuals usually have a subnormal IQ of around 50 (the normal population average is 100); however, they often exhibit auditory-social skills, including speech and music, which are well *above* their mental age (Panksepp and Trevarthen, 2009). Williams children are highly sociable, keen to communicate their feelings, and often have a notable aptitude for performance, dance, and music (Panksepp and Trevarthen, 2009).

There is much that can be drawn from the intriguing set of abilities exhibited by Williams syndrome individuals, not least of which is the obvious association between sociality, expressiveness, and musicality (including descriptive linguistic ability), and the dissociation between these skills and 'higher' functions such as reading, writing, mathematics, and visuospatial abilities. There appears to be a correlation between sociality, empathic, and other emotional capacities (or lack thereof) and the experience of music. This does not appear to affect musical *ability* (autistics can be technically excellent, and Williams no better than average), but it appears to have a profound influence on their *experience* of music, and consequent incentive to undertake such activities.

The mechanical lack of emotional expression exhibited by autistics and Asperger syndrome sufferers has interesting parallels with the findings reported by Jürgens (1992) concerning the anterior limbic cortex (see Chapter 7). Destruction of this area in humans results in an inability to produce *voluntary* joyful exclamations, angry curses, or pain outbursts. Without the use of the anterior limbic cortex voluntary vocal utterances, which would normally be highly emotionally communicative, sound more or less monotonous, with very flat intonation (Jürgens and von Cramon, 1982; reported in Jürgens, 1992). It is not known precisely the extent of neurological

---

[3] A well-known high-functioning autistic, Temple Grandin is a professor of animal science at Colorado State University, and was one of the subjects of Sacks's book *An Anthropologist on Mars*; the title is derived from her own description of how she feels around behaviourally 'normal' individuals.

damage in autistic individuals (although parts of the cerebellum are underdeveloped in fourteen of eighteen autistics; Davison and Neale, 1994), but it could be the case that the areas of the brain responsible for voluntary emotional exclamations also control voluntary emotional expression in music performance, and that these are affected.

Interestingly, Sacks (2008) reports several cases of musical activities facilitating social interactions in severely autistic individuals; it seems that music can facilitate engagement, 'drawing out' individuals into engaging with others, and alleviating self-orientated stereotyped movements and stress (Sacks, 2008, p. 319, and footnote). In other words, it appears that musical activities can provide a surrogate medium for social engagement, in the absence of an ability to generate such a medium spontaneously, and that it can actively supplant *self*-orientated stereotyped behaviours with *other*-orientated behaviours.

The role of the social context in which music is experienced

The associations between music and emotion considered so far have focused mainly on the direct relationship between the listener and the music. It has been noted that, in addition, stimuli from the listener's immediate context may feed into the emotional state experienced at the time, and thus come to be associated with the music itself. A contextual cue that can be very important in the personal experience of emotion is the emotion apparently being experienced by others in the immediate environment. We use other peoples' emotional reaction to a situation in a variety of ways to influence our own reaction, both consciously and unconsciously. As Sloboda and Juslin (2001) elucidate, infants frequently look to adults for a cue as to how to react to a fall, being far more likely to cry if the parent seems worried or upset than if they seem unconcerned. It seems likely that such cuing as to how to respond to a given circumstance occurs in other situations too, and is an important part of a child's learning the appropriate (or inappropriate, as the case may be) way to respond to events or interactions. As we get older we also may ask ourselves how a given person *in absentia* (e.g. a parent or other role model, such as a literary or screen character) would respond to the situation we are in and try to respond in the same way. At a less conscious level, emotions are also 'contagious' (Hatfield et al., 1994) in that we respond to a prevalent mood (Kraut and Johnston, 1979), which directly influences our own emotional state (Wild, Erb, and Bartels, 2001) (see Chapter 9).

As well as the benefits of a positive influence on our own emotional state, 'contagion' of emotional state can help us to understand the emotional state of others which, as discussed in Chapter 9, can have significant advantages in its own right—be that in terms of forming and maintaining important social and procreative relationships, or avoiding and preventing harm at the hands of others. These are very sensitive capabilities, and by drawing upon many of the same capacities as interacting with a person, including physical expression of emotion (as discussed in Chapters 7–9), music makes use of these abilities. In addition, there is the aforementioned role of social contextual cues in the experience of a response to a given stimulus.

Whilst the precise emotional reactions engendered by a given piece of music can be individual-specific, as discussed in the preceding sections, much of the value of listening to music in a communal situation (for example, at a concert or club) is that you are not listening in isolation. It is a shared experience. One knows, as a member of an audience, that even if one's own reactions to the music are quite personal and self-specific (and it is from this perception of them that they gain some of their value), the surrounding people are similarly experiencing such reactions. Thus the experience of a strong *individual-specific* reaction is also a *shared* experience, with the knowledge that the reaction in others, even if subtly different, is being elicited by the same stimulus. It is noteworthy that in the majority of cultures (including Western ones until relatively recent history) musical behaviour is a communal and participatory activity, rather than a solitary one, and in participating in musical activity, the process of entrainment also plays a very significant role in achieving a parity of arousal state between individuals (see Chapter 9).

It should be pointed out that this particular rationale is not specific to music, but applies to any communal situation involving the experience of strong emotions. For example, when watching a comedy show or riding on a rollercoaster one's reaction to the situation tends to be stronger, and more visibly and audibly expressed, in the presence of others, although the salience of emotion is individual-specific. Indeed, Kraut and Johnston (1979) found that people in situations that were likely to make them smile were more likely to do so in the company of other people than when they were on their own. This shared experience of an individual-specific and personal reaction to a common stimulus is also something that musical experience has in common with ritual and religious experience (for a further exploration of these parallels, see Morley, 2009).

PHYSIOLOGICAL, NEUROLOGICAL, AND NEUROCHEMICAL CORRELATES WITH THE EXPERIENCE OF EMOTION IN MUSIC

Suggestions that the form of emotion experienced when listening to music is not the same as a 'genuine' emotion are belied by physiological responses and by observations of neural activity during music listening. A wide variety of evidence now shows that listening to music activates cortical, subcortical, and autonomic systems involved in emotional processing in a similar way to other emotional stimuli (Trainor and Schmidt, 2003).

Whilst intrinsic emotional properties of music have to date received less laboratory examination than extrinsic properties, that which has been carried out suggests that there are genuine physiological reactions experienced in response to these stimuli within the structure of the music. Features such as syncopations, melodic appoggiaturas, and enharmonic changes create, maintain, confirm, or disrupt musical expectations (Sloboda and Juslin, 2001), and can elicit physiological reactions such as changes to breathing rate, heart rate, blood pressure, temperature, galvanic skin response (conductivity due to perspiration) (Krumhansl, 1997), 'chills' and 'shivers' (piloerection), and weeping (Sloboda, 1991, 1998; Panksepp, 1995; Panksepp and Bernatsky, 2002; Blood and Zatorre, 2001).

For example, Sloboda (1991, 1998) reports that specific properties of music seem to evoke particular emotional and physiological responses, regardless of the aesthetic type of the music (classical or pop, for example). He asked subjects to report at which points of a variety of pieces of music (from classical to contemporary pop and jazz) they experienced physical sensations such as racing heart, goosebumps or shivers down the spine, or tears. He found that there was a distinct correlation, regardless of type of music, between emotional-physiological response elicited and particular types of sequence in the music. Tears were correlated with melodic or harmonic sequences, downward harmonic movements, appogiaturas, and suspensions. Shivers were associated with harmonic, textural, or dynamic discontinuities, and racing heart with syncopation and 'other forms of accentual anticipation' (Sloboda, 1998, p. 27).

Nyklicek et al. (1997) have showed that respiration rate in listeners was directly affected by the *intensity* of a musical stimulus, regardless of whether the stimulus was positively or negatively *valenced* (i.e. expressing a positive or negative emotion). There does also seem to be a relationship between valence and respiration rate as well, although it is less clear: musical stimuli that were deemed to express either happiness (positive valence) or agitation (negative valence) were found to increase respiration rate, with shorter inspiration and expiration times. This was also found by Krumhansl (1997) to be the case with excerpts deemed to express fear, although 'sad' and 'fear' excerpts caused less of an increase in respiration rate than did 'happy' excerpts. The valence of a musical stimulus does seem to have a relationship with body temperature too: negatively valenced (emotionally negative) music can actually induce a drop in body temperature, whilst positively valenced music tends to have the opposite effect (McFarland and Kennison, 1989).

Are these physiological reactions to music paralleled by emotion-related neurological and neurochemical activity? Blood and Zatorre (2001) studied neurological responses to music that subjects had reported as resulting in the experience of 'chills'. They found that the same brain areas (left ventral striatum, dorsomedial midbrain, and paralimbic regions) experienced increased cerebral blood flow in response to these musical excerpts as are activated in other experiences of pleasant emotion, as well as euphoria and the administration of cocaine. Similarly, other areas (the amygdala, hippocampus, and ventral medial prefrontal cortex) experienced reduced blood flow in the same way as in the experience of intense emotion under other circumstances (Blood and Zatorre, 2001; Trainor and Schmidt, 2003). In other words, there is clear physiological and neurological evidence that the experience of emotional responses to musical stimuli is equivalent, in many ways at least, to the experience of emotional responses to other stimuli.

Chills are effectively induced by sustained high-pitched crescendos (Panksepp and Trevarthen, 2009), and it has been suggested that this may be because of the close resemblance of this stimulus to the cry that infants make when separated from their parents—a primal signal that solicits attention and social care, and which also results in a sense of lowered body temperature and desire for warming contact (Panksepp and Trevarthen, 2009). Blood and Zatorre (2001) showed through a PET study that cerebral blood-flow patterns when 'chills' are experienced in response to music showed considerable overlap with those witnessed in response to other reward stimuli (including food and sex) and euphoria-inducing drugs, as well as with representation of bodily states in response to emotional events. These effects seem to be related to deep-rooted opioid responses in the evolutionarily ancient emotional systems of the brain.

Opioids including oxytocin have an important role in emotional bond formation between adults as well as between parents and infants. Adult attractions are promoted by similar brain chemistries to those of mother–infant bonding, which include oxytocin systems as well as other opioids. They also are effective at reducing separation distress (Panksepp and Trevarthen, 2009). Both opioid and dopamine systems are active in peak musical experiences. Endogenous opioids (those produced by the brain itself), including oxytocin, may be responsible for the calming effects of listening to certain types of music (Panksepp and Trevarthen, 2009); indeed, opioid *blockers* diminish the emotional effects of music, including the incidence of 'chills' (Goldstein, 1980). The neurochemical dopamine has a slightly different role in emotional experience in music: 'Rewards of music may arise partly from the brain dopamine systems (Menon and Levitin, 2005) that integrate the search for and appreciation of reward (Alcaro et al., 2007)' (Panksepp and Trevarthen, 2009, p. 120). So musical stimuli can induce neurochemical reactions that are related directly to both reward and gratification, and to the positive emotional effects of social bond formation.

In addition to music being able to induce positive emotional reactions, certain qualities of musical stimuli can induce aversive reactions. One such example, which seems to be a human universal related to actual neurophysiological reactions to the stimulus, is the aversion to dissonant sounds, as opposed to consonant ones.

Tramo et al. (2003) conclude, on the basis of studies of neurological activity and neuropathology, that the human perception of consonance and dissonance is a product of the functioning of auditory nerve fibres. Perception of *harmony* is not simply a consequence of the absence of 'roughness' perceived with dissonance, but a positive product of the fact that the auditory nerve fibres fire in such a way that

> When a harmonic interval is played, neurons throughout the auditory system that are sensitive to one or more frequencies (partials) contained in the interval respond by firing action potentials. For consonant intervals, the fine timing of auditory nerve fibre responses contains strong representations of harmonically related pitches implied by the interval (e.g. Rameau's fundamental bass) in addition to the pitches of notes actually present in the interval. Moreover, all or most of the partials can be resolved by finely tuned neurons throughout the auditory system. (Tramo et al., 2003, p. 127)

This is in contrast to dissonant intervals, which result in nerve fibre activity that does not evoke implied related pitches, and which contain many partials that cannot be individually resolved, due to their proximity; these actually interfere with each other and cause fluctuations in the firing of auditory neurons, which results in the perception of roughness in the sound, and of dissonance (Tramo et al., 2003). This physiological-neurological response has a direct relationship with emotional response, with dissonant stimuli usually eliciting an aversive reaction. Neuropathological studies indicate that this preference for consonant stimuli involves the parahippocampal cortex, which is involved in emotional judgements. Gosselin et al. (2006) found that individuals with extensive damage to the parahippocampal cortex did not distinguish between consonant and dissonant music—or, at least, did not show the usual aversive reaction to dissonant music, which even neonates usually do.

Indeed, it appears that this preference for consonant, as opposed to dissonant, intervals is instinctive and innate. Infants exhibit more positive emotional responses, and greater attention, to consonant pieces than to those containing many dissonant intervals (Trainor and Heinmiller, 1998; Zentner and Kagan, 1998; Trehub, 2003). This preference appears not to be learned from speech sounds in the infant's immediate environment, as it is also exhibited by newborn infants with deaf parents who communicate only using sign language, but it may be an innate preference based on the preponderance of consonant rather than dissonant sounds in normal speech and environmental stimuli (Trehub, 2003).

In summary, the perception of dissonance is caused by actual interference in the firing of auditory neurons, whereas consonance (harmony) actually stimulates the experience of additional notes (other partials) that are not present in the sound itself. This is a product of the way in which the neurology of the human auditory system functions, before any 'interpretation' of the stimulus occurs. The emotional reaction to this physical-neurological effect—aversion to dissonance and preference for consonance—seems also to be an innate human universal, which involves the parahippocampal cortex in the emotional judgement.

This case is a good illustration that there are some biological factors—specifically properties of the auditory neurological system—that constrain the range of notes used in musics throughout the world, the ways in which notes are combined to form chords and intervals, and some of the emotional effects of these combinations.[4] Tramo et al. (2003) propose, for example, that the widespread appeal of Western popular music is a consequence of the fact that it taps into two universal tendencies/abilities:

> 1: universal competence in auditory functions needed to extract the pitch of a note and to analyse the harmonic relationships among different pitches, and
> 2: universal competence in cognitive functions that parse acoustic information and associate perceptual attributes with emotion and meaning. Experimental results suggest that similar perceptual attributes can be associated with similar emotions and social contexts across cultures (Gundlach, 1935; Balkwill and Thompson, 1999). (Tramo et al., 2003, p. 128)

Panksepp and Bernatsky (2002) suggest, in their review of their own neurological research, that 'the emotional impact of music is largely dependent on both direct and indirect (i.e. cognitively mediated) effects on subcortical emotional circuits of the human brain that seem to be essential for generating affective processes' (p. 137). Ultimately their research, and that of others, leads them to state that 'Our overriding assumption is that ultimately our love of music reflects the ancestral ability of our mammalian brain to transmit and receive basic emotional sounds that can arouse affective feelings which are implicit indicators of evolutionary fitness. In other words, music may be based on the existence of the intrinsic emotional sounds we make (the mammalian

---

[4] Note that this is not to say that musical forms are dictated by physiological universals, but that the potentially infinite variety of notes and combinations of notes (chords and intervals) is limited to a comparatively small number of forms to which the human auditory system can respond. To make an analogy, the actual structure of a building (stone, brick, wood, size of rooms, etc.) built on a particular piece of land is not *prescribed* by the nature of the ground on which it is built (steep, horizontal, flat, uneven, hard, soft, dry, wet, etc.), but the possibilities for that building's form are *limited* by the nature of the ground.

prosodic elements of our utterances), and the rhythmic movements of our instinctual/emotional motor apparatus' (p. 134).

As for the cognitive foundations of these prosodic elements, they hypothesize that 'music was built upon the prosodic mechanisms of the right-hemisphere that allow us affective emotional communications through vocal intonations' (Panksepp and Bernatsky, 2002, p. 136). This is all very much in keeping with the findings of the preceding chapters, with the proviso that the importance of these mechanisms is not their role as *indicators* of evolutionary fitness, but as a fundamental *component* of evolutionary fitness. That is to say that the production and processing of the prosodic elements of vocalization are social skills that are fundamentally important to individual and group survival. Music uses these same mechanisms.

CONCLUSIONS

Music elicits emotional responses in listeners for a variety of reasons. Musical stimuli can induce an emotional response both with and without conscious cognition of why it has done so, due to inherent properties of the music itself, and how it is processed. Some emotional responses can be elicited as a consequence of learned association with particular circumstances from our own experience, for example, others as a consequence of direct resemblance to ecological phenomena to which we have instinctive or conditioned responses.

A third form of emotional response occurs as a consequence of processing by, and stimulation of, auditory and kinaesthetic mechanisms associated with interpersonal interaction. Musical stimuli can be interpreted as having human-like properties, and can have similar effects to interacting with a person, through being processed using mechanisms related to the interpretation of meaning in interpersonal interaction. The contours of musical stimuli can have much in common with physical (including vocal) expression of emotional state, stimulating the interpretation of emotion across the other media that would normally be associated physically with that contour. The dynamic character of public physiological expression, and musical contour and tempo, are processed as part of the same system of expression with some of the auditory cues in music being interpreted *in the same way* as physiological and corresponding auditory expressions of emotional state.

We can react to such cues sympathetically, through recognizing those emotional cues, or empathically, feeling a shared experience with the emotion detected, if it elicits the same emotion as is being expressed. There are strong associations between sociality, empathic ability, expressiveness, and motivation to musicality, all being particularly prominent in Williams syndrome individuals, and often severely diminished in autistic individuals.

Properties of musical stimuli can elicit genuine physiological reactions equivalent to those elicited by emotional expression in other media, such as changes in respiration, heart rate, temperature, and tingling, and they are processed by many of the same mechanisms. Some of these reactions are caused by neurochemicals that are related to the formation of social bonds, reduction of separation anxiety, and seeking of reward and gratification.

The context in which we experience music is also very important in determining the emotion, and intensity of emotion, experienced. Especially important in this respect is the social context—the extent to which the experience is shared with others, and their reaction to the same stimuli, with emotions being 'contagious' and self-reinforcing. Musical experience can gain much of its value from a sense of a profoundly personal response coupled with the sense of shared experience; meanwhile, when practised alone, it can act as a surrogate for interaction and shared experience.

11

Rationales for Music in Evolution

> it isn't us playing our instruments that makes music happen; it's all those people caring about what's happening in this dank, dark room right now.... a whole roomful of people, happily freaking out together.... it becomes vividly clear that the song is the point and must be disappeared into.... no shame down there in nothingland, 'cause everyone's the same there—no me's, just us.
>
> Kristin Hersh (2010), p. 113

INTRODUCTION

This chapter looks at two different aspects of music in evolution: the first section concerns adaptive rationales for the practice of musical behaviours, the second considers how musical behaviours might fit within previous proposed models of cognitive evolution.

The preceding chapters have been concerned with the evolution of the capacities which form fundamental components of musical behaviours, and how and why these might have come to form the foundations for such behaviours. A separate question concerns the perpetuation of musical practices. Do musical behaviours only make use of functions and capabilities of the human anatomy and neurology that were already in place, or did the use of music, or early music-like activities, contribute to shaping these capabilities? In other words, are there, or were there at some point in the past, selective advantages to carrying out musical behaviours?

It has often been remarked that music is hardly essential for survival, the implication being that there is no reason to believe that musical behaviours could have been selectively important. However, this is to misunderstand the processes of evolution by selection, be it natural or sexual selection. To confer a selective benefit a behaviour or trait need not be *essential* for survival, it need only confer a slightly greater likelihood of survival to procreation, or a greater rate of procreation (thus perpetuating that trait) than would otherwise be the case.

Discussion of selection for musical behaviours can be concerned either with selection for carrying out musical activities, as a behavioural package, which could occur at a group or an individual level, or with selection for each of the foundations of musical abilities, which could have other selective pressures acting upon them as a factor of other functions that they fulfil.

The elements of the latter case, the evolution of capacities for music, and their interrelationships, have formed the focus of the preceding chapters, in which it has been argued that musical behaviours have particular effects on us because they use mechanisms that are socially valuable, grounded in and fundamentally important in interaction and the formation of social relationships. The fact that musical and proto-musical behaviours use mechanisms that are selectively important in contexts other than their use in music does not diminish the importance of musical behaviours, from an evolutionary perspective, if music is not simply *making use of* existing cognitive mechanisms that already existed (selected for already), but is *a development of* those mechanisms, fulfilling the roles of those mechanisms in an additional context. Whilst the foundations may initially have been selectively favoured as a consequence of their fulfilment of particular purposes, music developed within the context of those uses, and musical behaviours have the potential to fulfil some of those same purposes—potentially in even more effective ways.

Such a suite of related capacities could continue to develop in tandem, with interdependence increasing between them, whilst still fulfilling other functions. Such co-use of mechanisms could unite them functionally in this new behavioural manifestation: subsequently they could be selected for in tandem as part of a behavioural system, changes in one mechanism 'bootstrapping' changes in others. As Huron (2001) puts it, 'If music is an evolutionary adaptation, then it is likely to have a complex genesis. Any musical adaptation is likely to be built on several other adaptations that might be described as premusical or protomusical. Moreover, the nebulous rubric *music* may represent several adaptations, and these adaptations may involve complex co-evolutionary patterns with culture' (Huron, 2001, p. 44).

There are several ways in which behaviours, the capacities which support those behaviours, and evolution by selection may interact. In the case of musical behaviours, these can be characterized as the following selective processes:

1. Selection for *capacities* underlying musical behaviours because of their value in other circumstances. Musical practice then gains its efficacy (effects and wider emotional and social benefits) from its use of these capacities. This efficacy then may or may not itself have selective benefits, but this would be a separate process;

2. Selection for the *capacities* that support musical behaviours, through their use in music, because of benefits of *exercising those capacities together* for other aspects of life (i.e. the action of participating in musical activities itself *indirectly* facilitates individual survival and procreation);
3. Selection for the *capacities* that support musical behaviours in the context of their *efficacy in their use in music*—i.e. benefiting the practice of music itself, which is itself selectively advantageous for some reason (i.e. the action of participating in musical activities itself *directly* facilitates individual survival and procreation);
4. Sexual selection for musical *capacities* via musical *practices* because of the practices *indicating* fitness of participants due to potential survival benefits of the capacities that support them;
5. Cultural selection for particular musical *practices* (including capabilities to participate in those practices), which then may or may not have a biological effect via gene-culture co-evolution, through social functions of musical practice having knock-on effects on individual survival and reproductive rate.

Note that none of the five selective processes is mutually exclusive—any or all of them could potentially have been acting during evolutionary history, at different times or simultaneously. The distinctions between these different selective processes have not generally been made in discussions of music and evolution; as a consequence some proposed models have conflated various aspects of them, and others have favoured certain processes to the exclusion of others.

Process (1) has been discussed extensively in the preceding chapters; this includes rationales regarding the origins of musical behaviours in parent–infant interaction (e.g. Dissanayake, 2000) and in synchronous chorusing (Merker, 1999), for example. In addressing the question of selective roles for musical behaviours themselves, via processes (2), (3), (4), and (5), it remains to be asked whether there are genuine circumstances in which such mechanisms could have operated and, if so, how they would have influenced the development of musical behaviours. The following sections consider rationales that have been proposed for selective advantages to musical behaviours. These rationales frequently incorporate more than one of the selective processes in the list.

NON-ADAPTIVE MODELS OF MUSICAL ORIGINS

The idea that music could have evolutionary origins and selective benefits was widely speculated upon in the early part of the twentieth century, in the light of increasing bodies of ethnographic research and the development of Darwinian theory. However, this approach fell rapidly out of favour in the years after the Second World War, for social and political more than practical reasons, with the rejection of biological and universalist ideas in anthropological and musicological fields (Brown et al., 2000). As a consequence, views advocating a purely cultural basis for musical behaviours have been prominent for much of the last fifty or so years. Two influential voices have been an (evolutionary) psychologist (Pinker, 1997) and a psychologist of music (Repp, 1991). Repp takes a position typical of the dominant opinion of the post-war years. He states that

> There is no reason to believe that there is a universally shared, innate basis for music perception. Although the possible survival value of music has often been speculated about (e.g., Roederer, 1984), music has not been around long enough to have shaped perceptual mechanisms over thousands of generations. Clearly, music is a cultural artefact, and knowledge about it must be acquired. Moreover, in contrast to speech, this knowledge is acquired relatively slowly and not equally by all individuals of a given culture. (Repp, 1991, p. 260)

In this statement Repp makes several observations about music which, along with the aforementioned idea that music has no obvious immediate survival value, underlie the perspective that musical behaviours are purely a cultural product with no evolutionary foundation. However, in this he conflates selection for the capacities for the perception of music with selection for the practice of music as a learned cultural behaviour, and seeks to dismiss both. These are two very different propositions. Repp is arguing that music perception (and by extension musical practice) makes use of mechanisms that were selected for to fulfil other purposes, *and* that it has not had any influence on shaping those mechanisms, thus being exclusively a cultural product which is learnt during an individual's life time. In his use of the term 'music' he is clearly referring to the culturally specific aspects of musical practice, rather than the capacities for musicality, though he dismisses the possibility of evolution acting on capacities for musicality on the basis of the (in his opinion) short history of culturally specific musical practice, which conflates selection for the capacities with selection for the behaviours. By his assertion that 'music has not been around long enough', what he is actually arguing is that musical practices of the kind familiar to us from modern human cultures have not, in his opinion, been able to shape the perceptual mechanisms of which they make use (via processes (2), (3), (4), or (5) in the scheme outlined earlier) in the time available.

The basis of the explicit assertion that music has not been around for long is not clear. This is one of the issues that investigations into the origins of the behaviours would hope to address, although it is highly unlikely that there was ever a single moment at which something we would definitively categorize as 'music', as opposed to 'behaviours incorporating musicality', came into existence. Indeed, such distinctions are often very difficult to make in observing the musics of contemporary human societies. Even conservatively taking the first evidence for instrumental behaviour to be the first evidence for the existence of the capacities for music, the archaeological record discussed in Chapters 3 and 4 shows that the evidence dates back over 40,000 years (1,600–2000 generations)—plenty of time for selection to act further on the underlying capacities. But selection would have had to have acted on the capacities for these behaviours before they manifested themselves in this way anyway. With regard to the capacities for musical behaviour, assertions that there is no evidence for a universally shared, innate basis for music perception, and that music is clearly a cultural artefact about which all knowledge must be acquired, are contradicted by the evidence considered in Chapters 7–10. The ethnogenic elements of music, such as particular styles and instrumental skills, are only part of what defines musical ability and performance, and, as with specific languages and cultural traditions, shape the manifestation of what are biological capacities.

Pinker's position is somewhat different from that of Repp. Pinker is a strong advocate of adaptive foundations for many human behaviours, but takes the view that music simply makes use of functions and capacities of the human anatomy and neurology that were already in place, without itself having any selective value derived from that usage (Pinker, 1997). He argues that music is a 'technology' that we have created, to make stimulating use of existing innate faculties such as language, emotion, and fine motor-control, and that these faculties evolved independently through other selective pressures (i.e. that selection under process (1) occurred, but not under processes (2), (3), (4), and (5)).

However, as has been illustrated in the preceding chapters, whilst some or even all of the various mechanisms underlying musical activity have been selected for in the context of wider roles, many are functionally interdependent and the contiguous use of them could in itself result in their further development and the shaping of behaviours which came to be musical. Further, in terms of selective advantages to such behaviours, these may not be limited to the shaping of the mechanisms of perception and production: there is also the possibility that the use of certain faculties in the combinations required by musical behaviours might confer selective advantages in their own right.

SOME ADAPTIVE RATIONALES FOR THE USE OF MUSIC

A variety of adaptive benefits to musical behaviours have been advocated by other researchers, involving various combinations of the five selective processes outlined. These include roles for music in group bonding, group display behaviours, sexual selection, and the development of individual cognitive versatility.

Music and group cohesion

In reviewing the literature regarding the musical perceptual abilities of infants, Kogan (1997) concludes that almost all the skills of musical perception are present in children under the age of 1. He suggests consequently that these abilities are innate, and because they apparently co-occur, indicate that there is a single domain in the brain responsible for these skills which must have been selected for during the brain's evolution. The idea that human cognition consists of a 'Swiss Army knife' of functional modules (or domains), which have been naturally selected for, has been popularized in evolutionary psychology by Barkow et al. (1992) and other evolutionary psychologists subsequently. According to Kogan (1997) the music-dedicated domain that he proposes would likely have been selected for as a consequence of the ability of communal music-making to promote group morale and identity.

It should be pointed out that such a selective benefit, if being considered part of a single module, must be applied to all ancestral forms of musical behaviour, if one argues that the mechanisms of which music makes use were selected for to support musical activity from the outset. This means that, if they were selected for as domain-features specific to music, proto-musical behaviours must have had these social-bonding functions from their earliest occurrence.

Whilst also considering that music makes use of a specific cognitive domain or module, Sperber (1996) believes that this is one that music has 'hijacked', and that it was not originally selected for by music. In this he concurs somewhat with the view of Pinker (1997), although without diminishing the significance of musical behaviour itself in the way that Pinker does. He proposes that the original use of this domain was to process complex sound patterns discriminable by pitch variation and rhythm, which would have been used in early human communication when the vocal tract was limited. This is indeed what is suggested by the evidence considered in Chapters 8–10. That evidence also suggests, though, that it would be wrong to consider music to have 'hijacked' this set of capabilities, any more than modern language has, but that music and language are equally derived from this communicative capability that was a common ancestor to both.

It seems unlikely, from the evidence considered in the preceding four chapters, that musical behaviours are controlled by a dedicated 'music module'. In contrast, whilst the capacities that it uses are undoubtedly innate and finely tailored to that purpose to work in tandem, they serve other fundamentally important purposes too, from which the use of music has emerged, and upon which it has built. The 'module' might be said to exist in the sense of a set of interrelated functions that work together in particular—and perhaps unique—ways when supporting musical activities, but cannot be wholly separated from other roles that those functions fulfil in other, related, ways; however, having complementary functions these could come to be selected for in their use together, as a package of capabilities that complement each other.

Roederer (1984) was one of the first authors to reopen the field of investigation into possible selective benefits to the practice of music, suggesting several potentially selectively beneficial traits of music, which have been explored further by several other authors since. He pre-empts Papousek (1996a) and Dissanayake (2000) by suggesting that music may be developed from mother–infant communication. This type of interaction, he suggests, could have selective benefits by strengthening emotional bonds between mother and infant, and by providing a sort of pre-training for the ability to extract information from the musical components of speech, such as vowels, inflections, or pitch-cues found in oriental languages. The idea of the development of music from infant–mother interaction has been discussed at length in Chapter 8.

Roederer also suggests that music has an important ability to transmit emotional information to whole groups of people at once, equalizing the emotional state of the group, which results in a bonding effect between the group-members. The foundations, and interrelation, of musical vocalizations and rhythmic activity in interpersonal emotional interaction and fine muscular control have been explored in depth in Chapter 9, and the effect of experiencing such stimuli under larger group circumstances in Chapter 10. It should be evident from those investigations that the feelings and physiological changes that can be evoked by music draw upon, and stimulate mechanisms of, interpersonal emotional empathy, amongst other consequences. The potential benefits of these effects are explored further later in this chapter.

There is a variety of evidence from other disciplines to support the hypothesis that music can have direct effects on the nature and efficacy of social interactions. Ethnographic and ethological evidence shows that musical activity is predominantly social. As explored in Chapter 2, the vast majority of musical activity in modern hunter-gatherer societies, and other traditional societies, is communal; until recent history the same applied to Western musical activity too.

At a biological, rather than behavioural, level musical experience has been linked with the release and action of various hormones (see Chapter 10). Freeman (1995, 2000) reports that the hormone oxytocin aids in the formation of strong positive emotional memories, and in the supplanting of negative emotional memories. It is released in females whilst lactating, and in males and females following sexual orgasm, and fulfils an important role in interpersonal bonding (both pair-bonding between sexual partners and mother–infant bonding). In addition it is also involved in the reduction of separation anxiety, and in peak musical experiences (Panksepp and Trevarthen, 2009). Its effects seem to be strongest during major activations of the limbic system, such as during trauma or ecstasy. As mentioned in Chapters 9 and 10, the emotional limbic system seems to be particularly activated during learning, and during emotional and rhythmic motor activities. Given this, Freeman's (1995) suggestion that it has important effects in musical experience seems likely to be correct, and not only during the perception, but especially the production of musical behaviour; this provides a good neurological reason why music can fulfil important roles in the formation of social bonds, both in interpersonal interaction and in group musical activities, such as crowd chants (Huron, 2001; Freeman, 2000).

In addition to music clearly being able to regulate arousal in large group situations, Fukui (reported in Huron, 2001) has found that subjects listening to their own choice of music experienced reduced levels of testosterone; lower testosterone levels correlate with reduced aggression, competitiveness, and conflict. As Huron points out, further experiments that manipulate the music listeners are exposed to (rather than focusing on music that they prefer to listen to) would add considerably to the applicability of these findings. In fact, research into hormonal responses to music appears to be in its infancy at this stage, and is clearly an area of investigation that could prove most fruitful in the future regarding connections between musical and other interactive activities.

The idea of a social basis for music has also been proposed by Sloboda (1985). He observes that all cultures require organization for survival, and that 'modern' cultures have 'many complex artefacts that help us to externalize and objectify the organizations we need and value' (Sloboda, 1985, p. 267), but that in non-literate societies the organizational structures must often be expressed in terms of the ways that people interact with each other. He proposes that music can be used as a mnemonic framework for the knowledge of a community and as a way of expressing the structure of social relations.

Sloboda does not cite any specific ethnographic examples, but an example of the former is certainly provided by the Australian Aborigines' musical tradition, and by some of those of the Yupik, described in Chapter 2. It is also true that the musical performances observed in the Plains Indians and African Pygmies (also discussed in Chapter 2), whilst communal, do nevertheless tend to stress roles in the performance for different genders and individuals (e.g. hunters or medicine men) and so could have a role in reinforcing social organizations or roles. Whether these uses have a selective benefit for a group is difficult to test empirically, as it would be impossible to remove music from the peoples and observe the consequences. Nevertheless, it is clear that musical activities do fulfil these roles, and could likewise confer on the participants the advantages associated with such reiterations of social roles and relationships.

Music and dance as a coalition signalling system

Hagan and Bryant (2003) suggest that rather than music and dance *causing* social cohesion, they *signal* social cohesion that was achieved by other means. Their overall thesis is that 'for humans and human ancestors, musical displays may have... functioned, in part, to defend territory (and perhaps also to signal group identity), and that these displays may have formed the evolutionary basis for the musical behaviours of modern humans' (p. 27).

They justify this position, and reject other explanations, on a number of grounds. They argue that musical behaviours cannot contribute directly to the cohesion of a group, as they are not a good indicator of an individual's ability to contribute to the group's survival. They state that group cohesion exists on the basis of the perception of benefits to be offered and gained by individuals, and shared mutual goals. Musical behaviour, they say, cannot allow an individual to communicate their fitness interests or abilities (see Miller 2000a, 2000b, and later in this chapter, for a contrary position).

However, this views group cohesion purely in terms of perceived cost/benefit and the perceived value of group membership, but ignores factors of emotional bonding and loyalty engendered by mutual emotional experience, which could contribute to the perception of shared mutual goals. An individual may already have established their credibility within a group, in terms of their *ability* to contribute to its survival, but this provides no indication of their *likelihood* of doing so, or to whose survival they will contribute. Hagan and Bryant's criticism could be levelled equally at higher primate grooming behaviours, which do not have as their foundation an indication of the ability to hunt, for example, any more than does musical behaviour. The ability to instigate, express, and maintain emotional bonds and loyalty, however, is fundamentally important to the maintenance of the networks that exist within a group, be they reproductive, familial, parental, or relating to the hierarchical structure of the group as a whole. Any group which fragments under adversity, with individuals withdrawing their skills to themselves, will be less likely to survive than a cohesive group under similar conditions.

Hagan and Bryant (2003) also argue that music has no useful role in establishing group goals and coordinating group actions in the face of adversity (specifically, outside aggression—which is far from the only form of adversity that a group may face), a role which (they point out) language performs adequately. This may well be true, but apart from the fact that musical behaviours might well pre-date linguistic ability, what music *can* do is signal and strengthen ingroup (vs. outgroup) membership. Participating in such activities may allow individuals to identify themselves firmly with the particular group with which they perform, as a consequence of the shared experience. This (musical) experience is not merely a passive one, but is, in almost all societies, an actively participative one. Apart from the potential of any shared interactive activity to create a sense of inclusiveness, as has been illustrated in Chapters 7–10, musical behaviours themselves appear to rely on cognitive mechanisms grounded in emotional interaction. One would expect that their ability to perform such a role would be even greater than other mutual activities. The value of shared emotional experience in creating strong bonds of loyalty, especially in adversity, but also in positive circumstances, is well known. It is the basis of the loyalty and friendship associated with members of military units years after the events themselves, and is the rationale for initiation rites to any number of other groups in all societies.

Hagan and Bryant highlight that a major function of vocalizations in primates, both on an individual basis and, to a lesser extent, on a coordinated group basis, is territory defence. They point out that the possibility of territory defence having been an early role for song has been neglected relative to considerations of its possible roots in mate attraction, social- or pair-bonding, and individual or group recognition. Like these latter behaviours, territory defence is a role of the vocalizations of other animals, and Hagan and Bryant propose that territorial considerations may have been the driving role of musical development.

It seems likely that, once developed, complex vocalizations could have come to be used in all of these contexts in hominins. However, unlike social-interactive and bonding vocalizations, territorial and warning vocalizations require only limited tonal range and control to be effective (with the exception of birdsong, which often consists of complex sequences used for territorial purposes; this, however, does not occur at a group-cooperative level, so is not analogous to the situation being described by Hagan and Bryant). The characteristic feature of the human (and latter hominin) vocal tract, however, is its range and versatility, and the control over extended sequences of such vocalizations that seems to have emerged in tandem. In contrast to social-emotional vocalizations, such properties would be superfluous in warning and territorial vocalizations which in higher primates can (and still do) function most effectively in the form of abrupt vocalizations with little contour. Such vocalizations seem a poor candidate for the roots of the fine tonal control used by musical vocal behaviours.

So neither Hagan and Bryant's rejection of a social cohesion basis for musical interaction, nor their justification for a territorial basis of musical vocalization, seems wholly convincing. It seems unlikely that these factors formed the evolutionary basis for musical behaviours in modern humans, as they propose. This, however, is not to say that musical behaviours *could not* have come to be used in such a context, only that it seems rather unlikely that such a context was a selective force for their development. In other words, there is a distinction to be made between coalition-signalling being an 'evolutionary basis for musical behaviours' and being 'a reason why such behaviours may have been perpetuated', two distinctly different propositions. Whilst coalition signalling should be rejected as an explanation for the former, it may well be a good explanation for the latter. As Hagan and Bryant say, 'Territorial defense and alliance formation both require communicating credible information about group capabilities to non-group members, information that would deter intruders but attract allies' (p. 27).

They propose that music and dance act as good indicators of group stability and the ability to carry out complex coordinated actions. This seems a perfectly reasonable proposition; an example of it fulfilling exactly that role is provided by the New Zealand All Blacks famous *haka*. Hagan and Bryant propose that this is the case because the quality of the music and dance performance of a group corresponds to the amount of time they have in which to create and practise the performance, which would be a product of the group's stability and coordination—i.e. the quality of the coalition performing them.

In fact, such indirect rationalizing is unnecessary to explain this value of the behaviour. Such a display indicates a more direct and fundamental fact about the group—its level of cohesion and cooperation, and as Hagan and Bryant originally say, its stability and ability to carry out complex coordinated actions. It indicates these things because it both *relies on* them and *engenders* them. The ability of music to act as a forum for the practice of integrated, complex, coordinated group activities resulting in a powerful sense of membership and cohesion is one of its major values, and provides a very coherent explanation as to why such behaviours persisted at a group level. One of the manifestations of this role may indeed have been 'coalition signalling', and this may even have led to its perpetuation, but it seems more likely that any selective benefit associated with such activity would be conferred indirectly, thus via process (2) rather than (3). This potential of music to act as a forum for the execution of other social behaviours has been explored further by Cross (1999a, 2001).

Music and sexual selection

Darwin originally proposed that music in humans had its roots in courtship songs, as he believed that the vocalizations with greatest pitch changes made by apes tend to be produced by males when soliciting mates (Darwin, 1871). The courtship basis for musical performance has been echoed by a few authors since, including Atema (quoted in Ahuja, 2000), who created a reconstruction of the so-called 'Neanderthal flute' from Divje babe I (see Chapter 3 for an analysis and critique of this object). He has proposed that Neanderthals might have wooed their prospective partners with musical performances.

However, interestingly, there is actually little evidence for courtship songs in primate societies; most examples consist of homologues in distantly related species such as birds. There is some evidence, however, of correlations between 'singing' and pair bonding in some primate species. *Indri, Tarsius, Callicebus* and *Hylobates* (Geissman, 2000) are the only primate genera to have some members who sing, and in all cases both males and females sing. In most cases, duetting in bonded pairs also occurs. It is noteworthy that all known singing primate species are thought to have a monogamous structure, and this also applies to those bird species that duet (Geissman, 2000). To Geissman, 'This suggests that the evolution of singing behaviour in primates and of duet singing in general are somehow related to the evolution of monogamy' (p. 112); i.e. to the maintenance of a strong pair bond with a single other individual of the opposite sex. Because the four groups of primates that sing are not closely related, Geissman states that it is likely that this trait evolved four times, independently, and suggests that the same happened in hominins subsequently.

These singing behaviours do not seem to be used to 'woo' prospective mates, but occur between the members of an established mating pair. It does seem, however, that such activities *are* correlated with activities that increase pair bonding; Geissman and Orgeldinger (1998) observed that in ten siamang groups 'duetting activity was positively correlated with grooming activity and behavioural synchronisation, and negatively correlated with interindividual distance between mates' (Geissman, 2000, p. 111), suggesting that the activity is indeed related to the strength of pair bonds. In contrast, there is little evidence to suggest that our nearest relatives, the great apes, have specific sex-calls (Williams, 1980) and, whilst contexts of participation in musical activity can facilitate interpersonal bonding, including between members of the opposite sex, when music is used in relation to sex and relationships, it is usually within the context of the social relations of an entire social group (Sloboda, 1985).

Musical behaviours do, however, have various traits that could allow them to act as an indicator of fitness, a factor which may have shaped their development. A major advocate of a sexual selection hypothesis for the evolution of music in more recent years has been Miller (2000a, 2000b). Miller argues that musical behaviours can indicate sexual fitness for a number of reasons. Indicators of sexual fitness illustrate selectively important traits such as status, age, physical fitness, and fertility. Miller proposes that various features of music can act as good indicators of these traits, and that it is a good medium for their display. He suggests that dancing reveals aerobic fitness, coordination, strength, and health; voice control may reveal self-confidence and status;

rhythmic ability may indicate the 'capacity for sequencing complex movements reliably'; whist virtuosic performance per se 'may reveal motor coordination, capacity for automating complex learned behaviours, and having the time to practice', the latter of which may also indicate sexual availability, as it suggests lack of parenting demands. These properties of such displays could lead to aesthetic preferences for particular forms of those behaviours. This leads Miller (2000a) to propose that 'any aspect of music that we find appealing might also have been appealing to our ancestors, and if it was, that appeal would have set up sexual selection pressures in favour of musical productions that fulfilled those preferences' (p. 342).

It should be noted, however, that this logic does not necessarily follow—it implies that *any* musical trait for which there is any preference will subsequently be selected for by sexual selection. In fact there is an important qualifier that should be imposed: in the context of sexual and natural selection, selection for a particular trait can only occur if that trait can arise through biological change (mutation). This is not the same as the perpetuation of behaviours, which is cultural transmission. Also, if sexual selection was actually *responsible* for the various traits that humans find aesthetically appealing about music then we would expect a great deal of convergence in the form of musical behaviours, and which aspects of them are considered aesthetically appealing. In fact, whilst musical behaviours are universal and share features that are recognizable in all cultures, many (though not all) of the aesthetic preferences associated with them are culturally specific.

Miller admits that such rationales are speculative at this stage, and in fact his thesis is really a call for empirical testing of such ideas. Such research would indeed be a welcome addition to the field, as some of Miller's ideas of the fitness-display properties of music do seem intuitively to make sense. In terms of explaining the foundations of the appeal of musical stimuli, consideration of the biological, neurological, and evolutionary evidence discussed so far in the preceding chapters would seem a more fruitful line of investigation; however, they do provide legitimate mechanisms by which *musical behaviours* may have become refined, perpetuated, and spread. His theory is more an explanation of why they might have been developed into the forms we know and then propagated across the species, rather than an explanation of the form and appeal of the foundations of musical behaviours (i.e. the rationale deals with selective processes (4) and (5), rather than the others).

Miller says 'If one can perceive the quality, creativity, virtuosity, emotional depth and spiritual vision of somebody's music, sexual selection through mate choice can notice it too' (p. 355). This is true, but it relies upon the existence of value judgements on creativity, virtuosity, and emotional depth to such a behaviour. To say that sexual selection is responsible for these value judgements begs the question—there has to be a reason why such values are held, and held as applicable to music, before selection can act upon their manifestation.

Miller cautions against equating behaviours that are made *in* groups with behaviours that are made *for* groups; i.e. just because a behaviour is carried out in groups does not mean that it is carried out for the benefit of the group. He points out that the fact that music is most often performed in groups has led to the assumption that music's most important function was at the group level, when this need not be the case. This is a salient observation, and the attribution of any explanation should be treated with rigour; nevertheless, Miller's objections to the attribution of group selection benefits to musical behaviours have some flaws themselves. He claims that authors applying group selection models to specific situations make two errors. The first, he claims, is that they are often under the impression that it is a more friendly and 'humane' form of selection than individual selfishness, so people are inclined to favour it as an explanation. Miller argues that, in fact, group selection is no more 'humane' than individual selection, as the friendly groups are still outcompeting and exterminating other groups, replacing 'the logic of murder with the logic of genocide'. In fact, group selection mechanisms do not necessarily rely on neighbouring groups exterminating each other; group selection only has to rely on a group performing better in its own environment than other groups perform in their environment. The adverse conditions that they find themselves overcoming with the aid of a behaviour need not have anything to do directly with another group; they only have to survive when other groups do not.

The second error that Miller (2000a) believes authors make in attributing group selection as an explanation is a failure to consider 'free-riding' by members of the group who benefit from a group's behaviour without participating in it themselves, and its attendant costs. He says that if music had group-selective benefits, the situation could arise in which 'wallflowers' emerge among the group. They benefit from the group-level advantages of the dancing/music behaviour that is undertaken, but do not 'pay the enormous time and energy costs of dancing all night' (p. 352). Thus, the wallflower mutation would spread throughout the group within a few generations, and music/dance would be lost. There are two problems with Miller's objection. The first is that it presupposes that for music/dance behaviours to be beneficial at a group level they involve 'enormous time and energy costs'. This need not be the case. The second problem is that, if music did have a group-selective advantage, a group that became a community of 'wallflowers' would die out eventually, whilst the groups that survive would be those in which the balance between wallflowers and dancers has been, through selection at an individual level, kept at an optimum balance for the group, whatever that balance may be. This is not necessarily to advocate a group-selective benefit for musical behaviours, but to point out that Miller's objection does not undermine the possibility of group selection for music; in contrast, it provides an explanation for a way in which it might operate, if musical behaviours *are* advantageous at a group level.

In fact, Miller (2000a) points out that 'if music did have individual-level benefits, such as courtship benefits under sexual selection, it may be possible for group selection to reinforce them with group benefits' (p. 352). Essentially, a skill of interpersonal emotive-tonal vocalization used for forming and maintaining networks and loyalty is one that is beneficial at an individual level; the foundations for such individual behaviour, namely the ability to produce such vocal signals, and sensitivity to their content, may also allow a group activity using those behaviours to be beneficial at a communal level as a consequence of simultaneous contiguous emotional experience. This is in addition to the potential benefits inherent for an individual in having cohesive internal networks that support the survival and procreation of that individual.

In addition, musical activities have the potential to facilitate the fostering of the necessary emotional bonds between participating individuals that would facilitate procreation. As discussed in Chapter 9, two individuals dancing to the same musical stimulus both simultaneously cooperatively entrain their movements with each other whilst both entraining subserviently to music being played—the rhythmic concordance, physical cooperation and proximity, and more general achievement of commonality of experience required have much in common with the performance of sexual relations. Music could thus facilitate sexual selection in this way, which would be via process (3), rather than process (4) (display of capabilities) as Miller (2000a) describes.

Music and group selection

The potential role of music in selection at the level of the group has been mentioned in the context of the preceding models. The nature and existence of mechanisms of selection at the group level have been the subject of considerable debate, and so require some more detailed consideration here.

Shennan (2002) has carried out a comprehensive evaluation of models of selection as they apply to human prehistory, and discusses how selection can occur at numerous levels, including that of the group. It is, of course, always the individual who is the perpetuator (or otherwise) of genes, and who thus forms the crux of selection, but the likelihood of this occurring can depend upon the nature and perpetuation of the cultural group. Models of individual self-sacrifice for the benefit of the group or species as a whole have been shown to be unworkable. Within groups, competition between individuals is generally far more important, selectively, than cooperation: 'Mathematical analysis [has] demonstrated that in most circumstances, selection for interests at the individual level will override interests at the group level' (Shennan, 2002, p. 29), though altruism can potentially be selected for if directed towards kin, or others who reciprocate.

This mechanism of self-sacrifice in favour of group-level interests is not the same, however, as 'group selection' in the sense that it has been used in the preceding sections, which is in terms of group behaviours having an effect on the cultural environment in which individuals are living, feeding, and breeding. As Shennan puts it, 'all theoretical schools, including those that are sceptical about other levels of evolutionary process than that of individual inclusive fitness, recognize that such [individual] interests may often be served by cooperating rather than competing with other individuals of the same species' (p. 213). Gene theory, and the famous 'Prisoners' dilemma', shows that if individuals have similar aims they are frequently better off modifying them slightly (if necessary) in order to cooperate with others, than to compete with them. They do so on a tit-for-tat basis making only as much self-sacrifice as they perceive, from experience, that they are likely to receive in turn. Furthermore, these behaviours are likely to conform to 'pro-social norms' that emerge within a population, as a consequence of frequent interaction with the same people; adherence to these norms can have beneficial effects on the well-being of members of the group in terms of the recompense that they receive for other behaviours that they undertake (Bowles and Gintis, 1998). To put it another way, behaviours which constitute the optimal behaviour for an individual can be determined culturally as well as ecologically. In a social species the likelihood of an individual surviving to procreation, or of having a high rate of procreation, depends upon their 'cultural fitness' too—the way they behave in relation to their surrounding culture (e.g. Boyd and Richerson, 2005; Richerson and Boyd, 2005). In other words, the cultural environment is a part of the selective niche that is occupied by those individuals (e.g. Laland et al., 2000; Laland et al., 2010).

The potential value of behaviours that contribute to 'group cohesiveness', as musical behaviours could, is that they may make cooperative behaviours more likely to occur within the group than would otherwise be the case. This can benefit reproductive success at an individual level, and thus also increases the likelihood that the culture practising them will perpetuate them subsequently too. Bowles and Gintis (1998) have also demonstrated that 'populations whose interactions are structured in such a way that coordination problems are successfully overcome will tend to grow, to absorb other populations, and to be copied by others' (Shennan, 2002, p. 216). It can be seen that the emergence of musical behaviours as a pro-social norm which thus leads to the overcoming of coordination problems within a group could lead to the growth of such groups and the spread of those behaviours (selection via process (5)). Furthermore, not only could musical behaviours constitute a behavioural norm in their own right, but because of their foundations in social cognition, those individuals with well-developed musical capacities could also be those best at identifying and conforming to other norms of social interactive behaviour (in which case selection could operate via process (2)).

Hull (1980) has used the terms 'replicator' and 'interactor' to refer to genes and the individuals in which those genes reside, respectively. An interactor is defined as 'an entity that interacts as a cohesive whole with its environment in such a way that replication is differential' (Shennan, 2002, p. 65). These terms can also be applied to selection at the level of the group, with the 'replicators' now being the individuals and the 'interactor' being the group in its environmental context and in contact with other groups. In this circumstance, 'the replicators whose propagation is affected will be genetic and cultural traits at the individual level, as well as group-level traits' (Shennan, 2002, p. 253). Thus, if a group develops or adopts a structural and/or behavioural innovation which results in proliferation of that group (as described earlier), individuals within the group may gain greater reproductive success as member of that group than they would otherwise (despite possibly having to pay some cost associated with the innovation). The group and individual stand or fall together in terms of reproductive success.

Musical behaviours could fit well within these models of selection at the individual and group level, and these have the potential to provide a model for the development and spread of musical behaviours.

Music's multiple meanings and cognitive development

Whilst emphasizing that musical capacities appear to be based on universal and innate human biological capacities, Cross (1999a, 2001, 2003c) highlights the importance of the fact that the manifestation of musical behaviours based upon those foundations has a great deal of cultural specificity. Although the cultural elements (for example our learning and interactions with people and objects during infancy, childhood, and adolescence) may themselves be *constrained* by evolutionary forces, they are not *determined* solely by such factors. Music is different things to different peoples in different places and at different times—it has 'transposable intentionality' (Cross, 2003b). It is from this quality that Cross believes it derives its principal biological value.

In terms of contributing directly to survival, Cross (2001) observes that 'in itself, music does not seem capable of being a material cause of anything other than a transient hedonic encounter. It seems to be inefficacious' (p. 99) (i.e. musical practice cannot have a selective benefit via process (3)). However, precisely because of its 'non-efficaciousness and multiple potential meanings', Cross believes that 'music can be both a consequence-free means of exploring social interaction and a "play-space" for rehearsing processes necessary to achieve cognitive flexibility' (p. 99) (i.e. musical practice could have a selective benefit via process (2)). It should be noted that Cross's conception of *efficaciousness* is in terms of immediate and direct survival benefit as a consequence of the action (process (3)), and the efficaciousness of longer-term consequences of the activity is not outside his model (process (2)). It seems, from the evidence considered in Chapters 7–9, that musical activities do not simply provide an opportunity for carrying out social interaction, but actually use the mechanisms of social interaction and expression—thus they could be valuable not simply as a means of exploring social interactions, but actually as a vicarious stimulus and exercise of those capacities, in a more abstract way. The fact that musical or proto-musical production and perception function within the media of social interaction actually *gives* them efficaciousness— their value can come from stimulating and exercising those mechanisms; i.e. a transient hedonic encounter may itself be efficacious and ultimately have survival benefits.

As for the multiple meanings of music, Cross suggests that this property of music can allow for the development of cognitive flexibility in human infants because musical or proto-musical behaviour has the potential to make use of several domains of intelligence at once. It thus provides the opportunity to practise the integration and control of biological, psychological, social, and physical systems, and what is more, can do so in a context of limited risk. So not only can musical behaviour provide the opportunity to develop 'capacities for flexible social interaction', building upon innate social intelligence, it can also contribute to the development of cross-domain cognitive flexibility.

It can be seen that these are actually two separate functions that can be fulfilled by music, and the former does not depend upon the latter. The value of proto-musical behaviours in providing a forum for the stimulation of social interactive capacities could pre-date any requirements for cognitive flexibility, whilst musical activities have the *potential* to use several domains of intelligence at once, due to their non-specificity of meaning.

Cross (2001) suggests that 'If evolution has shaped the human mind, it has probably selected at the level of infant predispositions, and culture can be thought of as particularising—shaping into specific and distinct forms—the expression of those predispositions' (p. 98). So should we consider music to be solely the perpetuation into adulthood of a behaviour that is selectively useful in infancy and childhood? This is not Cross's implication, and the weight of the evidence considered so far strongly suggests otherwise. Music is built upon behavioural foundations that are useful at all stages of life history, and at all levels of the group. That selection may occur at the level of infant predispositions does not mean that behaviours are selected for because they manifest themselves usefully in infants. This is not selection for infant behaviours (contra Falk and others)—it is selection for infant capacities which give a predisposition to develop certain behaviours in adulthood. For example, infants are not born with linguistic behaviours intact—they cannot speak language—however, they appear to be born with the mechanisms and predisposition to develop this important behaviour into adulthood. That the capacity to develop an adult behaviour is being selected for does not mean to say that the developing skill is not useful in the meantime, of course. So these considerations do not undermine Cross's suggestion that the existence and development of musical capacities in infants and children may themselves be selectively important. As Cross (2001) points out, however, whilst these capacities and predispositions have been biologically selected for, the forms that the behaviours take when they are developed are culturally determined.

This means that 'In effect, the mature musics of a culture are, in part, constituted through a persistence into the adult world of childhood patterns of thought and behaviour' (Cross, 2003c, p. 82). Cross suggests that this trait of musical behaviours may be connected with the process of altricialization in hominins. Altriciality refers to the amount of time spent by a species in a pre-reproductive state, relative to its lifespan. Humans are especially altricial, and it appears that this trait correlates with neocortical development in other primates too, and with social complexity. Joffe (1997) has suggested that a greater time spent in a juvenile state allows for the fine-tuning of the social skill repertoire required for increasingly complex societies, as well as maintenance of access to flexible patterns of thought and behaviour which may confer benefits in terms of behavioural adaptability (Cross, 2003c). Despite the fact that such a process must confer considerable survival disadvantages as well, such as an increased duration of dependence and greater likelihood of death before procreation, evidently just such a process has occurred in hominin evolution; between *Homo habilis* and *Homo ergaster* (Foley and Lee, 1991), there appears to be an increase in the duration of the pre-reproductive state, and a considerable marked increase in juvenile duration between *Homo erectus* and Neanderthals and modern humans (Dean et al., 2001). The advantages conferred must have exceeded these potentially high costs. In reviewing this evidence, Cross (2003c) suggests that this process may have allowed for the perpetuation and development into adulthood of proto-musical behaviours too. Given that the evidence considered in Chapters 8 and 9 strongly suggests a foundation for proto-musical behaviours in social-emotive vocalization and vocal, corporeal, and manual gesture (useful in mediating increasingly complex social relations), it indeed seems likely that an increased period of development of these capabilities could lead to the development and refinement of proto-musical behaviours.

As for the 'transposable meaning' of musical behaviours, Cross suggests that this may have implications for human cognitive evolution as a whole. He suggests that as well as stimulating cognitive flexibility ontogenically, it may lie at the root of the development of cognitive flexibility in phylogenic terms, and thus the development of what is probably the defining characteristic of human cognition over that of other animals.

given that it seems feasible that music plays a role in the development of cognitive and social flexibility for modern human infants, it could be that the emergence of proto-musical behaviours and their cultural actualisation as music were crucial in precipitating the emergence of the cognitive and social flexibility that marks the appearance of *Homo sapiens sapiens*. (Cross, 2001, p. 100)

How music might fit into models of the emergence of cognitive flexibility and cross-domain intelligence in the evolution of modern humans is the subject matter of the next section.

MUSIC AND COGNITIVE EVOLUTION

The preceding chapters have explored the evolution of the capacities that underlie musical behaviours, and the first section of this chapter has looked at possible reasons why musical behaviours may have been advantageous. This section examines how the development of musical capacities fits within various models of hominin cognitive evolution that have been proposed. Writings on cognitive evolution have typically focused on the emergence of language and symbolic behaviour as the defining characteristics of modern humans; these discussions have generally neglected to mention musical behaviours at all, or if considered they are subsumed under the umbrella of art and symbolism. The evidence in the preceding chapters suggests that symbolic and 'artistic' associations with musical behaviours, although having the potential to add to musical experience, are not fundamental to those behaviours and that key properties of musical behaviours are derived from social interaction capabilities. However, the symbolic and social structures that have emerged in humans have clearly shaped the forms of musical activities that we recognize in human societies today.

Cognitive modularity and symbolic thought

Mithen (1996) argues that the creation of art forms requires that the 'barriers' between specific functional cognitive domains (such as Social Intelligence, Technical Intelligence, and Natural History Intelligence) be broken down. He suggests that this occurred in the so-called 'creative explosion' of around 40,000 years ago (at the start of the Upper Palaeolithic in Europe), with the advent of general-purpose language, and along with it evidence of visual art and representation. As we have seen in Chapter 3 this is also the time from which the earliest known musical instruments date. Leaving aside the question of when such capabilities actually first emerged (a 40,000-year date is now no longer accepted by most scholars; see Mellars et al., 2007, for example), this hypothesis raises the question of whether cross-domain fluidity (and with it symbolic thought) would have been required for the production of music.

The convention of attributing symbolism to music, and treating music as an art form, may be a consequence of our own modern use of it. When we hear music it evokes emotional response, and with it often comes associated imagery. This may, indeed, be an exclusive product of the cognitive fluidity of the modern human mind; as a result music has come to be *used* to evoke particular imagery. However, imagery and symbolic associations are not *inherent* properties of music. Whilst music is frequently attributed with symbolic properties, such properties, and the attribution of them, are not necessary to its production. Indeed, the ethnographic evidence discussed in Chapter 2 describes several examples of cases where music is viewed as having no symbolic associations or content at all. It is probably largely the modern Western usage and categorization of the 'creative arts' which has come to result in music being seen as *requiring* symbolic thought.

Further, as discussed in Chapters 8–10, the evocation of an emotional response by music is not necessarily reliant upon symbolic associations, but rather upon properties of our perceptual system: music can affect our moods and create emotional response without any direct symbolic associations being made at all. A good example of this is represented by the aforementioned findings of Watt and Ash (1998) that music has an action on the individual similar to the actions of a person. This results in listeners most readily assigning to music human-like attributes such as trait-qualities (e.g. age and gender) and state-qualities (e.g. emotions), but this need not be as a consequence of a *symbolic* association being made by the listeners between the music and a person; it is as a consequence of the human perceptual system processing properties of the musical signal in the same way as interpersonal interaction. To use Peircean semiotic terms, the music has inherent *iconic* rather than *symbolic* properties (see Chapter 10). These findings are consistent with the idea that melody in music could have developed from the use of a form of interpersonal vocalizations, rather than having as its foundation abstract symbolic associations.

Despite the fact that abstract symbolic associations are not necessary to the production and perception of musical activity, such activity can exercise several domains of intelligence at once. As discussed earlier in this chapter, Cross (1999a, 2001, 2003c) has suggested that, since musical activities appear to provide an excellent medium for the development of multi-domain cognitive and physical behaviour in infants, they may also lie at the root of the development of cross-domain flexibility in *Homo sapiens*. Music can play a role in the development of cognitive and social flexibility in modern human infants because it is built on foundations that fulfil that process; for protomusical behaviours using those mechanisms to develop in modern humans may simply have required that they be a useful manifestation of those mechanisms for the development of cognitive and social flexibility.

On top of these reactions to/experiences of music, it can be attributed further meanings on the basis of the emotional responses elicited (see Chapter 10) and symbolisms that are culturally and personally derived. This means that the specifics of its effects on people and the interpretation of and reaction to a musical stimulus can be very personal. In Cross's words, music has 'floating intentionality'—beyond the anthropomorphic properties of music it has symbolic ambiguity—but it does not require an ability to have cross-domain flexibility, in the sense of transferring knowledge and expertise from one domain to another and making analogical connections. Having had its foundations in the social intelligence domain, music has the potential to mean different things to different people, on the basis of the emotions and interactions stimulated. It can also gain meaning from specific associations, and from abstract associations, but does not need to do so in order to be valuable to its participants. So music does not (have to) have specific symbolism associated with it, and much of its current value comes precisely from its non-specificity. Thus the emergence of this form of proto-musical behaviour does not have to post-date the emergence of 'symbolic thought' (analogical, cross-domain thought where something can *stand for* something else), as is generally thought to be the case for linguistic and visual artistic behaviour, with direct or abstract symbolism. But it does require the cognitive flexibility for the associated skills not to be firmly rooted in a single domain. Whilst it has social intelligence as its foundation and it is in social context that it is most efficacious, its execution and perception are not limited to being literally experienced as social interaction; they allow interaction to take place at several levels, vicarious and literal, and provide a medium for the exercise of such interactions.

It seems reasonable therefore to suggest that ancestral forms of music using only rhythm and/or voice need not have required any domain of cognition other than the social, if they were used as a group or interpersonal activity rather than as an individual's own piece of creative performance. Subsequent developments of musical behaviour may have benefited from increasing cross-domain flexibility, and as Cross (2001) suggests, musical behaviours may even have engendered some aspects of such cognitive flexibility. Mithen (1996) suggests that a more sophisticated social intelligence (involving a theory of mind) was probably well developed as early as *Homo* (or *Australopithecus*) *habilis*, allowing it to live in relatively large social groups, although the increase in encephalization of *H. ergaster*, as well as the increase in its habitat range and adaptability might make this a better candidate for greater complexity of social organization. Those foundations of musicality which have their roots in capacities for social mediation must have become increasingly sophisticated from this time.

Mimesis, culture, and cognition

An alternative model of human cognitive evolution, not based on modular models of cognition but focusing on developments of mechanisms of cognition related to memory, representation, and interaction, has been proposed by Donald (1991, 2001). Donald argues that hominin cognitive evolution occurred in three main transitions, between four stages that he calls *episodic*, *mimetic*, *mythic*, and *theoretic*. Theoretic culture describes the most recent developments in human cognition, which are a consequence of the storage of large quantities of information in external symbolic media; this development Donald places around 40,000 years ago in *Homo sapiens*, on the basis of evidence from Upper Palaeolithic Europe. The first stage, episodic culture, describes the mode of cognition of apes, in which their behaviours are unreflective and largely responsive to the prevailing situation—'their lives are lived in the present, as a series of concrete episodes' (p. 149), albeit with the potential to be mediated by episodic memories of past events. Episodic cognition and culture are constrained largely by memory systems.

There appear to be three main systems of memory in animals (including humans), *procedural*, *episodic*, and *semantic*, with modern humans apparently being the only creatures to possess a semantic memory system, according to Donald. Procedural memory is concerned with learning sequences of actions, such as catching an object in flight. It preserves the general principles for action, rather than the specifics of a given situation. This type of memory system is possessed by most animals. In contrast, episodic memory seems to be unique to birds and mammals, and appears to be more evolved in higher primates than in other species, in its subtlety. Episodic memory contrasts with procedural memory in that it is concerned with particular episodes in the life of the organism, specific events and their associated stimuli. Episodic and procedural memory appear to use separate neural systems and to have evolved separately, as one can be damaged without detriment to the other. It is this system which is the foundation, and constraint, of higher primate cognition and social systems. Semantic memory is, according to Donald, possessed solely by humans, and is concerned with symbolism, analogical associations, and mental manipulations, and 'knowledge' in the form of explicit facts and propositions. Note that humans retain, and use, all three systems, but the question is which of them constitutes the defining mode of cognition for their culture. Wynn and Coolidge (e.g. 2005, 2010) emphasize instead the importance of *working memory*, or working attention, for human cognition, in drawing upon and integrating information from these different memory systems—this idea is discussed further later in the chapter.

As noted, apes have a highly developed form of episodic memory, but apparently semantic memory capability is not developed. They are highly adept at event perception and episodic storage, have some self-perception and problem-solving skills, and have social structures that require the maintenance of complex relationships. What they lack, however, is an ability for representation: 'Animals excel at situational analysis and recall, but cannot re-present a situation to reflect on it, either individually or collectively.... Semantic memory depends on the existence of abstract, distinctively human representational systems' (Donald, 1991, p. 160). At some point in hominin evolution, the transition from one governing memory system and associated type of culture to the other occurred.

For Donald (1991, 2001) there is an important intervening stage between ape and modern human cognition, the 'missing link' in this transition. This is what he calls 'mimetic' culture. Mimesis is the consequence of a cognitive development which allowed voluntary control and self-awareness of the expression and mimicry of action. This includes corporeal, gestural, vocal, and facial expression of emotion, and action. As discussed in Chapter 9, these are intimately related systems in many animals, and evidence suggests that there is a single controller responsible for expression in these different media; the difference, for mimesis, is the emergence of *voluntary control* over this system. Evidence discussed in Chapter 7 (e.g. Jürgens, 1992; Davis et al., 1996) indeed suggests that the emergence of the neurological systems for such voluntary control, of vocalization in particular, occurred some time between our common ancestor with chimpanzees (*c*.6 million years ago) and the common ancestor of Neanderthals and modern humans (*c*.600,000 years ago).

The emergence of mimesis has several implications, and four main by-products in terms of the types of behaviours it allows: mime of emotion; imitation of action; development of skill and understanding through rehearsal and integration of action; and gesture-like mime, expressing emotion as an intentional communicative act.

This development of conscious control over action allowed the development of play-acting, precise imitation, gesture, and body language. It also allowed, through the ability to imitate action and to rehearse and understand skills, the refinement of many skills, such as throwing, cutting, making stone tools, and the ability to pass such information and skills on, i.e. pedagogy and practice. Finally, it allowed the making of intentional, modulated vocal sounds, i.e. voluntary control over vocal prosody 'including deliberately raising and lowering the voice, and producing imitations of emotional sounds' (Donald, 2001, p. 261). For Donald, the fundamental factor at the root of the mimetic repertoire of behaviours is 'the invention of culture as a collective means of accumulating experience and custom' (Donald, 2001, p. 263).

Donald places this development with early *Homo erectus* (*Homo ergaster*), for a number of reasons. Not least, this species shows a great increase in encephalization (as discussed in Chapter 6), and it is with this species that we see the first evidence for widespread omnivory. *Homo ergaster* also appears to be the first hominin to show the ability to occupy vastly differing ecological niches and expand its range throughout the globe. Recent finds in Dmanisi, Georgia, of a relatively small-brained hominin, which has been tentatively identified as a *Homo habilis*-like *Homo ergaster* (Vekua et al., 2002), now suggest that *H. erectus* was not the first hominin to expand its geographical range outside Africa, although *H. erectus* was probably still the first to occupy significantly different environments.

Donald's (2001) description of *H. erectus* as having 'left behind toolmaking sites, seasonal hunting camps and continuously occupied fire sites' (p. 261) has to be treated with caution, however, as there is much debate regarding the interpretation of the evidence for hunting camps and fire use (see, for example, Klein, 2009), and it represents a viewpoint that is far from universal in palaeoanthropology. Nevertheless, we can at least say that hominins were forming groups with wider ranges, had the ability to behaviourally adapt to new environments, to develop some novel skills and to pass them on, and to coordinate group activities such as hunting (or scavenging). It seems likely that stable groups with strong social organization would be required for these activities to be possible.

The physiological evidence considered in Chapter 5 also indicated that it was indeed with *Homo ergaster* that the ability emerged to voluntarily moderate breathing sufficiently well to produce vocalizations controlled for pitch, intensity, and contour, even if it lacked the ability to produce extended vocalizations. It was also with *Homo ergaster* that the first developments in vocal *range* appear to occur; if it is indeed at this time that mimetic ability emerged, this explains the continued development of both vocal control and range over the subsequent speciations.

The evidence discussed in Chapters 7-9 concurs strongly with Donald's model of the emergence of mimesis, even though the majority of it is not cited by Donald himself. The capacities that Donald describes, their interdependence, and their foundations in increased social complexity form a coherent system, the constituent parts of which must have emerged over the course of hominin evolution. Although Donald's own evidence for placing this initial development with *Homo erectus* is sometimes less than consensual, the other evidence considered here concurs with the attribution to *Homo erectus* of significant advances in these capacities, and their culmination with 'archaic *Homo sapiens*' (the descendants of *Homo erectus* that were ancestral to Neanderthals and modern humans, such as *Homo heidelbergensis*).

What Donald's 'mimetic' stage highlights is the interrelationship of the capacities for self-aware voluntary vocal and corporeal control, and the apparently inevitable consequences of such abilities for social interaction, expression, the potential for social complexity, and the practice and pedagogy of skills. It also leads to more self-directed focus of attention, towards corporeal stimuli and control, expression and action, and away from external stimuli.

This, according to Donald (2001), has particular implications for the integration of rhythmic corporeal movement (including facial, gestural, and vocal rhythmic control) with rhythmic stimuli, using many motor systems in a unified way, and constantly appraising them during the activity and in review. It is the repertoire of abilities that constitute mimetic cognition that is responsible for our ability to regulate our own emotional responses, on the one hand, yet also to be moved by and participate in mimetic displays such as national funerals, military marching, group dancing, and chanting at a sports game. It results in a type of cultural interaction based on conventional non-verbal expression, through body language, eye contact, facial expression, and vocal tone. It is also, in the same way (though Donald makes little mention of it), the foundation that underlies musical behaviours, rhythmic and vocal, perceptive and participative.

Note that this cognitive development is not one that in itself allows for great innovation in technology, but it does allow for the adaptability to accommodate new situations, ecological and social. In a sense, Donald's model of the importance of the development of mimesis is not inconsistent with ideas of modularity, as it falls largely within the domain of social intelligence. However, one of the main implications of Donald's model is that this social foundation had great ramifications for the emergence and propagation of all manner of other cognitive and behavioural developments. The behavioural and cognitive capacities inherent in mimesis themselves form the foundations for the integration and refinement of cognitive skills previously restricted in their development, including the evolution of the capacities for complex vocalizations and ultimately language.

In terms of the timing and importance of these developments in social intelligence, Donald's (2001) and Mithen's (1996) models concur; it is in terms of the domain specificity that they differ. Donald himself puts it: 'Note that this is all domain-general training. Virtually any cognitive module might prove relevant to cultural life, and thus the habits of social interaction tend to engage the generalist aspects of brain function' (Donald, 2001, p. 257). Increasing complexity of cultural life and the demands placed on the attendant social behaviours may themselves select for cognitive flexibility. For an individual, the social environment is as much a part of the selective environment as any other part. The way an individual interacts with it—responds to it and influences it—can be as important to their survival as the way they interact with the rest of the natural environment.

The next stage in Donald's model builds on this foundation. He calls it, somewhat poetically, the 'mythic' stage, and this is characterized by the emergence of language, between archaic and modern humans. The emergence of fully linguistic behaviours has traditionally been given the most credence as the defining step in human cognition, responsible for much or all other symbolism, aesthetics, and characteristic aspects of human cognition. For

Donald, this is not the case. Whilst language is clearly hugely important in the manifestation and transference of knowledge, and allowed the externalization of knowledge into the 'distributed cognition' of a community (Donald's final—'theoretic'—transition, with modern humans in Europe), linguistic complexity was not itself the driving force: 'The priority was not to speak, use words, or develop grammars. It was to bond as a group, to learn to share attention and set up the social patterns that would sustain such sharing and bonding in the species' (Donald, 2001, p. 253); in other words, the language capacity was initially shaped by the social world, rather than vice versa; increasingly complex societies led to increasingly complex behaviours, initially—although there would also be feedback, of course.

This, in turn, led to the development of narrative thought, abstract symbolism, and external representation of knowledge in the 'mythic' and 'theoretic' stages of Donald's model; whilst the implications of these developments are very great, and there is an accordingly large body of literature to that effect, Donald's thesis regarding mimesis is that we should not underestimate the importance of this earlier stage of cognitive development and what it allowed: 'The great divide in human evolution was not language but the formation of cognitive communities in the first place', and 'symbolic cognition could not self-generate until those communities were a reality' (p. 254).

It is evident that it is in this first major cognitive development (of *mimetic* capacities) that the greatest proportion of the capacities for musical behaviours emerged (and were integrated) in hominins. The latter developments of full language, symbolic representation, and externalized knowledge would certainly have shaped those abilities in ways that would allow the development of the diversity of forms and roles of music that we see today, but these forms and roles do not themselves constitute *capacities* for music. The attribution of symbolic associations to music, and the integration of lyrical content to add specificity to its emotional content or add to its information-carrying potential, are sophisticated add-ons, using new skills, supported by the edifice but not the central edifice itself.

Enhanced Working Memory

Wynn, an archaeologist, and Coolidge, a psychologist, have, over the last decade, published a series of studies on the significance of working memory in higher cognitive functions (e.g. Coolidge and Wynn, 2007; Wynn and Coolidge, 2003, 2004, 2005, 2006, 2007, 2010; Wynn et al., 2009). Working memory, in the influential model originally proposed by Baddeley and Hitch (1974), and developed and tested extensively since, allows the ability to hold information in mind, and to process and integrate additional information at the same time. This includes current sensory information and information and skills already a part of procedural and declarative ('factual') memory. Coolidge and Wynn argue that a development in working memory capacity, or Enhanced Working Memory (EWM), through increasing the capacity to focus attention and to process and integrate information, allowed the development of many of the higher cognitive functions which, psychological research shows, are uniquely developed in humans today. These include, for example, complex goal-directed actions, flexibility in problem solving, generation of innovative solutions to problems, analogical reasoning, planning over long distances or time, and aspects of linguistic capabilities, symbolic thinking, and theory of mind. Coolidge and Wynn make the important point that hypotheses about the evolution of human cognition ought to be grounded in genuine cognitive functioning, and that such developments ought also to be identifiable in the material products of the behaviours that they allow.

Many of the day-to-day behaviours of hominins, including modern humans, although complex, do not necessarily require a working memory capacity enhanced beyond that of earlier ancestors, which means that it can be difficult to definitively identify points at which enhancements to elements of working memory such as the visuo-spatial sketchpad, phonological loop, and the central executive occurred. Wynn and Coolidge (2005) discuss four key domains of human behaviour in which archaeological evidence would have the potential to indicate development in EWM: technologies whose creation and/or use involve action remote in time and/or space and planning for future contingencies (e.g. traps and possibly complex weapons), managed foraging systems involving planning for future contingencies (e.g. resource management, storage, and agriculture), algorithmic information processing (using rules and/or devices) and, possibly, marine navigation over large distances.

According to Wynn and Coolidge (2005), a conservative interpretation of the archaeological record for these behaviours allows the identification of tasks that clearly required EWM only in the late Upper Palaeolithic; less conservative interpretation encompasses the record of the last 100,000 years or so, but only that associated with anatomically modern *Homo sapiens*. It is important to note that the absence of evidence for such behaviours earlier does not necessarily mean that the capacities described were *definitely* not present, or that earlier versions of memory functioning were not highly sophisticated and successful (see, for example, Wynn and Coolidge, 2004, regarding Neanderthals' expert intelligence), but they do argue that the lack of evidence for these capabilities is *most likely* to indicate a lack of EWM before *Homo sapiens*.

The implications of working memory capacity for 'higher' cognitive functions are substantial. The identification of developments in these capabilities, in terms of behavioural corollaries, is open to debate, though, both in terms of the criteria and the record for them, though this is not the place for an extended analysis. What is important in this context is to ask which aspects of the model might have implications for understanding the evidence for the prehistory of music, and which of the evidence associated with the prehistory of music might have implications for the model.

The manufacture of musical instruments specifically for deliberate sound production in particular future circumstances and for particular purposes would seem a good candidate for the type of technologies which Wynn and Coolidge describe as evidence for EWM. As discussed in Chapters 3 and 4, at present the earliest evidence for such objects comes from the Upper Palaeolithic. The performance of specific types of music for specific circumstances (i.e. as part of a formal tradition) would seem likely to require the use of 'algorithms' of those musical structures. The act of performance itself, of formally structured 'pieces' at least, requires complex feedback between attention and procedural and declarative memory, but there could be much variation in these requirements and the relative interaction of procedural and declarative memory with the attentional aspects. If the capacity to simultaneously appraise and integrate numerous sources of sensory and stored information via working memory is limited to *Homo sapiens*, then this would limit these formal aspects of musical activity to *Homo sapiens* too—though this would not rule out other forms of musicality in the absence of this capacity. The capacities that allow the communicative *efficacy* of musicality, explored in the preceding chapters, i.e. the cognitive foundations that allow the creation and extraction of social-emotional information from rhythmic and melodic stimuli, would appear to build upon much more deep-rooted and ancient foundations. In other words, some specific forms of their use, in formalized form and contexts, would require EWM capacities, but the efficacy of musicality and its access to those cognitive foundations do not themselves rely on an EWM capacity, and less formalized musical behaviours would be possible without.

CONCLUSIONS

Chapters 5–10 were much concerned with the evolution of underlying capacities for musicality; this chapter has discussed whether there are any reasons, selective or otherwise, why musical behaviours may have become developed and perpetuated within hominin (and/or human) groups. Past discussions of the potential values of musical behaviours with regard to selection have often failed to distinguish between rationales for the evolution of musical capacities, and rationales for the perpetuation of developed musical behaviours; further, there has often been a failure to make a clear distinction between biological selection for musical behaviours, and non-biologically selective reasons why they may be perpetuated. It is suggested here that there are five main ways in which musical behaviours, the capacities which support those behaviours, and evolution by selection may interact, and that these are not mutually exclusive—any or all of them could potentially have acted at different times or simultaneously in evolutionary history. What remains is to clearly model likely ecologically-based scenarios in which such mechanisms might have operated, and to then test their likelihood against the available evidence. Such rationalizing is at an early stage.

It is possible that developed musical behaviours could provide a good medium for the use and display of various traits related to fitness, and that aspects of those behaviours might be 'fine-tuned' by sexual preferences exhibited under such circumstances. However, this mechanism cannot account for those preferences' existence, or for their being applicable to music to start with. These require other explanations. Further, a distinction needs to be made between cultural sexual selection and biological sexual selection: *behaviours* could conceivably be perpetuated as a consequence of sexual preference, but unless the behaviour is a consequence of a biological trait, which is then itself selected for, such perpetuation will be through social emulation and not Darwinian sexual selection.

One way—with a biological basis—in which music could be a particularly potent expression of reproductive fitness lies in its roots in the ability to communicate emotionally and to empathize, bond, and elicit loyalty. An individual who is talented in these respects may well be more appealing to the opposite sex, because they are more likely to be able to form strong social alliances, and strong pair and family bonds. Good musical ability may vicariously indicate such abilities, as the cognitive capacities relied upon are in many respects shared.

The fact that people are frequently strongly drawn to same-gender music groups actually does not undermine these appeals of music, as forming strong alliances is not a gender-specific activity; music can also fulfil the role of engendering strong feelings of empathic association and group membership (and thus loyalty and cooperativity are both more likely to be offered and reciprocated). Assertions that music can only *indicate* group cohesion are mistaken in this respect: whilst musical behaviours may not directly indicate an individual's ability to contribute to a group in subsistence terms, such a criticism might equally be levelled at grooming activity. This fails to account for any benefit music might have in the respects discussed earlier, namely as an indication of an individual's *likelihood* of contributing to a group or to specific individuals, as indicated by their networks and loyalty, and as a mechanism for actually stimulating and maintaining those networks and loyalties.

Musical vocalizations gain much of their value from their complexity and pitch range, whereas most territorial and warning vocalizations gain their value from precisely the opposite traits, so the latter make unlikely candidates for the selective driving forces for the evolution of vocal complexity. Group musical activities may nevertheless form an important signal (and stimulus) of group cohesion, cooperation, and loyalty, and a forum for the coordination of complex interactions; they still frequently fulfil this role. A distinction again needs to be made between a 'selective basis for the foundations of musical behaviours' and 'reasons why such behaviours may have been perpetuated'. Coalition signalling is not a strong candidate for an explanation for the former, but it may well be a good explanation for the latter. Group displays of musical behaviours can indicate group stability and the ability to carry out complex coordinated actions precisely because they can *engender* these things.

So musical behaviours could be valuable not only as means of exploring social interactions, but actually as a vicarious stimulus and exercise of those capacities, in a more indirect way. Musical or proto-musical behaviours also have the potential to make use of several domains of intelligence at once, relying on the integration and control of biological, psychological, social, and physical systems; furthermore it gives the opportunity to practise and develop these integrated skills in a context of limited risk. Mechanisms of group and individual selection (Shennan, 2002; Bowles and Gintis, 1998; Boyd and Richerson, 2005) may prove fruitful in generating models of mechanisms through which musical behaviours may developed and spread, through these advantages conferred on a group and individuals within it.

This may seem like an attempt to assimilate all the possible explanations for music's value in social groups, and an unwillingness to reject any. This is not the case—there are problems with various aspects of the evolutionary rationales proposed to date. However, this does not mean that the value that music is argued to have in these models is itself unimportant. Whilst the different *rationales* for those selective roles of music have often been mutually exclusive, the roles themselves are not. What actually seems to be the case is that all of the rationales discussed here manifest advantages that are facets of the social interactive value of musical behaviours, as this role, ultimately, underlies them all.

So how do the various models of cognitive evolution relate to the considerations of the possible selective advantages of musical behaviours, and their emergence as behaviours in their own right? Donald's account includes elements that most directly complement the subject matter of the preceding chapters, concerning the evolution and integration of the various capacities for musical behaviours, focusing as it does on action, embodied interaction, and imitation. With regard to the evidence examined in Chapter 5, it puts these integrated foundations in place with *Homo erectus*, and then developing between *Homo erectus* and the last hominins that were ancestral to both Neanderthals and modern humans, as part of a continuity of the development of cognitive capacities that mediated increased social complexity in hominin evolution. The implications of the evidence discussed in Chapters 5 and 6 were that such capacities, and their physiological concomitants, had become highly developed by *Homo heidelbergensis*. It is likely that many of the potentially useful roles of proto-musical behaviours, discussed in the first section of this chapter, could have been fulfilled at this time by these integrated capacities. It also seems likely that at this time we could see exhibited various behaviours that, whilst they may not fit our common perception of *music*, we would recognize as *musicality*, consisting of fundamental elements of musical behaviour and fulfilling some of the roles that musical activities provide today. At this stage there would still have been considerable overlap between the roles fulfilled by proto-musical vocalization and by social-emotional proto-linguistic vocalizations. The behaviours manifest from these capacities would have been essentially socially functional and visceral—part of the extended continuum of social 'grooming' activities. But the contiguous use of these interdependent capacities could have itself helped to instigate the integration of wider, 'new' cognitive abilities in the descendants of *Homo heidelbergensis*.

The diversity of cultural forms of musical behaviours that we see today depends upon the capacities for analogical thinking and external symbolic representation that are uniquely developed in *Homo sapiens*, descendents of the African *Homo heidelbergensis*. It is only with these capabilities that behavioural innovation and consequent social and cultural diversity could shape those proto-musical behaviours into the diversity of forms that we recognize as music today. Cognitive fluidity any greater than that inherent in social intellectual capacities is not a necessary prerequisite of musical behaviours, even today, but the development of such analogical thinking and symbolic external representation would undoubtedly have subsequently had a great influence on the shaping of the *manifestation* of musical behaviours and their cultural and personal associations. It would be under these circumstances that we would say that developed linguistic vocal and musical behaviours first truly diverged from their shared heritage. However, the preceding shared heritage of linguistic and musical vocalizations, in terms of physiology, neurology, production, perception, development, and affective content, is still very much in evidence today.

12

Conclusions

> If you ask music a question, it answers and then just keeps getting louder and louder, never shuts up...Do you want your face grabbed and shouted at? Probably not; at the very least, it's irritating. But now that it's happened to me, I know that music is as close to religion as I'll ever get. It's a spiritually and biologically sound endeavour—it's healthy.
>
> Kristin Hersh (2010), pp. 10–11

> Many cultures have provided colorful stories describing how humans acquired the capacity for music. A few bold or reckless scholars have ventured to offer biological, psychological, social, cultural or religious accounts of possible origins. Most scholars have wisely steered clear of the issue of music's origins, since clearly the enterprise is patently speculative. At its worst, proposals concerning music's origins are fiction masquerading as scholarship.
>
> David Huron (1999), p. 1

The wealth of different types of evidence now available that contribute to our understanding of musical behaviours in humans means that we can be bold without, it is hoped, being reckless in our investigation of their origins. Rather than an issue of which to steer clear (in fact, Huron's words of caution precede his own writings on the subject), it is one that should be explored in more depth as fundamental to any consideration of the emergence of human cognition and social behaviour. Investigation of the development and interrelationships between musical capacities need be no more speculative than any other study in human origins, and where speculation is necessary it can be well informed and, increasingly, have testable implications.

Frequently attempts to examine the 'origins' of things generate a desire to identify 'the moment' when the species or behaviour in question 'became' what we recognize today. Music is no exception—there is a tacit desire to pinpoint a time when what we think of as a *separable* activity (as we often think of our music) can be identified, existing in its own right as distinct from other activities. But the fact is that if we were looking directly at another culture, today or in the past, this would not be the main criterion for identifying whether something was musical or not. Plentiful ethnographic examples indicate that the activity 'music' does not need to be perceived by its practitioners as being distinct from other activities and elements of life in order for it to be identifiably musical. And in fact, as much of the preceding analysis has illustrated, music today is not *genuinely* separable from other aspects of our lives and interactions. Although there is a sum-total version of it that is commodifiable, its use, form, and elements in production and perception are integrally related to other aspects of our existence. So we cannot expect to be able to identify a 'moment' when music appeared, because there was no such single moment, but we can hope to understand how activities we would recognize as musical emerged. One starting point is identifying certain properties that musical activities possess across all cultures, and certain capabilities that are required to be able to participate in such activities.

THE NATURE OF MUSIC

To reiterate the discussion in Chapter 1, musical activities rely on the ability to voluntarily produce sequences of sounds moderated for intensity and/or pitch and/or contour, generated by metrically-organized muscular movements co-ordinated (entrained) with an internally or externally perceived pulse, plus the ability to process and extract information from such sounds. When tonal content is incorporated, musical behaviours amongst all humans involve the encoding of pitches into between three and seven pitches which are unequally separated across the scale, including the perfect fifth, favouring consonance and harmony over dissonance, and organizing sequences of sounds so that they have a deliberate temporal relationship to each other. These common features are a consequence of properties of our biology, though the forms and roles of musical behaviours that build upon these foundations are hugely varied. This investigation into the evolutionary prehistory of music has therefore been concerned with the roots and developments of these capabilities, how they are used together, and how they relate to other capabilities that we possess.

Despite the great variation between cultures in the forms and purposes of music, it is clear that the foundations of musicality help to shape the roles that musical behaviours take in cultural contexts. Amongst the diverse hunter-gatherer groups whose music was examined, musical activity is almost always communal and accompanied by rhythmic dancing, social (including ceremonial) use of music is very important, and music is also performed on occasion as a communal activity purely for pleasure. In each case there is increased performance of ceremonial and communal music and dance during periods when the groups meet up with fellow communities, during their most difficult subsistence season. These commonalities exist in spite of the populations under consideration being very widely geographically spread and occupying very different environments.

Music and dance can have important uses in engendering group cohesion and altering mood, and can facilitate group interactions and communality, within and between groups. Music can also take a role in teaching dance and as a mnemonic framework for information about the world. Learning such information through embodied gestural activity can be more effective (and important) than doing so via other means, especially in the absence of stored written records. In the majority of instances examined, the music itself has no inherent symbolism, and its creation does not rely on capacities for symbolic behaviour, but it can be used to accompany symbolic activities, and has the *potential* to be attributed with considerable symbolic importance. All of the groups discussed believe themselves to have come from the land, to be akin to the other fauna of their environment, and use music to try to influence the world around them.

The music of the Amazonian Pygmies and the Native Americans of the Plains, and much of that of the Native Americans of the Arctic, contains minimal lyrical content, using emotive tonal *vocables* instead of lexical words to add to the emotional effect of the music. In these cases the efficacy of the music in eliciting communality is not dependent upon language or, in fact, any obvious abstract symbolism. The music can nevertheless have wider significances (such as its perceived influence on the world) which are predicated on symbolizing, but this would appear not to be necessary for many of the significant forms and effects of the activity. Some of the Arctic Native American music, and most of that of the Western Desert Australians, in contrast, carries much symbolic significance in lyrical content, using the music as a mnemonic structure for critical cultural and ecological information.

So many of the significant roles of musical activities in these communities are predicated on the ability of those activities to engender communality, group cohesion, commonality of mood, and to facilitate interaction, and these effects are not themselves predicated on symbolizing, but symbolic significance tends to be attributed to such activities too, as part of the wider body of symbolic cultural structures of the societies.

In each of the populations considered, the vast majority of melodic content is provided not by instrumentation but by the human voice, and much of the percussion is provided by the human body. When instrumentation is used, in most cases it is percussive and is made from readily occurring natural resources with relatively little modification, such as skins, wood, cane, or other vegetable matter such as gourds. In archaeological contexts such materials would be unlikely to survive and, were they to do so, in many cases evidence of human modification would be difficult to detect.

We would not, therefore, be justified in expecting musical activities to rely on symbolizing for much of their efficacy, or to be predominantly instrumental. The human body constitutes very effective instrumentation, and much of the significance of musical activities is derived from its modification of arousal, soliciting of cooperation, and induction of shared experience. Significant symbolic significance is attributable to those effects, such that they can take on much additional meaning, but they are not predicated upon that symbolic significance. This means that for activities incorporating musicality to be effective they need not rely on a symbolizing capacity; it is clear, though, that the presence of a symbolizing capacity would add much to the development and diversification of such activities, as a product of cultural and symbolic significance.

These observations have implications for our interpretation of the archaeological record for musical behaviours. They mean that we should not expect that the first incidence of instrumentation, however significant in other ways, indicates the first emergence of musical behaviour. It also highlights the fact that there are numerous types of sound-producer that have probably been lost to the archaeological record, and that some types, which do survive, may require especially keen analysis to recognize them as such. The common characteristics between the different hunter-gatherer groups considered, particularly the importance of tonal vocalization in adding emotional content, and the use of musical activities to engender cohesion, cooperation, and shared arousal and emotional state, would suggest that such traits of musical activities are either the products of a long shared heritage or characteristics of musical activities that naturally emerge as a consequence of the nature of musicality's foundations when coupled with the demands of those subsistence circumstances. Either of these scenarios would lead us to expect musical activities amongst earlier *Homo sapiens* hunter-gatherers also to have exhibited these traits.

While instruments can provide evidence of musical activities, it is clear that the ability to produce tonally varied sounds and rhythmic sequences does not rely on instrumentation—musical instruments are tools that we can use achieve this, but we already possess excellent tools to fulfil these roles within our bodies themselves. Even when instruments are used, the finely-controlled, temporally-organized manual, gestural, and breathing control required uses the same systems as are used when instruments are not used; the body is another instrument or, more correctly, manufactured instruments are extensions of the use of the body as an instrument.

Humans possess the ability to produce vocalizations more varied in tone, and of greater duration than other higher primates are able to produce. We also have a uniquely developed ability to *voluntarily* plan and execute *novel* vocalizations of varied structure and complexity. Although the degree of voluntary control of these vocalizations, and their complexity, are greatly increased in humans compared to our nearest relatives, our vocalizations nevertheless still rely on direct input from mechanisms which organize reflex-like vocalizations that communicate emotional state and arousal.

These capabilities are essential for producing melodic vocalizations—what we typically call singing—but tonally controlled vocalizations can have roles beyond this. The use of vocalizations of increased duration, tonal range, and with increasingly fine control over articulation, clearly has important roles in speech too. Studies of the neurological relationships—and dissociations—between language functions and musical functions in the brain can provide some insight into the interdependence and specialization of these capabilities, and thus the ways in which their evolutionary developments are likely to have been related, and to have diverged.

Whilst the specialized linguistic structures of lexicon, syntax, and grammar allow the transmission of enormous quantities of meaning, in normal speech these occur in association with variations in tonal contour, rhythmicity, and intensity which themselves transmit highly significant information. These latter modes of information transmission seem to have a far older evolutionary heritage, neurologically and in terms of related capacities possessed by our primate relatives, than the linguistic structures of lexicon, syntax, and grammar.

It is not the case that 'music is in the right hemisphere' and 'language is in the left hemisphere'; both music and language functions use both left- and right-hemisphere structures. Certain sub-functions of music and language seem to be shared, whereas functional lateralization does seem to be the case for others. In particular, areas in the right hemisphere appear to be responsible for processing and production, in both melody and speech vocalization, of prosodic melody, pitch control, tonality of singing, timbre processing, and voice recognition. Left hemisphere regions appear to be implicated in production and processing of semantic verbal meaning and syntactic sequences, as well as rhythmic production and perception, and some aspects of conscious auditory analysis.

The *production* of both vocal melodies and speech contours expressing emotion and intention (speech prosody) draws upon related structures, including deep-rooted and evolutionarily ancient structures involved in tonal-emotional expression in other primates. Similarly, the *processing* of tonal content in both speech and music seems to rely on the same structures as each other: it appears that the ability to discriminate intonation patterns in speech (prosody) uses the same pitch discrimination mechanism as is used for pitch processing in music. However, the use of this mechanism by music is more refined than modern linguistic speech requires. This involves not only cortical neurological structures that have become especially refined for the task, but also evolutionarily ancient subcortical structures in the right hemisphere which are also used for processing emotional content in other media.

So vocal tonal production and the processing of tonal information each involve a combination of both evolutionarily ancient structures involved in primate emotional vocal signalling, and other structures which, whilst also used for other forms of communication, have become finely tuned to the demands of musical activity. Whilst certain of the neurological structures used in the production and processing of musical stimuli are highly specialized for this purpose, it remains to be demonstrated that any are exclusively dedicated to this function. The mechanisms used also frequently have important roles in the processing of other types of auditory (and physical) stimuli. These mechanisms may be used in *combinations*, however, that are unique to music processing and production—the combination of neurological structures, and the interaction between them in producing and processing musical signals, constitute a specialized use of those mechanisms.

The specialized functions involved in producing and comprehending *linguistic* verbal meaning have emerged later than these systems for producing and perceiving emotional tonal content, apparently building upon some of the same structures in the left hemisphere that are required for the performance of planned sequences of complex muscular movements, including both vocalization (through laryngeal and orofacial muscular control) and rhythmic behaviours. Marin and Perry (1999) propose that linguistic speech may have evolved from an already complex system for the voluntary control of vocalization, and that the brain functions that are specialized for singing and speech represent hemispheric specializations that are a later evolutionary development building upon this shared foundation.

Koelsch and Siebel (2005) conclude from studies of the development of speech and singing in human infants that at an early age, the human brain treats language as a special case of musical stimulus. Human infants are born with abilities fundamental to musical processing, including the perception of frequency, timing, and timbre, and are able to extract different emotional content from vocalizations, on the basis of tone and rhythm alone, before any understanding of language exists. Such emotional content to vocalizations is strongly correlated with facial expression, which has direct effects on tonal quality. This leads to universal correlations between aspects of tonal quality and emotion expressed, and we use facial affect (emotional expression) and vocal affect to inform about the content of each other, interdependently within both production and perception.

Vocalizations directed towards infants (ID speech) communicate in a non-linguistic way, communicating emotion, modulating arousal, and eliciting attention and emotional response. They do so through properties that are also shared with music, including variable pitch contour, high rhythmicity, and repetitive motifs. These characteristic features of ID vocalization are also characteristics of the non-linguistic elements of adult-directed (AD) speech, and they apparently share the same foundations and roles in emotional expression.

The music-like characteristics of our vocalizations act upon cognitive-perceptual mechanisms that respond emotionally to emotional cues, and the characteristics of musical stimuli act upon the same mechanisms, because the *function* of these mechanisms is to respond in this way to these cues in *all* vocalizations, infant-directed and adult-directed. This emotional response to emotional cues is the foundation of empathy and successful interpersonal interaction, and musical stimuli act upon the mechanisms responsible as a reified form of the cues inherent in human emotional interaction.

Manual gesture, body language (corporeal gesture), facial expression (facial gesture), and vocalization (vocal gesture) should all be considered as elements of a single system of gestural expression and perception of emotional state. Affective (emotional) content can apparently be interpreted equivalently across multiple modes of expression, in visual, auditory, and kinaesthetic media, each of which can represent tension, release, and particular emotions, underlining the cross-modality of affective expression and interpretation. The production of particular facial expressions and particular body postures actually causes us to experience some emotional response as a consequence, with feedback between bodily expression and emotional experience. Furthermore, perception of these facets of gesture in others also directly invokes responses in our own emotional-gestural system.

This means that in producing, and even to an extent in perceiving, a particular sound (and other gestural action) we generate some emotional response in ourselves due to the kinaesthetic feedback from the physiology and neurology required to produce that gesture. As well as being able to observe such cues in others, we can empathically experience something of their emotional state in mirroring them with our own bodies. Musicality and rhythmic movement involve deliberate control and sequencing of this system, and doing so cooperatively with others requires us to adopt equivalent forms of these gestures between individuals, which leads to some parity in arousal level and sharing of emotional state. These capabilities also constitute critical elements of the capacity for mimesis (see Donald, 2001).

A variety of evidence suggests that it is not the case that rhythmic and tonal elements of sound production have come to be used together over the course of the development of musical capabilities in humans, but that they are, at a deep level, interdependent. Although the production and perception of rhythm and melody involve some neurologically specialized and distinct areas of the brain, it would appear that at a fundamental level rhythmic and tonal production share important mechanisms. Production of complex vocalization relies on priming of the whole motor system, and the instigation and execution of complex muscular movements involve a cognitive rhythmic motor coordinator irrespective of the musculature that is used, resulting in the coordination of the complex patterns of muscular movement of gesture (finger, hand, arm, shoulder, and joint musculature) and vocalization (orofacial, laryngeal, and respiratory musculature). This is rhythmic and cyclical, and coordinates with tonal rhythm rather than lexical elements of vocalization, and early infant gestures are emotive and rhythmic (coordinating with prosodic rhythm rather than lexical elements). Coordination between bodily gesture and vocalization is innate, deep-rooted, evolutionarily ancient, and prelinguistic.

Participation in musical activities requires not only the production of vocal and corporeal gesture, but also the coordination of those gestures with an internally or externally generated pulse (entrainment). Changes in frequency of that underlying pulse, in both production and perception, have direct effects on arousal level (and consistent frequency can effectively moderate level of arousal); such effects of beat frequency on arousal are innate and evolutionarily ancient. The process of entraining to a beat may be cooperative or subservient, or some combination of the two, and has the potential to allow both 'losing oneself' in the stimulus and/or a profound sense of physical cooperation, and synchronization of arousal level between individuals entrained to the same beat. Many musical experiences feature a powerful combination of both these effects, as individuals cooperatively (symmetrically) entrain their movements with each other whilst both entraining subserviently (asymmetrically) to music being played. In terms of the rhythmic concordance, physical cooperation and proximity, and more general achievement of commonality of experience required, music, dancing together, and sexual relations share much in common—and it is perhaps not surprising that they are often associated. The value of this experience of cooperative subservient entrainment is also related to the physicality of bodily gesture—the physical expression of emotional state—as well as to the direct overlaps in the mechanisms for the perception and production of musical stimuli. Entrainment in the experience of music makes systematic use of our systems for understanding the emotional states and intentions of others through physical gesture, and is an inherent part of all musical experience. Musical behaviours have their foundations in a system of vocal-gestural expression and comprehension of emotional state.

Music can elicit emotional responses for a wide variety of reasons, not just through properties of the musical stimuli themselves, but also as a consequence of the circumstances in which they are experienced, and previous personal experience. These responses can be both consciously and unconsciously evoked. Some can be a consequence of the resemblance of aspects of the music to ecological phenomena to which we have instinctive responses. The context in which we experience music is also very important in determining the emotion, and intensity of emotion, experienced. Especially important in this respect is the social context—the extent to which the experience is shared with others, and their reaction to the same stimuli, with emotions being 'contagious' and self-reinforcing—strengthened by shared experience of them. The potential of particular musical stimuli to become strongly associated with particular circumstances in our own prior experience is itself likely to be due to the ways in which we store memories of sensory experiences with high emotional salience.

There is no doubt that the emotional responses to musical stimuli are genuine, and in many cases equivalent to emotional reactions elicited under other circumstances, being processed by many of the same mechanisms. Musical stimuli can elicit genuine physiological responses such as changes in respiration, heart-rate, temperature, and tingling; some of these reactions are caused by the production of neurochemicals in response to music that are related to the formation of social bonds, reduction of separation anxiety, and seeking of reward and gratification, including sexual gratification.

Musical stimuli can be interpreted as having human-like properties, and can have similar effects to interacting with a person, through being processed using mechanisms related to the interpretation of meaning in interpersonal interaction. The contours of musical stimuli can have much in common with physical (including vocal) expression of emotional state, stimulating the interpretation of emotion across the other media of expression that would normally be associated physically with that contour. The dynamic character of public physiological expression and musical contour and tempo are processed as part of the same system of expression, with some of the auditory cues in music being interpreted *in the same way* as physiological and corresponding auditory expressions of emotional state. Indeed, there are strong associations between sociality, empathic ability, expressiveness, and motivation to musicality. When practised alone, musical experience has the potential to act as a surrogate for interpersonal interaction; meanwhile, when (as is more commonly the case) practised communally, the elicitation of a profoundly personal emotional response and the establishment of a sense of shared experience has the potential to be an especially potent combination, producing a powerful sense of commonality between the individuals sharing the experience.

CONCEIVING THE FOUNDATIONS OF MUSIC

It was noted earlier that musical behaviours have their foundations in a system of vocal-gestural expression and comprehension of emotional state. I have previously (Morley, 2002a, 2003) argued that specialized melodic and linguistic behaviours grew out of a shared precursor, a form of emotional communication in which the tonal vocal elements are part of a multi-modal system of auditory, visual, and kinaesthetic expression of emotional content, whose efficacy is rooted in the development of the mimetic systems that allow deliberate control over this system and our empathic understanding of its meaning in others. This system still plays critical roles in paralinguistic communication and in parent–infant vocalizations, as well as being central to music and dance. The evidence discussed earlier allows us now to say more about this system and its relationship to music and dance, such that we can seek to define the pan-cultural 'core' of musical behaviours, upon which cultural variations then build. Musical behaviours do not merely represent a *use of* this communicative system but a specialized *version of* it:

Music-dance is deliberate metrically-organized gesture
(and deliberate metrically-organized gesture is music-dance)

- *Gesture* constitutes embodied expression of emotion, vocal and orofacial and/or corporeal. i.e. gesture is motor action incorporating input from emotion-controlling systems (this thus excludes other metrically-organized movement—such as hammering—from the definition).
- *Metrically-organized* refers to the coordination of the gestures with an internally- or externally-generated temporal pattern.
- *Deliberate* refers to being under conscious control (this excludes from the definition of music-dance unconscious metrically-organized gesture, although this might be said to possess 'musicality'. Note that undertaking musical activity involves numerous unconscious processes, but the act itself is volitional).

It will be noted that this definition is of music-dance as a single phenomenon. Whether the deliberate metrically-organized gesture is experienced as music or dance, or both, will depend upon the sensory media over which the gestural signal is predominantly experienced. This definition removes from the equation issues of the intention of the producer of the gesture and the context in which it is produced which, although having a critical influence on the experience of music, are clearly very varied in modern human musics. It also encompasses possible ancestral forms of musical behaviours, as well as all musical behaviours as we know them today, without the need to seek to make a distinction between the two—i.e. a point where one became the other (if, indeed, such a point could be said to have existed). It can be applied to all experiences of music-dance, be they 'live' and participatory or recorded and solitary. It deliberately avoids cultural specificity, and whilst it does not explicitly stipulate specific properties of music which are universal, it describes the universal underlying causes of those properties. Whilst it is broad enough to encompass any musical activity of any culture and era, it is sufficiently precise such that any activity conforming to this definition would be considered to be musical (music or dance). When human ancestors possessed, and used, the ability to produce deliberate metrically-organized gesture, then they were participating in activities that we would recognize as music-dance.

Dance is here viewed as a part of a broadly defined music, and music as part of a broadly defined dance. Music and dance can be distinguished in the perception of the beholder, but in each case the signal itself is deliberate metrically-organized gesture. Within this definition:

> Music is the auditory product of deliberate metrically-organized gesture;
> Dance is the physical product of deliberate metrically-organized gesture.

Product here refers to the sensory stimuli experienced by gesturer and observer. It should be noted, however, that in reality the auditory and physical products are interdependent and rarely experienced in isolation, never by the producer of the signal, and rarely by the perceiver of it. The auditory signal always requires physical gesture, and will be in many ways experienced physically—not just in that it is a physical signal but in that we process that signal in a way that overlaps with the physical experience of producing it, even when we do not directly witness its production. These are just two different sensory elements of the same underlying cause (deliberate metrically-organized gesture) that may each be to a greater or lesser extent emphasized in the signal—as heat and light are two different sensory elements of the same underlying cause (radiation) that are nevertheless experienced simultaneously in differing proportions. In our analogy, we are interested in the experience of both the 'heat' and the 'light', but in order to fully understand these we must understand the nature of the 'radiation'.

Musicians undertaking the act of creating sound are performing a form of dance in doing so—it is impossible to produce a musical signal without embodied gestural action. Pure vocalization (singing), in the form of vocal/orofacial gesture, might be considered an exception to this statement, but this is because we tend to classify dance as a corporeal activity, and do not include vocal/orofacial gesture in this classification. Both are, nevertheless, products of the same emotional-gestural systems; in this sense, and in the sense that it is possible to produce an embodied gestural action without producing an auditory signal, musical sound-making could be considered to be a subcategory of dance.

This definition is certainly reductionist, in that there is much about the *experience* of music that it does not describe, but the fundamental, *essential* elements of our musical experience are derived from the fact that what we are experiencing is metrically organized gesture.

Deliberate metrically-organized gesture is made possible by the collection of deep-rooted and selectively-beneficial capacities for mediating complex social interaction and relationships, relying on interdependent systems of tonal, corporeal, and rhythmic expression, comprehension, and moderation of emotional state. These 'innate human abilities that make music production and appreciation possible (Blacking, 1969/1995)' (Malloch and Trevarthen, 2009a, p. 4) can be collectively termed *musicality*.

'Music' should thus not be considered to be a stand-alone activity that *makes use of* a set of otherwise unrelated mechanisms that fulfil other purposes, but instead is a specialized category of the use of a set of related and interdependent mechanisms that underlie a broader capacity, *musicality*, which itself serves a wide variety of selectively important roles. Musical activities derive many of their important effects from this heritage. Musicality involves, and is involved in, cultural learning of information (including emotional information) and establishment and communication of shared understanding (including establishing sympathetic movement, recollection, and collaborative planning, in temporally organized narratives) (Malloch and Trevarthen, 2009a). Many of these properties of musicality are closely related to those of mimesis, as outlined by Donald (2001); the development of the mechanisms for understanding the emotional states and intentions of others through imitating their physical actions—and the development of deliberate control over this system—would have been a critical element of the development of musicality and its wider benefits related to cooperation and social learning.

Some of the capacities that support musicality were selected for reasons other than to support the culturally specific forms of music—those capacities support the broad repertoire of behaviours that depend upon corporeal emotional gestural expression and imitation that we call musicality—but they were selected for reasons that were related to the roles that musical behaviours can now fulfil.

Whether the roles fulfilled by musical behaviours could themselves have selective benefits is a separate issue, though the distinction has not always been clearly made in the past. Several rather different models have been proposed in the past for ways in which musical activities, or ancestral versions of them, could have selective benefits for their practitioners. In each case, however, the proposed manifest advantages of musical behaviours are products of the social interactive foundations of musical behaviours, as this role ultimately underlies them all. With the exception of the use of musical structures for the storage and transmission of information, none of these potentially selectively-important uses is predicated on a capacity for symbolic behaviour, which has frequently been argued to be unique to *Homo sapiens*. The ability to carry out deliberate metrically-organized gesture pre-dates the emergence of our species by many millennia. It is proposed here (Chapter 11) that there are five different mechanisms by which the practice of musical behaviours and biological selective processes could be related, which are not mutually exclusive, and that any or all of them could potentially have acted at different times or simultaneously in evolutionary history. It remains to clearly model the range of ecologically based scenarios in which such mechanisms might have operated, and to then test their likelihood against the available evidence. In any case, it is clear that participation in collective musical activities can have powerful effects on participants, establishing concordance in level of arousal and emotional state, stimulating the release of neurochemicals for emotional bonding, and thus resulting in a profound sense of commonality and shared experience.

A TIMELINE FOR THE EMERGENCE OF MUSICALITY

Vocalizations in our primate relatives are, as in humans, frequently geared towards soliciting care, facilitating friendly interactions and instigating reconciliatory interactions. What emerged in the course of the evolution of pitched-contoured vocalizations in human ancestors was an *increased range and increased control* of pitch contour, as part of the wider system of gestural communication outlined earlier, allowing greater vocal versatility, expressiveness, and thus efficiency, in paralinguistic affective communication. This would have built upon the type of limited pitch control already used for emotive-tonal-social vocalization amongst higher primates.

The fossil record of human ancestors indicates that australopithecines, the bipedal but rather ape-like predecessors of the *Homo* genus, possessed a vocal anatomy and neurological structure that was not significantly different from that of other higher primates today. It appears that the first major changes with regard to vocal physiology and control occurred around the time of the emergence of our own genus, *Homo*. Approximately 2 million years ago, *Homo habilis* exhibited developments in left-hemisphere neuroanatomy related to fine manual control and associated areas controlling the planning and execution of fine muscular sequences, including those of vocal and orofacial musculature. In *Homo habilis* this development was initially small. This species would probably have had the ability to make discrete vocal utterances of limited tonal range and duration, for emotional expression, parent–infant communication, warning, and signalling, similar to those exhibited by modern higher primates. These would have been generated predominantly by neurological mechanisms that were right-hemisphere and subcortically localized, but with a small increase in the ability to plan and execute such sequences, controlled by this left-hemisphere development.

By the time of the emergence of *Homo ergaster*, around 1.8 million years ago, a number of further changes had taken place. The shift to a fully upright bipedal posture and increases in brain size led to changes in cranial morphology that impacted on both the vocal tract and inner ear. The position of the larynx altered to be somewhat lower in the throat, increasing the size of the supralaryngeal soundspace, potentially allowing a greater range of vocal sounds to be made. Whether this occurred initially as a by-product of the physiological changes associated with bipedalism, or directly as a consequence of advantages associated with increased versatility of vocal communication, is impossible to assert; however, the fact that the increase in supralaryngeal soundspace was accompanied by an increase in cervical vertebral innervation, allowing finer control over the larynx itself, suggests that increased vocal control was indeed an important selective force at that time. In any case, the latter benefits would have soon become selectively advantageous in the situations of increased social complexity contiguous with the move to diverse environments, especially with the expansions of *Homo erectus* into new ecologies in and beyond Africa.

The production and perception of tonal prosodic sounds of vocalization were (and are) handled predominantly by the right hemisphere and subcortical emotional-limbic systems, but the left-hemisphere mechanisms that began to emerge with *Homo habilis*, responsible for the planning and muscular execution of those sounds, became increasingly developed. In particular, early in the emergence of our genus, this is likely to have included the development of the neurological pathway from the periaqueductal grey matter to the nucleus ambiguus, which is responsible for fine laryngeal, orofacial, and expiratory control. This allowed *voluntary* control over the structure and complexity of vocalized sequences, hitherto impossible. The instigation of vocalizations was, as it is today, handled by a rhythmic motor coordinator also responsible for complex patterns of muscular movement of corporeal gesture (finger, hand, arm, shoulder, and joint musculature) as well as vocalization (orofacial, laryngeal, and respiratory musculature), with gestural movements being associated with the prosodic rhythm of vocalization. The increasingly sophisticated vocalizations thus produced were predominantly prosodic, emotional-tonal, and used in social interaction, a form of vocal social grooming. The evidence thus suggests that *Homo ergaster* and *Homo erectus* had developed the capability to produce an increased range of vocal sounds, controlled for pitch and intensity, and coupled with gestural emotional expression. The range of sounds producible by the larynx, although increased, is unlikely at this point to have been as great as that of which modern humans are capable, and it appears that initially these vocalizations would have been of a limited duration relative to our own.

The physiology and neurology on which these abilities depend appear to have undergone selective modification over the next million years or so, between *Homo ergaster* and the last common ancestor of Neanderthals and *Homo sapiens*. It would appear that over this period there was a strong selective pressure for the development of increased tonal range, duration, and control of vocalization, with vocalization playing an increasingly important role within an ever more complex system of gestural communication. Specimens of *Homo heidelbergensis*, from 300,000–600,000 years ago, possess a vocal anatomy, and neurological connections to it that allow for control over both vocal range and duration, which are virtually indistinguishable from our own. Increased development of the ability to plan and execute complex vocal and orofacial muscular sequences would need to be coupled with this, and indeed, there was further disproportionate development in the temporal region of the left hemisphere over this time. Since these *Homo heidelbergensis* specimens include examples that are ancestral to Neanderthals as well as specimens ancestral to *Homo sapiens*, it is likely that these features were also possessed by the immediate predecessor of *Homo heidelbergensis*, the last common ancestor of Neanderthals and *Homo sapiens*.

Amongst human adults today, the majority of our vocal communication remains focused on exchanging social information and mediating social relationships. It would seem that social-affective content was initially the most important component of vocal communication, one element of a system of gestural expression and comprehension of emotional state, and that this remains a very important component of our vocalization behaviours today. Non-verbal vocal utterances act as expressions of personal state and relationships between the self and other. Such utterances, which are also coupled with equivalent body language and facial expression, can express personal state and reactions (wellbeing, approval, disapproval, disgust, etc.) and can also solicit such information from another individual. Both the expression and solicitation of such information, as well as the ability to share in the experience of such states, are obviously important and advantageous skills for forming and managing social relationships, including cooperative, mating, and parent–infant relationships. Individuals most effectively able to establish and maintain pair-bonds and alliances through this ability would have a significant selective advantage over less able fellows.

In this system, the production and perception of the gestural (vocal, orofacial, and corporeal) expression of emotion is a consequence of priming of the rhythmic-motor and emotional systems, and results in further priming of those systems. Rhythmic sequences, and the prosodic content and rhythm of tonal sequences, prime this system, and each other. This results in a multimodal relationship between rhythmic and emotive-prosodic content in the form of vocal, auditory, visual, and kinaesthetic expression of emotion. It would seem that what emerged over the course of the evolution of *Homo* was the ability to *deliberately* use this system, with increasing control over the form, range, and duration of these deliberate metrically-organized gestures.

This system of vocal and kinaesthetic communication of emotion constituted the foundation for vocal communication out of which later emerged culturally-shaped melodic, rhythmic musical behaviours and semantic, lexical linguistic capabilities. Whilst this communicative system would not be described as 'music' in the sense of the discrete culturally-mediated behaviours that we see in human societies today, it contained all of the core elements of musical behaviours—the capacities for *musicality*—in an integrated, interdependent system, and would certainly be described as musical. It is proposed that infant-directed vocalizations in modern humans are one specialized perpetuation of this sophisticated form of non-linguistic interpersonal interaction which was earlier used between all individuals; similarly, modern human musical practices are another specialized perpetuation of the use of this system.

At some point in or after our last common ancestor with Neanderthals specializations in vocalization form began to occur, in terms of linguistic and melodic vocalizations becoming discrete entities, with syntactic structure and lexical associations emerging in relationships between vocal and gestural action in some circumstances. Other authors have proposed points in the emergence of our species at which these came to dominate the structure of vocalizations, in the form of lexical syntactic language. In terms of the emergence of culturally-formed musical behaviours, their foundations in social interaction, emotional expression, and fine control and planning of corporeal and vocal muscular control lends them extremely well to exercising the integration of important cognitive skills. The exercise of musicality could become increasingly important, and beneficial, at both an individual and group level, with increasing social complexity within and between groups.

Because music production and perception are handled by the brain in ways that are complex and related to interpersonal interaction and the formation of social bonds, they stimulate many functions thus associated. It seems that musical participation, even without lyrics or symbolic associations, can act on the brain in ways that are appealing to humans because of their vicarious stimulation of fundamentally important human interactive capacities— musical production and perception stimulate parts of the cognitive system concerned with interpersonal interaction, empathy, and expression. Not only do they do so on an individual basis, but they also thus add to the potency of activities which are, in any case, social and interactive. Consequently, musical behaviours could be instrumental in fostering in-group cooperation and relations, the advantages associated with which could lead to the rapid spread of such traditions within and between groups.

These are by no means music's only appeals, or only way of eliciting emotional response, both of which have many associations in the other activities and symbolisms that, in modern humans, can be culturally built up around music. However, the underlying social-emotional factor is a fundamental element of music's action, *irrespective* of those other activities and symbolisms which may be 'built on' afterwards. The traditional perception of musical behaviours as necessarily being part of a repertoire of symbolic behaviours is probably due largely to Western conceptions of music, and to the association of music with *other* activities with which music in most cultures is now contiguous. It also possibly owes something to the misconception that musical behaviour is always carried out 'for its own sake', and with no direct benefit or consequence. Having said this, the emergence of symbolizing capabilities in modern humans would have added immeasurably to the diversity of circumstances and forms in which musical behaviours were manifested, as well as adding many more layers of meaning and association for participants in such activities.

There is every reason to believe that the earliest *Homo sapiens* possessed both the physical and the neurological capacities necessary to engage in recognizable musical behaviours, and that the capacities for, and wider benefits of, musicality would have been evident much earlier than this. Because of the melodic capabilities of the vocal tract, we can expect that melodic instrumentation would have been a relatively late development in the prehistory of music. We can also expect that rhythmic percussive activity would have accompanied complex vocalization from an early time, because of the fundamental link between the production of planned complex sequences of prosodic vocalizations and the instigation of corporeal motor control, described earlier.

The archaeological evidence that we have for deliberate sound production in the later Palaeolithic currently comes solely from Europe. However, the earliest material comprises sophisticated objects and dates to the arrival in Europe of *Homo sapiens*, around 40,000 years ago, suggesting that they were a product of a longer tradition. The majority of those artefacts that are widely interpreted as musical instruments are flutes or pipes made from bone, and it may be that the use of large avian bones at this time for instrument manufacture marks an innovation based on the use of a new subsistence resource. The dominance of this sort of instrument in the record is likely to be mainly a product of the fact that bone preserves where other suitable raw materials do not, and that flutes are more easily recognized as instruments than other types of sound-producer—particularly percussive ones, which form such an important part of the musics of the world's cultures today. There are other artefacts that may constitute instruments. Bones and stones that appear to have been deliberately struck to produce percussive sound, as well as possible rasps and bullroarers, may also constitute evidence of the musical activities of these early modern humans in this part of the world. Use-wear analysis on these objects will help to resolve whether or not they were indeed used for deliberate sound production, as well as increasing the likelihood of the identification of such objects in the future. It is likely, then, that flutes represent just the tip of the iceberg of what was actually produced and, in turn, that the instruments produced represent the tip of the iceberg of musical activities carried out.

The auditory environment is likely to have been very significant to Upper Palaeolithic populations, and there is evidence to suggest that they made deliberate use of acoustic properties of their surroundings. Such evidence seems, on the surface of it, more intangible and difficult to discuss than the visual aspects of their activities, and this has led to less discussion of auditory phenomena than visual representation. In some cases, however, it seems that the visual and auditory worlds were deliberately related to each other. The development of sophisticated and innovative investigative techniques has already added to our understanding of the form of such activities in the Palaeolithic, and if we are to ensure that we look in the right places for relevant evidence, we need to maintain a very open conception of the types of contexts in which musical sound production might have occurred and what artefacts might have been used in that process.

Precisely reconstructing the contexts of use of many of the known artefacts is not possible, due to the lack of detail in excavation records for the numerous finds recovered in the earlier part of the twentieth century, though more recent finds have the potential to provide much more detailed information. Analysis of existing finds can also reveal important information about their manufacture and some aspects of their use; some of the archaeological evidence for musical activities comes from sites that are known to have been used for aggregations of large numbers of people, and shared traditions of manufacture and decoration suggest long-distance contact over large areas. It is clear from the other evidence discussed in this book that activities incorporating musicality draw upon aspects of cognition that are fundamentally related to social interaction, and that such activities can be particularly efficacious in social contexts.

In some cases the choice of raw material used for the instruments seems to have been particularly significant to their users, and this sort of additional significance would certainly rely on symbolic frameworks of beliefs about the world. Although many aspects of musical activity do not rely on a capacity for symbolism to have profound effects on the participants, those effects can clearly take on significant extra meaning when interpreted symbolically, and this extends the importance of musical activity from the purely social and emotional experience into drawing meaning from, and giving meaning to, other aspects of life. For this reason the contexts in which musical activity occurs in symbolic *Homo sapiens* populations are hugely diverse. For example, amongst the modern hunter-gatherers discussed, musical activities become increasingly important in times of subsistence stress, and are frequently viewed as supernatural in origin, powerful, and capable of influencing the people and the world around.

There are fundamental parallels between musical practice and ritual practice, at both non-symbolic and symbolic levels (for a discussion of these parallels, see Morley, 2009, and Dissanayake, 2009), meaning that these activities very often co-occur, often to the extent of being interdependent. Instruments have the potential to produce sounds which, whilst being interpreted using the same mechanisms as are used for perceiving other, natural, sounds (including the voice), go beyond what is naturally producible by the body—this can lend to them the powerful combination of being perceived *in the same way* as a natural sound, whilst being alien to the normal world and thus being attributed special significance. A visual parallel to this combination is provided by images that combine natural features in ways that do not occur in nature, such as the lion-man from Hohlenstein-stadel (an ivory figure with the body of a human and head of a lion), which whilst combining natural elements creates a product which is supernatural. Hohlenstein-stadel is in the Lone valley, very near the Ach valley home of the earliest Palaeolithic flutes, and the lion-man dates to the same period as those flutes. The representational imagery that survives from the Palaeolithic clearly illustrates that such perceptual interpretations occurred amongst these populations, and it seems very likely that such significance would also have been attached to the auditory medium too.

Musical activities were certainly not trivial or incidental aspects of Palaeolithic life, and there can be little doubt that the musical behaviours undertaken in the European Upper Palaeolithic were ones that we would recognize today. The ubiquity of musical behaviours amongst human cultures throughout the world today—and indeed, through recorded history and later prehistory— illustrates that music and dance are integral to all varieties of modern human society. The diversity, in location and period, of evidence from the Upper Palaeolithic does nothing to disavow us of this view of human society at that time; the implications of that diverse evidence for the specific natures of those activities are something we can hope to elaborate more in the coming years, with increasingly sophisticated excavation and analysis.

Musicality has a long evolutionary heritage, developed from foundations present in our primate ancestors and undergoing particular refinement over the course of the million years or so between the emergence of our genus, *Homo*, and our last common ancestor with Neanderthals. Increasingly refined musicality has supported increasingly complex social interactions through the mediation, expression, and comprehension of emotional state and arousal in other individuals. Behaviours that *deliberately* seek to moderate emotional state and arousal through these mechanisms, behaviours that we would describe as musical, can have significant personal and social benefits, and would not have been limited to those human ancestors for whom the use of a symbolizing capacity was important. However, in *Homo sapiens*, for whom so much of culture and cognition is built upon symbolizing, such deliberate metrically-organized gesture could take on a vast array of additional significances and effects, such that the universal biological capacity for musicality became the recognizable music that now inextricably permeates so much of our social world. Music is not just a pervasive feature of human society, musicality is a pervasive fact of our humanity; as Andersson and Ulvaeus (1978) imply, without a song or a dance we would hardly be human.

Appendix

Appendix Table 1. Inventory of Palaeolithic reputed pipes and flutes

Cross-reference	Period	Origin	Described as	Material (if known)	Details of Age	Description	Status & Comments	Location (if known)	Main References
	Late Middle or Early Upper Palaeolithic	Ilsenhohle, Germany	Flute	Diaphysis of a 'mammal species'	No stratigraphic record	Originally five holes. 'Details not known' (Turk and Kavur, 1997)		?	Hülle, 1977; Turk and Kavur, 1997
Swabian Alb 1	Aurignacian	Geissenklösterle, Germany	Flute	Swan radius	Upper Aurignacian; 43,150–39,370 cal BP Aurignacian II split bone points in same layer (Archaeological Horizon II)	Geissenklösterle Flute 1 3 holes preserved (diameter 5.3 × 3.4 mm, 3.5 × 3.0 mm, 2.8 × 2.4 mm), distance between holes 30–40 mm. Originally 18–19 cm long; now 12 cm, with the largest bore being 8–9 mm	Certainly sound-producer. Reconstructed from 23 pieces excavated 1990, published 1995	Württembergisches Landesmuseum, Stuttgart	Hahn and Münzel 1995 Turk and Kavur, 1997; Richter et al., 2000; Goldbeck, 2001; Münzel et al., 2002; d'Errico et al., 2003; Higham et al. 2012
Swabian Alb 2	Aurignacian	Geissenklösterle, Germany	Flute	Swan (radius?)	Upper Aurignacian; 43,150–39,370 cal BP Aurignacian II	Geissenklösterle Flute 2 1 hole remaining	Certainly sound-producer. Reconstructed from 7 pieces excavated 1973, published 1995	Württembergisches Landesmuseum, Stuttgart	Hahn and Münzel, 1995 Turk and Kavur, 1997; Richter et al., 2000;

(*Continued*)

Appendix Table 1. Continued

Cross-reference	Period	Origin	Described as	Material (if known)	Details of Age	Description	Status & Comments	Location (if known)	Main References
					split bone points in same layer				Goldbeck, 2001; Münzel et al., 2002; d'Errico et al., 2003; Higham et al., 2012
Swabian Alb 3	Aurignacian	Geissenklösterle, Germany	Flute	Mammoth ivory	Upper Aurignacian; 43,150–39,370 cal BP Archaeological Horizon II	Geissenklösterle Flute 3 Length: 187 mm Numerous finely-carved notches along edges of the two halves to facilitate binding with resin	Certainly a sound-producer. Complicated manufacture involving splitting ivory, hollowing, and then rebinding. Appears to replicate bird bone form. Reconstructed from 31 pieces excavated 1974–9, published 2004	University of Tübingen collections	Conard et al., 2004; Conard et al., 2009; Higham et al., 2012
Swabian Alb 4	Aurignacian	Hohle Fels, Germany	Flute	Radius of *Gyps fulvus* (Griffon vulture)	Basal Aurignacian deposits of Layer Vb; initial Upper Palaeolithic occupation, c.40,000 ya; certainly >35,000 cal BP	Hohle Fels Flute 1 Length: 218 mm Diameter: 8 mm 5 holes preserved. 2 V-shaped notches cut into proximal end	Reconstructed from 12 pieces excavated 2008, published 2009	University of Tübingen collections	Conard et al., 2009

					Proximal end of radius = proximal end of flute. Length of unmodified radius c.340 mm			
Swabian Alb 5	Aurignacian	Hohle Fels, Germany	Flute fragment	Mammoth ivory	Hohle Fels Flute 2 Length: 11.7 mm Width: 4.2 mm Thickness: 1.7 mm Portion of finger hole preserved. Striations on internal and external surfaces	Basal Aurignacian, Feature 10, Base of Archaeological Horizon Va	Excavated 2008, published 2009	University of Tübingen collections	Conard et al., 2009
Swabian Alb 6	Aurignacian	Hohle Fels, Germany	Flute fragment	Mammoth ivory	Hohle Fels Flute 3 Length: 21.1 mm Width: 7.6 mm Thickness: 2.5 mm Incised lines on outer surface and 9 notches on edge	Basal Aurignacian, lowest Aurignacian unit of Archaeological Horizon Vb; initial Upper Palaeolithic occupation, c.40,000 ya;	Excavated 2008, published 2009 Greater thickness and dimensions suggest not part of same object as Flute 2	University of Tübingen collections	Conard et al., 2009

(Continued)

Appendix Table 1. Continued

Cross-reference	Period	Origin	Described as	Material (if known)	Details of Age	Description	Status & Comments	Location (if known)	Main References
						Striations on internal and external surfaces	certainly >35,000 cal BP		
Swabian Alb 7	Aurignacian	Vogelherd, Germany	Flute fragment	Bird bone	From reworked contexts, but majority of finds from site are securely dated to Aurignacian; pre-date 30,000 ya (Conard et al., 2009)	Vogelherd Flute 1 Length: 17.5 mm Width: 5.8 mm Thickness: 1.8 mm 1 partly preserved finger hole, 7 small notches along edge	Reconstructed from 3 fragments, excavated 2005, published 2006	University of Tübingen collections	Conard and Malina, 2006; Conard et al., 2009
Swabian Alb 8	Aurignacian	Vogelherd, Germany	Flute fragment	Mammoth ivory	From reworked contexts, but majority of finds from site are securely dated to Aurignacian; pre-date 30 kya (Conard et al., 2009)	Vogelherd Flute 2	Excavated 2008, published 2009	University of Tübingen collections	Conard et al., 2009
	Aurignacian	Abri Blanchard, France	Flute	Bird	Aurignacian I (?)	4 holes on front side of bone, 2 on rear		?	Harrold, 1988 (p. 177); Jelinek, 1990 (p. 18).
	Aurignacian		Flute?		Aurignacian II	1 hole		?	Bayer, 1929; Horusitzky,

	Bukovácer Höhle, Lokve, Croatia		Cave bear (*Ursus spelaeus*)		1955; Brade, 1975; Albrecht et al., 1998; Holderman and Serangeli, 1998a.
Aurignacian?	Liegloch, Austria	Flute	Juvenile cave bear tibia (*Ursus spelaeus*)	4 holes in a zigzag arrangement; distance between distal holes similar to those of Divje babe I (Turk and Kavur, 1997)	Mottl, 1950, p. 22; Albrecht et al., 1998; Holderman and Serangeli, 1998a and b
				Age uncertain	?
				Aurignacian? Age uncertain	
Aurignacian	Potočka Zijalka, Slovenia	?	Cave bear femur (*Ursus spelaeus*)	?	Brodar and Bayer, 1928; Brade, 1975; Scothern, 1992
				Aurignacian II	
Aurignacian	Spy, Belgium	3 bone tubes	?	Incised, 3 specimens, 52, 60, and 70 mm long	Otte, 1979; Scothern, 1992
				Aurignacian. 32–17 kya	L'Institut Royal des Sciences Naturelles de Belgique (IRScN), Brussels
Aurignacian	Spy, Belgium	Bone tube	?	Incised, 70 mm long	Otte, 1979; Scothern, 1992
				Aurignacian. 32–17 kya	L'Institut Royal des Sciences Naturelles de Belgique (IRScN), Brussels
Aurignacian	Spy, Belgium	Bone tube	?	Incised, 37 mm long	Otte, 1979; Scothern, 1992
				Aurignacian. 32–17 kya	L'Institut Royal des Sciences Naturelles

(*Continued*)

Appendix Table 1. Continued

Cross-reference	Period	Origin	Described as	Material (if known)	Details of Age	Description	Status & Comments	Location (if known)	Main References
Buisson's fig. 4.1 (Scothern 1)	Aurignacian	Isturitz, France (Salle de Saint-Martin)	Pipe fragment	Right ulna of diurnal raptor, probably vulture (distal end)	IA designates Passemard's Layer A. This refers to Layer Aω in Salle Saint-Martin, typologically dated to *Aurignacien typique*	IA Sup 1921 77142 Length: 74.0 mm Diameter: 15.0–20.0 mm Bore: 14.0–15.3 mm 3 holes on anterior, diameters 5.0 × 8.0 mm and 4.0 × 8.5 mm, third hole broken. Distal end heavily worked and may form embouchure, proximal end broken in region of third hole. Several traces of abrasive scraping, 9 short deep parallel incisions between second and third hole	This is the oldest of the pipes from Isturitz, the only one to come from Aurignacian contexts and the only one to be found in Salle Saint-Martin (Salle Sud). This was the first pipe to be published by Passemard (in 1923), although he had excavated the Magdalenian examples from Salle d'Isturitz earlier	Collection Passemard, Musée des Antiquités Nationales, Saint-Germain-en-Laye, Paris [de Belgique (IRScN), Brussels]	Passemard, 1923; Passemard, 1944; Buisson, 1990; Lawson and d'Errico, 2002

Buisson's fig. 2(i) (Scothern 16)	Gravettian	Isturitz, France (Salle d'Isturitz)	Pipe fragment	Left ulna of diurnal raptor, probably *Gypaetus barbatus*—bearded lammergeier/vulture	IF3a designates Passemard's Layer F3 in Salle d'Isturitz, typologically dated to Gravettian. Corresponds with Saint-Périer layer IV	Periosteum (bone lining) still present, so not worked internally (Scothern, 1992, p. 94) IF3a 1914 75252 A3	Virtually complete pipe reconstructed from two separate finds by Buisson (1990). Initial analysis and description carried out by Buisson (1990). Detailed analysis of piercings and markings, and reconstruction, carried out by Lawson and d'Errico (2002)	Collection Passemard/ Collection Saint-Périer, Musée des Antiquités Nationales, Saint-Germain-en-Laye, Paris	Buisson, 1990
Buisson's fig. 2(ii) (Scothern 2)	Gravettian (originally designated 'Final Aurignacian (Gravettian)' by Saint-Périers)	Isturitz, France (Salle d'Isturitz)	Pipe fragment		Ist. III 83888 designates items from Saint-Périer Layer III in Salle d'Isturitz (corresponding with Passemard Layer C, immediately above Layer F3); all recent chronologies agree Gravettian cultural horizon	Ist. III 1939 83888 (a) Combined dimensions: Length: 212.0 mm Diameter: 11.0–14.0 mm Bore: 10.8–11.8 mm 4 holes on posterior, diameters 5.0 × 8.0 mm and 7.0 × 5.5 mm	Highlights stratigraphical issues associated with the site, with one part coming from a layer equivalent to Ist. IV and the other from layer Ist. III		Saint-Périer and Saint-Périer, 1952; Buisson, 1990; Lawson and d'Errico, 2002

(Continued)

Appendix Table 1. Continued

Cross-reference	Period	Origin	Described as	Material (if known)	Details of Age	Description	Status & Comments	Location (if known)	Main References
					(see Goutas, 2004). The original designation 'Final Aurignacian' derived from Breuil's typology (Lawson and d'Errico, 2002)	One end finely finished and complete, the other end slightly damaged. Longitudinal abrasive scraping and numerous fine incised parallel lines perpendicular to length. Means of blowing possibly V-notched tongue duct (Scothern, 1992)	(a) designation added by IM to differentiate from items illustrated in Buisson (1990) figs. 4.7, 4.8, and 4.9 (described later)		
Buisson's fig. 3.3 (Scothern 10)	Gravettian	Isturitz, France (Salle d'Isturitz)	Pipe fragment	Left ulna of diurnal raptor, probably vulture (proximal end)	Ist. IV designates Saint-Périer's layer IV in Salle d'Isturitz, typologically dated to Gravettian. Corresponds to	Ist. IV 1936 83889 Length: 56.5 mm Diameter: 13.6–15.0 mm Bore: 10.8–12.0 mm 2 holes on anterior,		Collection Saint-Périer, Musée des Antiquités Nationales, Saint-Germain-en-Laye, Paris	Saint-Périer and Saint-Périer, 1952; Buisson, 1990

Buisson's fig. 3.4 (Scothern 11)	Gravettian	Isturitz, France (Salle d'Isturitz)	Pipe fragment	Right ulna of diurnal raptor, probably vulture (proximal end)	Ist. IV designates Saint-Périer's layer IV in Salle d'Isturitz, typologically dated to Gravettian. Corresponds to Passemard's Layer F3	Ist. IV 1939 83889 (a) Length: 67.0 mm Diameter: 15.0–18.0 mm Bore: 12.3–13.0 mm 1 hole on posterior, 1 hole on anterior, diameters 4.0 × 6.0 mm. Breaks at the extremities, one in the region of the posterior hole. Traces of abrasive scraping; fine incisions perpendicular and oblique to the longitudinal axis on the surfaces opposite the holes	(a) designation added by IM	Collection Saint-Périer, Musée des Antiquités Nationales, Saint-Germain-en-Laye, Paris	Saint-Périer and Saint-Périer, 1952; Buisson, 1990

diameters 5.0 mm. Breaks at the extremities. Traces of abrasive scraping

Passemard's Layer F3

(*Continued*)

Appendix Table 1. Continued

Cross-reference	Period	Origin	Described as	Material (if known)	Details of Age	Description	Status & Comments	Location (if known)	Main References
Buisson's fig. 4.3 (Scothern 17?)	Gravettian	Isturitz, France (Salle d'Isturitz)	Pipe fragment	Ulna of small bird	Ist. IV designates Saint-Périer's layer IV in Salle d'Isturitz, typologically dated to Gravettian. Corresponds to Passemard's Layer F3	Ist. IV 1939 83889 (b) Length: 91.0 mm Diameter: 7.0 mm 2 holes on posterior, diameter 2.5 × 3.0 mm. Distal end heavily worked, proximal broken in the region of one of the holes. Worked end shows incisions that are traces of the removal of the distal epiphysis. Traces of abrasive scraping on surface	(b) designation added by IM. Unpublished before Buisson (1990)	Collection Saint-Périer, Musée des Antiquités Nationales, Saint-Germain-en-Laye, Paris	Buisson, 1990
Buisson's fig. 5.1(a)	Gravettian	Isturitz, France (Salle d'Isturitz)	Pipe fragment	Right ulna of a diurnal raptor, probably vulture	Ist. IV designates Saint-Périer's layer IV in Salle d'Isturitz,	Ist. IV 1939 86757	Virtually complete pipe reconstructed from three separate finds by Buisson (1990). Initial	Collection Favre, earliest Saint-Périer collection, Musée des Antiquités Nationales, Saint-	Saint-Périer and Saint-Périer, 1952; Buisson, 1990

(Scothern 13)	(proximal end)		Means of blowing possibly V-notched tongue duct (Scothern, 1992)	analysis and description carried out by Buisson (1990). Detailed analysis of piercings	Germain-en-Laye, Paris
Buisson's fig. 5.1(b)	Pipe fragment	typologically dated to Gravettian. Corresponds to Passemard's Layer F3	Ist. IV 1939 (a)	and markings, and reconstruction, carried out by Lawson and d'Errico (2002). (a) and (b) designations added by IM. On the basis of	
Buisson's fig. 5.1(c)	Pipe fragment		Ist. IV 1939 (b) Combined dimensions: Length: 165.0 mm Diameter: 12.0–24.0 mm Bore: 11.0–17.0 mm	4 regular holes on anterior surface, diameters 6.0 mm × 7.0 mm and 7.0 mm, hole edges polished	Buisson's (1990), fig. 5 it would appear that the dimensions of the three component parts of the pipe are as follows: Ist. IV 1939 86757: L: 87 mm D: 12–14 mm 2 complete holes (d: 7 × 6 mm and 7 × 7 mm), 2 partial holes

(Continued)

Appendix Table 1. Continued

Cross-reference	Period	Origin	Described as	Material (if known)	Details of Age	Description	Status & Comments	Location (if known)	Main References
(Scothern 4?)						Breaks at the extremities, one of which impinges on the body of the pipe to the region of one of the holes, the other of which affects the proximal epiphysis. The damage in the region of the hole closest to the proximal part of the bone is recent and probably caused by an excavation tool. Several traces of longitudinal scraping visible on diaphysis. Fine rectilinear and wavy short incised lines perpendicular to main axis on flattest surface;	Ist. IV 1939 (a): L: 44 mm D: 4–10 mm Bone fragment/splinter Ist. IV 1939 (b): L: 70 mm D: 18–25 mm No holes evident		

Buisson's fig. 6	Gravettian	Isturitz, France (Salle d'Isturitz)	Bone fragment	Detached epiphysis removed from bird ulna	Ist. IV designates Saint-Périer's layer IV in Salle d'Isturitz, typologically dated to Gravettian. Corresponds to Passemard's Layer F3	Ist. IV 1942 Detached epiphysis of bird bone ulna, removed by cutting a groove and then snapping off; suggests on-site manufacture of bird bone pipes (Lawson and d'Errico, 2002)	Collection Saint-Périer, Musée des Antiquités Nationales, Saint-Germain-en-Laye, Paris	Buisson, 1990; Lawson and d'Errico, 2002
Buisson's fig. 3.6 (Scothern 8)	Gravettian	Isturitz, France (Salle d'Isturitz)	Pipe fragment	Left ulna of diurnal raptor, probably vulture (proximal end)	Ist. IV designates Saint-Périer's layer IV in Salle d'Isturitz, typologically dated to Gravettian. Corresponds to Passemard's Layer F3	Ist. IV 1946 83889 Length: 93.0mm Diameter: 15.0–27.0 mm Bore: 12.0–17.0 mm 1 hole on posterior and 2 on anterior, diameters 4.0 mm, one of which very close to another. Breaks at the extremities one	Collection Saint-Périer, Musée des Antiquités Nationales, Saint-Germain-en-Laye, Paris	Saint-Périer and Saint-Périer, 1952; Buisson, 1990

(Continued)

Appendix Table 1. Continued

Cross-reference	Period	Origin	Described as	Material (if known)	Details of Age	Description	Status & Comments	Location (if known)	Main References
						of which close to the opposing holes. Traces of intense longitudinal abrasive scraping, clear fine parallel incisions perpendicular to the longitudinal axis on posterior surface. Possibly originally block & duct (Scothern, 1992)			
Buisson's fig. 5.2	Gravettian	Isturitz, France (Salle d'Isturitz)	Pipe fragment	Right ulna of large bird, possibly golden eagle (proximal end)	Ist. IV designates Saint-Périer's layer IV in Salle d'Isturitz, typologically dated to Gravettian. Corresponds to Passemard's Layer F3	Ist. IV 1946 86756 Length: 88.4 mm Diameter: 10.8–20.5 mm Bore: 8.6–13.0 mm 3 holes on posterior, diameters 4.0 × 6.0 mm and 5.0		Collection Favre, earliest Saint-Périer collection, Musée des Antiquités Nationales, Saint-Germain-en-Laye, Paris	Saint-Périer and Saint-Périer, 1952; Buisson, 1990

Buisson's fig. 3.7	Gravettian	Isturitz, France (Salle d'Isturitz)	Pipe fragment	Left ulna of diurnal raptor, probably vulture (proximal end)	Ist. IV designates Saint-Périer's layer IV in Salle d'Isturitz, typologically dated to Gravettian. Corresponds to Passemard's Layer F3	Ist. IV 83889 (a) Length: 108.0 mm Diameter: 13.0–17.0 mm Bore: 11.0 mm 1hole on posterior, 1 hole on anterior, diameters 4.0 mm. Breaks at the extremities, one in the region of the posterior hole. Several traces of × 6.5 mm. Both ends fractured, one break straight and in the region of a hole, the other irregular and in the region of the proximal epiphysis, due to crushing. Traces of scraping or fine abrasion	(a) designation added by IM. Unpublished before Buisson (1990)	Collection Saint-Périer, Musée des Antiquités Nationales, Saint-Germain-en-Laye, Paris	Buisson, 1990

(Continued)

Appendix Table 1. Continued

Cross-reference	Period	Origin	Described as	Material (if known)	Details of Age	Description	Status & Comments	Location (if known)	Main References
						abrasive scraping; fine parallel incisions perpendicular to the longitudinal axis, on posterior and lateral surfaces			
Buisson's fig. 4.4	Gravettian	Isturitz, France (Salle d'Isturitz)	Pipe fragment	Bird bone, otherwise unidentifiable	Ist. IV designates Saint-Périer's layer IV in Salle d'Isturitz, typologically dated to Gravettian. Corresponds to Passemard's Layer F3	Ist. IV 83889 (b) Length: 52.0 mm Diameter: 7.5 mm Bore: 6.0 mm 3 holes, one of which is quadrangular, diameter 2.0 mm, other two on opposite surface at broken ends. Broken at both ends in the region of the holes that are on the same face; traces of	(b) designation added by IM. Unpublished before Buisson (1990)	Collection Saint-Périer, Musée des Antiquités Nationales, Saint-Germain-en-Laye, Paris	Buisson, 1990

| Buisson's fig. 4.2 (Scothern 6) | Gravettian | Isturitz, France (Salle d'Isturitz) | Pipe fragment | Ulna of diurnal raptor, probably vulture (distal end) | IF3a designates Passemard's Layer F3 in Salle d'Isturitz, typologically dated to Gravettian. Corresponds to Saint-Périer's layer IV | IF3a 1914 75253 A Length: 74.5 mm Diameter: 14.0–15.0 mm Bore: 12.5–13.2 mm 1 hole on posterior, diameter 3.5 mm. One end is heavily worked and may form embouchure, the other is broken in the region of the hole. Numerous traces of abrasive scraping; fine incised wavy lines along length on lateral and anterior surfaces | abrasive scraping on surface | Collection Passemard, Musée des Antiquités Nationales, Saint-Germain-en-Laye, Paris | Passemard, 1944; Buisson, 1990 |

(Continued)

Appendix Table 1. Continued

Cross-reference	Period	Origin	Described as	Material (if known)	Details of Age	Description	Status & Comments	Location (if known)	Main References
Buisson's fig. 4.6 (Scothern 7)	Gravettian	Isturitz, France (Salle d'Isturitz)	Pipe fragment	Ulna of diurnal raptor, probably vulture (diaphysis only)	IF3β designates Passemard's Layer F3 in Salle d'Isturitz, typologically dated to Gravettian. Corresponds to Saint-Périer's layer IV	IF3β 21 75253 B Length: 51.0 mm Diameter: 13.0 mm Bore: 12.0 mm 2 holes, one in posterior surface, one in anterior. Broken at both ends in region of holes; several traces of abrasive scraping		Collection Passemard, Musée des Antiquités Nationales, Saint-Germain-en-Laye, Paris	Passemard, 1944; Buisson, 1990
Buisson's fig. 4.7	Gravettian (originally designated 'Final Aurignacian (Gravettian)' by Saint-Périers)	Isturitz, France (Salle d'Isturitz)	Pipe fragment	Right ulna of diurnal raptor, probably vulture (diaphysis only)	Ist. III 83888 designates items from Saint-Périer Layer III in Salle d'Isturitz (corresponding with Passemard Layer C, immediately above Layer F3); all recent chronologies agree Gravettian	Ist. III 1939 83888 (b) Length: 74.0 mm Diameter: 15.0 mm Bore: 14.0 mm 3 holes, two on anterior surface, one of which complete and has diameter 4.5 × 5.2 mm, one hole on posterior	(b) designation added by IM to differentiate from item illustrated in Buisson (1990), fig. 2 (described earlier), and figs. 4.8 and 4.9 (described later)	Collection Saint-Périer, Musée des Antiquités Nationales, Saint-Germain-en-Laye, Paris	Saint-Périer and Saint-Périer, 1952; Buisson, 1990

			cultural horizon (see Goutas, 2004). The original designation 'Final Aurignacian' derived from Breuil's typology (Lawson and d'Errico, 2002)	Broken at ends in region of anterior and posterior holes. Several traces of abrasive scraping; fine short parallel incisions, perpendicular or slightly oblique to longitudinal axis appear on all surfaces	
Buisson's fig. 4.8	Gravettian (originally designated 'Final Aurignacian (Gravettian)' by Saint-Périers)	Isturitz, France (Salle d'Isturitz)	Pipe fragment	Bird bone, otherwise unidentifiable	Ist. III 83888 designates items from Saint-Périer Layer III in Salle d'Isturitz (corresponding with Passemard Layer C, immediately above Layer F3); all recent chronologies agree Gravettian cultural horizon (see Goutas, 2004). The original	Ist. III 1939 83888 (c) Length: 68.4 mm Diameter: 12.5–13.0 mm Bore: 10.5 mm 1 hole on flattest face, diameter 5.5 mm. Broken at the extremities, of which one break in the region of hole	(c) designation added by IM to differentiate from items illustrated in Buisson (1990), figs. 2 and 4.7 (described earlier), and 4.9 (described later)	Collection Saint-Périer, Musée des Antiquités Nationales, Saint-Germain-en-Laye, Paris	Saint-Périer and Saint-Périer, 1952; Buisson, 1990

(Continued)

Appendix Table 1. Continued

Cross-reference	Period	Origin	Described as	Material (if known)	Details of Age	Description	Status & Comments	Location (if known)	Main References
					designation 'Final Aurignacian' derived from Breuil's typology (Lawson and d'Errico, 2002)				
Buisson's fig. 4.9	Gravettian (originally designated 'Final Aurignacian (Gravettian)' by Saint-Périers)	Isturitz, France (Salle d'Isturitz)	Pipe fragment	Bird bone, otherwise unidentifiable	Ist. III 83888 designates items from Saint-Périer Layer III in Salle d'Isturitz (corresponding with Passemard Layer C, immediately above Layer F3); all recent chronologies agree Gravettian cultural horizon (see Goutas, 2004). The original designation 'Final Aurignacian' derived from	Ist. III 1939 83888 (d) Length: 104.0 mm Diameter: 8.5 mm Bore: 10.0 mm 2 holes, one on one surface, diameter 2.0 × 3.0 mm, the other on opposite surface, diameter 2.0 × 3.0 mm. Broken at the extremities, of which one break is close to a hole. Several traces of intense abrasive	(d) designation added by IM to differentiate from items illustrated in Buisson (1990), figs. 2, 4.7, and 4.8 (described earlier)	Collection Saint-Périer, Musée des Antiquités Nationales, Saint-Germain-en-Laye, Paris	Saint-Périer and Saint-Périer, 1952; Buisson, 1990

| Buisson's fig. 4.5 (Scothern 5) | Gravettian (originally designated 'Final Aurignacian (Gravettian)' by Saint-Périers) | Isturitz, France (Salle d'Isturitz) | Pipe fragment | Bird bone, otherwise unidentifiable (ulna?) | Ist. III 83888 designates items from Saint-Périer Layer III in Salle d'Isturitz (corresponding with Passemard Layer C, immediately above Layer F3); all recent chronologies agree Gravettian cultural horizon (see Goutas, 2004). The original designation 'Final Aurignacian' derived from Breuil's typology (Lawson and d'Errico, 2002) | Ist. III 83888 Length: 34.5 mm Diameter: 7.8–8.3 mm Bore: 6.6 mm 3 holes, one on one surface, two on opposite surface at broken ends. Broken at both ends in the region of the holes that are on the same face; several traces of abrasive scraping on surface scraping along length | Originally published (Saint-Périer and Saint-Périer, 1952) in more complete form than now exists. Complete version pictured in Saint-Périer and Saint-Périer (1952), plate IV, top left. Scothern gives original length as 125 mm, presumably on basis of Saint-Périer plate IV | Collection Saint-Périer, Musée des Antiquités Nationales, Saint-Germain-en-Laye, Paris | Saint-Périer and Saint-Périer, 1952; Buisson, 1990 |

(Continued)

Appendix Table 1. Continued

Cross-reference	Period	Origin	Described as	Material (if known)	Details of Age	Description	Status & Comments	Location (if known)	Main References
Buisson's fig. 3.2	Solutrean	Isturitz, France (Salle d'Isturitz)	Pipe fragment	Ulna of diurnal raptor, probably vulture (diaphysis only)	Ist. IIIa designates Saint-Périer's layer IIIa in Salle d'Isturitz, typologically dated to *Solutréen supérieur*	Ist. IIIa 1939 83887 (a) Length: 61.5 mm Diameter: 13.0 mm Bore: 12.3 mm 2 holes, one on posterior, diameter 4.2 mm, one on anterior at point of break. Broken at both ends. Longitudinal abrasions	(a) designation added by IM	Collection Saint-Périer, Musée des Antiquités Nationales, Saint-Germain-en-Laye, Paris	Saint-Périer and Saint-Périer, 1952; Buisson, 1990
Buisson's fig. 3.8	Solutrean	Isturitz, France (Salle d'Isturitz)	Pipe fragment	Right ulna of diurnal raptor, probably vulture (proximal end)	Ist. IIIa designates Saint-Périer's layer IIIa in Salle d'Isturitz, typologically dated to *Solutréen supérieur*	Ist. IIIa 1939 83887 (b) Length: 74.0 mm Diameter: 15.0–20.0 mm Bore: 14.0–15.3 mm 1 hole on anterior surface, diameter 3.5 mm and possible additional hole on same surface	(b) designation added by IM	Collection Saint-Périer, Musée des Antiquités Nationales, Saint-Germain-en-Laye, Paris	Saint-Périer and Saint-Périer, 1952; Buisson, 1990

| Buisson's fig 3.1 (Scothern 15?) | Magdalenian | Isturitz, France (Salle d'Isturitz) | Pipe fragment | Ulna of diurnal raptor, probably vulture (diaphysis only) | IE*a* designates Passemard's layer E*a* in Salle d'Isturitz, typologically dated to *Magdalénien moyen* | IE*a* 1914 P2 77153 Length: 68.0 mm Diameter: 14.4 mm Bore: 13.0 mm 2 holes, one on posterior, diameter 3.0 mm, one on anterior at point of break. Broken at both ends. Longitudinal abrasive scraping and numerous fine incised parallel lines perpendicular to length evidenced by trace of cupule at one end. Breaks at the extremities, one in the region of supposed second hole. Traces of abrasive scraping | Collection Passemard, Musée des Antiquités Nationales, Saint-Germain-en-Laye, Paris | Passemard, 1944; Buisson, 1990; Lawson and d'Errico, 2002 |

(Continued)

Appendix Table 1. Continued

Cross-reference	Period	Origin	Described as	Material (if known)	Details of Age	Description	Status & Comments	Location (if known)	Main References
Buisson's fig. 3.5 (Scothern 9)	Magdalenian	Isturitz, France (Salle d'Isturitz)	Pipe fragment	Left ulna of diurnal raptor, probably vulture (proximal end)	IEa designates Passemard's layer Ea in Salle d'Isturitz, typologically dated to *Magdalénien moyen*	IEa 1914 P1 77153 Length: 111.0 mm Diameter: 15.0–23.0 mm Bore: 13.0–15.0 mm 1hole on posterior, 1 hole on anterior, diameters 3.5 mm. Breaks at the extremities, one in the region of the posterior hole. Several traces of intense abrasive scraping, and fine parallel incisions perpendicular and oblique to the longitudinal axis on anterior, lateral, and		Collection Passemard, Musée des Antiquités Nationales, Saint-Germain-en-Laye, Paris	Passemard, 1944; Buisson, 1990; Lawson and d'Errico, 2002

| Scothern 3 | Gravettian | Isturitz, France (Salle d'Isturitz) | Flute fragment | Bird (p. 89) | 'Final Aurignacian' (Gravettian) | posterior surfaces. Ist SP52. [May be Lawson and d'Errico, 2002, plate I.6] Length: 52 mm Width: 12 mm. Blow hole diameter: 12 mm | SP52 numbers are Scothern's (1992) own, designating artefacts featured in the Saint-Périer and Saint-Périer (1952) publication but now apparently missing from the collections. MAY be Lawson and d'Errico (2002), plate I.6 (dimensions match closely), though this is a Passemard find so unlikely to have been published in Saint-Périer & Saint-Périer (1952) as Scothern's label suggests | Lost | Saint-Périer and Saint-Périer, 1952; Scothern, 1992; Turk and Kavur, 1997 |
| Scothern 4 | Gravettian | Isturitz, France (Salle d'Isturitz) | Flute fragment | Bird (p. 89) | 'Final Aurignacian' (Gravettian) | Ist SP52A. [May be Ist. IV 1939 (b), Buisson (1990), fig. 5.1 (c)] Length: 76 mm. Width 15–16 mm. No finger holes. Blow hole diameter: 8 mm | MAY correlate with Ist. IV 1939 (b), illustrated by Buisson as fig 5.1 (c)—part of a virtually complete pipe reconstructed by Buisson (fig. 5.1) | Lost | Saint-Périer and Saint-Périer, 1952; Buisson, 1990; Scothern, 1992; Turk and Kavur, 1997 |

(*Continued*)

Appendix Table 1. Continued

Cross-reference	Period	Origin	Described as	Material (if known)	Details of Age	Description	Status & Comments	Location (if known)	Main References
Scothern 12	Gravettian	Isturitz, France (Salle d'Isturitz)	Flute fragment	Bird (p. 89)	Gravettian. 28–22 kya	Ist SP52C. Length: 132 mm. Width: 17–27 mm. Bore: 15 mm. Blow hole diameter: 1.6 mm. Finger hole diameter: 8 mm. Possibly originally block & duct (Scothern, 1992)		Lost (Scothern, 1992, p.92), although in inventory Scothern lists location as Saint-Germain, Paris	Saint-Périer and Saint-Périer, 1952; Scothern, 1992
Scothern 14	Gravettian	Isturitz, France (Salle d'Isturitz)	Flute fragment	Bird (p. 89)	Gravettian. 28–22 kya	Ist SP52E. [Saint-Périer and Saint-Périer, 1952, plate VII, bottom left] Length: 94 mm. Width: 12–21 mm. Finger hole diameter: 6 mm. Bore ? (p. 92). Retains distal epiphysis (p. 94)	Listed in Scothern, 1992, table 7, p. 92, but not listed in Scothern's inventory. Scothern's text describes this as having 2 complete and 1 truncated finger hole (p. 94), whereas table 7 describes as left	Lost (Scothern, 1992, p. 92)	Saint-Périer and Saint-Périer, 1952; Scothern, 1992

Scothern 17	Gravettian	Isturitz, France (Salle d'Isturitz)	Flute fragment	Bird (p. 89)	Gravettian. 28–22 kya [83889 is Saint-Périer layer IV/Passemard Layer F3]	Ist 1939. [May be Ist. IV 1939 83889 (b), Buisson, 1990, fig. 4.3] Length: 89 mm. Width: 9–10 mm. 3 finger holes, diameters: 6 mm, 7 mm, 8 mm	Means of blowing possibly V-notched tongue duct (Scothern, 1992) This correlates with artefact pictured in Saint-Périer and Saint-Périer, 1952, plate VII, bottom left. This was missing also when Buisson examined the collection	Saint-Germain, Paris (Scothern, 1992, p. 92), although in inventory Scothern lists location as lost	Buisson, 1990; Scothern, 1992
						Two items with the code Ist 1939 are listed by Scothern (1992) in inventory and table 7 (p.92). One is Ist. III 1939 83888 (see earlier), and this MAY be Ist. IV 1939 83889 (b). (Buisson, fig. 4.3). Different width and finger hole diameters given in Scothern's table 7, p. 92: 0.6 cm, 0.8 cm, and 0.8 cm			
+ (Lawson and d'Errico, 2002, plate I.6)	Gravettian?	Isturitz, France (Salle d'Isturitz)	Pipe fragment	Ulna of diurnal raptor, probably vulture (distal end)	From Passemard 75252 collection (catalogued in 1929) so probably from the same excavation and	75252 A1 Worked bird-bone tube fragment not published by Buisson (1990) (Lawson and d'Errico, 2002)	MAY be Scothern's Ist. SP52 (Scothern 3) (dimensions match closely), though this is a Passemard find not a Saint-Périer and Saint-Périer (1952)	Collection Passemard, Musée des Antiquités Nationales, Saint-Germain-en-Laye, Paris	Lawson and d'Errico, 2002

(Continued)

Appendix Table 1. Continued

Cross-reference	Period	Origin	Described as	Material (if known)	Details of Age	Description	Status & Comments	Location (if known)	Main References
					context as IF3a 1914 75252 A3, i.e. Passemard's Layer F3 in Salle d'Isturitz, typologically dated to Gravettian.	Length: c.50 mm Diameter: c.8.5–11.5 mm (On basis of Lawson and d'Errico, 2002, plate I.6)	find as Sothern's label implies		
+ (Lawson and d'Errico, 2002, plate I.7)	Gravettian	Isturitz, France (Salle d'Isturitz)	Pipe fragment	Ulna of diurnal raptor, probably vulture (diaphysis only)	Ist. IV designates Saint-Périer's layer IV in Salle d'Isturitz, typologically dated to Gravettian. Corresponds to Passemard's Layer F3	Ist. IV 1935 Worked bird-bone tube fragment not published by Buisson (1990) (Lawson and d'Errico, 2002). Length: c.22.5 mm Diameter: c.10 mm (On basis of Lawson and d'Errico, 2002, plate I.7)		Collection Saint-Périer, Musée des Antiquités Nationales, Saint-Germain-en-Laye, Paris	Lawson and d'Errico, 2002
	Gravettian/ Perigordian	Goyet, Belgium	Whistle	Bird (p. 89)	Perigordian. 28–22 kya	Length: 105 mm. Central, crescent-shaped sound hole		L'Institut Royal des Sciences Naturelles de Belgique (IRScN), Brussels	Otte, 1979; Sothern, 1992

Gravettian/ Perigordian	Lespaux, Gironde, France	Flute fragment	Bird bone (p. 99)	Perigordian. 28–22 kya	Reconstruction produces B. Has definite sound-window, still intact (Scothern, 1992, p. 87)	Krtolitza Collection, Bordeaux	Couste and Krtolitza, 1961, 1965; Roussot, 1970; Scothern, 1992
					Length: 60 mm. Width: 10 mm. 1 intact central finger hole, diameter: 4 × 4.5 mm. 2 incomplete holes at proximal and distal ends of the bone (p. 99). All holes show clear evidence of tool use (Scothern, 1992).	Inconsistently with the text (Scothern, 1992, p. 99), as left, Scothern's inventory describes this artefact as having 2 finger holes, one with 20 mm diameter, other fragmentary. However, a 20 mm hole seems unlikely to be a functional finger hole	
Gravettian/ Perigordian	Maisieres Canal, Belgium	Bone pipe/ whistle?	?	Perigordian. 28–22 kya	Length: 42 mm	Questioned in Scothern's (1992) inventory, although in text, Scothern says this is 'more likely' to be an intentional sound-producer	L'Institut Royal des Sciences Naturelles de Belgique (IRScN), Brussels	Otte, 1979; Scothern, 1992
Gravettian/ Perigordian	Pair-non-Pair, Gironde, France	Flute	?	Upper Perigordian/	Length: 127 mm			Daleau, 1884–5,

(Continued)

Appendix Table 1. Continued

Cross-reference	Period	Origin	Described as	Material (if known)	Details of Age	Description	Status & Comments	Location (if known)	Main References
					Gravettian. 28–22 kya	Width: 107–24 mm. 3 finger holes, 2 intact, diameters: 4.5 and 6 mm diameters, the other damaged		Daleau Collection, Musée d'Aquitaine, Bordeaux	1885, 1896; Seewald, 1934; Malvesin-Fabre, 1946; Pequart and Pequart, 1960; Cheynier, 1963; Bourdier, 1967; Roussot, 1970; Scothern, 1992.
Gravettian/Perigordian		Pair-non-Pair, Gironde, France	Whistle?	?	Gravettian. 28–22 kya		Questioned	?	Da Veiga Ferreira and Cardosa, 1975; Scothern, 1992.
Gravettian/Perigordian		Pekarna, Moravia, Czech Republic	2 pipes/whistles?	Bird epiphyses	Gravettian. 28–22 kya	2 specimens.		?	Absolon, 1936; Scothern, 1992.
Gravettian/Perigordian		Pekarna, Moravia, Czech Republic	Whistle	?	Upper Palaeolithic/Gravettian? 28–22 kya?	Length: 140 mm. Produces pitch A.		?	Absolon, 1936; Buchner, 1956;

	Solutrean	Le Placard, France	?	?	Solutrean. 21–19 kya	?	Scothern, 1992.
	Solutrean?	Liceia, Barcarena, Portugal	Flute fragment	?	Solutrean? 21–19 kya?	Age questioned	Seewald, 1934; Scothern, 1992
	Upper Palaeolithic	Grubgraben bei Kammern, Austria	Flute	Medium ungulate right tibia	c.19 kya	Several examples from various levels	Da Veiga Ferreira and Cardosa, 1975; Scothern, 1992
						Length: 83 mm. Width: 13 mm. 2 finger holes, diameters: 3 and 6 mm	
						Broken distally and proximally. Length: 165.3 mm. 3 perforations, diameters: 51–5 mm	Einwögerer and Käfer, 1998; Käfer and Einwögerer, 2002
Scothern 18	Magdalenian	Isturitz, France (Salle d'Isturitz)	Flute fragment	Bird ulna (Scothern, 1992, p. 124)	Magdalenian. 18–12 kya	1st SP52F. Length: 175 mm. Width: 10–15 mm. Notch diameter: 9x2 mm (although inventory says 'blow hole, 0.2 cm diameter'). Bore?	Lost	Capable of being a sound-producer. Appears not to feature in Buisson's analysis, perhaps because no finger holes	Saint-Périer, 1947; Scothern, 1992
	Magdalenian?		Whistle	?	Magdalenian?		?	

(Continued)

Appendix Table 1. Continued

Cross-reference	Period	Origin	Described as	Material (if known)	Details of Age	Description	Status & Comments	Location (if known)	Main References
		Bolinkoba, Vizcaya, Spain				Not in inventory, though mentioned in text (Scothern, 1992, p. 114). Central blow hole. Resembles other later whistles (Scothern, 1992, p. 114)	Age questioned; may be as late as Neolithic		Barandiaran, 1973; Scothern, 1992
	Magdalenian	Fontarnaud, Gironde, France	Pipe	Bird	Final Magdalenian	Length: 70 mm	Shows definite cut-marks. Not demonstrably a sound producer (Scothern, 1992)	Ferrier Collection	Roussot and Ferrier, 1970; Scothern, 1992
	Magdalenian	Garrigue, France	Flute	?	Magdalenian Superior	Length: 68 mm. Width: 13–18 mm. 2 finger holes, diameters: 5 mm. Fragmentary. Very worn. No indication of mode of sound production. Holes tapered,		Musée d'Aquitaine, Bordeaux	Vircoulon and Deffarge, 1977; Scothern, 1992

Magdalenian	Gudenushöhle, Austria	Whistle	?	Magdalenian. 18–12 kya	however, characteristic of burin-type tool No. 22.387. Length: 40 mm. Width: 12 mm. Central hole	Natural History Museum, Vienna	Seewald, 1934; Scothern, 1992
Magdalenian	Le Placard, France	Notched whistle-flute	Eagle	Magdalenian IV–VI, 15,000–13,000 BP	LP.1. (Scothern's own denomination). Length: 140 mm. Width: 15–25 mm. Blow hole notch on posterior, diameter: 15 × 3 mm. Engraved, series of parallel striations creating arced lines. Similar to next entry, same date	'Undoubtedly a sound-producer' (Scothern, 1992, p. 125) Originally in Piette collection, Musée des Antiquités Nationales, Saint-Germain, Paris. Now in Poitiers University	Piette, 1874, 1875, 1907; Seewald, 1934; Marshack, 1970; Scothern, 1992
Magdalenian	Le Placard, France	Whistle-flute	Eagle	Magdalenian IV–VI, 15,000–13,000 BP	LP.2. (Scothern's own denomination). Length: 110 mm. Fragmentary. Engraved same as previous entry	Lacks notch or sound window. Marshack (1970) considers it too fragile to be functional. However, typologically very similar to previous entry, and lack of sound window/notch may be due only to fragmentary nature	Originally in Musée des Antiquités Nationales, Saint-Germain, Paris, now in Poitiers University	Piette, 1907; Seewald, 1934; Marshack, 1970; Scothern, 1992

(Continued)

Appendix Table 1. Continued

Cross-reference	Period	Origin	Described as	Material (if known)	Details of Age	Description	Status & Comments	Location (if known)	Main References
Magdalenian?	Magdalenian?	La Placard, France	Notched whistle-flute	?	Magdalenian? 18–12 kya? 'Also [as above examples] dated to mid-Magdalenian' (Scothern, 1992, p. 126)	LP.3 (Scothern's own denomination). Exact dimensions unknown	Age questioned. (Not questioned in text, only in inventory, see left)	? Location unknown	Piette, 1907; Seewald, 1934; Scothern, 1992
Magdalenian	Magdalenian	Le Placard, France	Notched whistle-flute	?	Magdalenian. 18–12 kya (Mid-Magdalenian)	54931. Length: 72 mm. Width: 10 mm. Bore: 8 mm. Blow hole diameter: 7 mm. Deeply engraved		Saint-Germain	Scothern (1992) (previously unpublished)
Magdalenian	Magdalenian	Le Placard, France	Flute fragment	?	Magdalenian. 18–12 kya	54931Pl. Length: 84 mm. Width: 14 mm. Bore: 10 mm. Two finger holes, diameters: 4 mm and 5 mm. Blow hole diameter: 15 mm	'Resemblant of Isturitz bones' of the Gravettian (Scothern, 1992, p. 124), although 'The significance of this artefact remains uncertain' (Scothern, 1992, p. 129)	Saint-Germain	Scothern, 1992
Magdalenian	Magdalenian	Le Placard, France	Notched whistle-flute	?	Magdalenian. 18–12 kya	54967. Length: 80 mm		Saint-Germain	Scothern (1992)

Magdalenian			(Mid-Magdalenian)	Width: 14–15 mm. Blow hole diameter: 6 mm. Deeply engraved.	Saint-Germain	(previously unpublished)
Magdalenian	Le Placard, France	Notched whistle-flute	?	Magdalenian. 18–12 kya (Mid-Magdalenian)	54967PL. Length: 110 mm. Width: 13–15 mm. Blow hole diameter: 8 mm. Deeply engraved	Scothern's (1992) inventory lists an artefact 54964PL; it seems likely that this is a typographical error, and refers to this artefact, as there is no 54964PL in the text (p. 124) or table 12 (p. 125), only 54967PL, and the dimensions described are identical in each case	Saint-Germain	Scothern (1992) (previously unpublished)
Magdalenian	Le Placard, France	Notched whistle-flute	?	Magdalenian. 18–12 kya (Mid-Magdalenian)	55185. Length: 94 mm. Width: 20–5 mm. Bore: 13 mm. Blow hole diameter: 10 mm. Deeply engraved		Saint-Germain	Scothern (1992) (previously unpublished)
Magdalenian	Le Placard, France	Flute fragment	?	Magdalenian. 18–12 kya	IFB321. Length: 73 mm.	'Fragmentary... with little indication of	Saint-Germain	Scothern, 1992

(*Continued*)

Appendix Table 1. Continued

Cross-reference	Period	Origin	Described as	Material (if known)	Details of Age	Description	Status & Comments	Location (if known)	Main References
						Width: 14–15 mm.	sound-production' (Scothern, 1992, p. 124). Scothern's (1992) inventory lists an artefact IFB312; it seems likely that this is a typographical error, and refers to this artefact, as there is no IFB312 in the text (p. 124) or table 12 (p. 125), only IFB321		
Magdalenian	Magdalenian	Le Placard, France	4 bone pipes	?	Magdalenian. 18–12 kya	4 unequal lengths of bone pipe. 'Possibly Pan pipes' (Mortillet, 1906; Scothern, 1992)	Age uncertain. This entry refers to four of the above specimens with matching decoration and features of design. Which four is not specified, in inventory, but most likely 54931, 54967, 54967PL and 55185, judging by text (p. 130). The bones show no remaining traces of having been attached to each	?	Mortillet, 1906; Megaw, 1968; Scothern, 1992

Magdalenian	Le Roc de Marcamps, Gironde, France	Bone pipe	Bird	Magdalenian V–VI	Length: 70 mm. Width: 10 mm. Bore: 7 mm. Blow hole diameter: 7 mm (although inventory says 0.3cm notch in upper bone)	other, so 'pan pipe' interpretation is unsupported, although they are typologically similar Position and form of the sound-holes suggests a block and duct or tongue duct principle of sound production (Scothern, 1992, p. 133). However, Scothern (1992) goes on to say that 'the Marcamps pipes … are difficult to reconstruct using modern tools due to their small dimensions, which almost rules out the possibility of being interpreted as effective, functional sound producers' (p. 134). She then goes on to say, though, that 'the Marcamps whistles … are the earliest notched whistles to be found as a series of multi-	Collection Maziaud, Musée de l'Aquitaine, Bordeaux	Roussot, 1970; Scothern, 1992
Magdalenian	Le Roc de Marcamps, Gironde, France	Bone pipe	Bird	Magdalenian V–VI	Length: 49 mm. Width: 9–11 mm. Bore: 8 mm. Blow hole diameter: 7 mm (although inventory says 0.3cm notch in upper bone)		Collection Maziaud, Musée de l'Aquitaine, Bordeaux	Roussot, 1970; Scothern, 1992
Magdalenian	Le Roc de Marcamps, Gironde, France	Bone pipe	Bird	Magdalenian V–VI	Length: 43 mm. Width: 6–7 mm. Bore: 6 mm. Blow hole diameter: 5 mm (although inventory says		Collection Maziaud, Musée de l'Aquitaine, Bordeaux	Roussot, 1970; Scothern, 1992

(Continued)

Appendix Table 1. Continued

Cross-reference	Period	Origin	Described as	Material (if known)	Details of Age	Description	Status & Comments	Location (if known)	Main References
	Magdalenian	Le Roc de Marcamps, Gironde, France	Bone pipe	Bird	Magdalenian V–VI	Length: 37 mm. Width: 7–9 mm. Bore: 6.5.00 mm. Blow hole diameter: 6 mm (although inventory says 0.3cm notch in upper bone)	0.3 cm notch in upper bone). stopped pipes' (p. 135)	Collection Maziaud, Musée de l'Aquitaine, Bordeaux	Roussot, 1970; Scothern, 1992
	Magdalenian	Les Roches, Sergeac, Dordogne, France	Flute fragment	?	Magdalenian. 18–12 kya	Length: 115 mm. Width: 19 mm. 2 finger holes, diameters: 4 mm and 5 mm, centrally located on posterior. No mouthpiece/blow hole. No remaining evidence of mode of sound production, although holes 'definitely worked' deliberately		British Museum, London	Collins, 1975; Scothern, 1992

Magdalenian	Lussac, France	Flute fragment	?	Magdalenian Superior	(Scothern, 1992, p. 119) No.38-23-37008. Length: 59 mm. Width: 19 mm. Upper section of artefact only remaining	Musée de l'Homme, Paris	Scothern, 1992
Magdalenian	Mas-d'Azil, France	Notched flute	?	Magdalenian. 18–12 kya	Length: 85 mm. Width: 5–8 mm. Notch diameter: 2 × 3 mm	May be too small to produce sound (Scothern, 1992, p. 123)	?	Pequart and Pequart, 1960; Scothern, 1992
Magdalenian	Padtberg, Munzingen, Germany	Signal pipe?	Alpine hare	Magdalenian. 18–12 kya		?	Seewald, 1934; Scothern, 1992
Magdalenian	Pas du Moir, France	Flute	A limb bone	Magdalenian, 18–12 kya	4 holes on the foreside, 2 on rear. Possibly bored. Ends sawn or cut and broken off straight. Notched mouthpiece-type opening (Marshack, 1990, quoted in Turk and Kavur, 1997)	?	Marshack, 1990; Turk and Kavur, 1997

(Continued)

Appendix Table 1. Continued

Cross-reference	Period	Origin	Described as	Material (if known)	Details of Age	Description	Status & Comments	Location (if known)	Main References
	Magdalenian	Pekarna, Moravia, Czech Republic	Whistle	Bird bone with sawn holes	(Magdalenian?) C14 dated to 12,940 ± 250 or 12,670 ± 80 BP (Svoboda et al., 1994, quoted in Turk and Kavur, 1997)	Length: 70 mm		Brno	Buchner, 1956; Scothern, 1992; Svoboda et al., 1994
	Magdalenian	Peyrat, Près de Terrason, Dordogne, France	Flute	?	Magdalenian IV	Fragmentary. Length: 60 mm. 4 holes, diameters: 5–6 mm, 2 of which at ends of bone fragmentary	'Possible transverse flute' though difficult to confirm (Scothern, 1992)	?	Cheynier, n.d.; Fages and Mourer-Chauviré, 1983; Scothern, 1992
	Magdalenian	Raymonden, France	4 bone fragments	Bird	Magdalenian IV–VI	Lengths: 39 mm, 41 mm, 132 mm, and 147 mm	Show definite cut-marks. Not demonstrably sound producers (Scothern, 1992, p. 121)	Musée du Périgord, Périgueux	Scothern, 1992
	Magdalenian	Rond du Barry, France	Whistle flute	Radius of whooper swan	'Final Magdalenian'	Median section missing. Signs of polishing at both ends. Engraved with 'fine striae and hatched oval designs' (Scothern, 1992, p. 121)	There remains no actual evidence of sound production, though clearly deliberately worked	?	de Bayle des Hermens, 1974 Bergounioux and Glory, 1952; Fages and Mourer-Chauviré, 1983;

'Middle or Upper Palaeolithic'	Goyet, Belgium	'Flute'	'Limb bone'	Undated	Goyet 132. Both epiphyses sawn off; 2 holes, 1 at end of one side, 1 near centre of other. Latter made by 'sawing the compact bone tissue' (Turk and Kavur, 1997)	No stratigraphic information; typology of lithics from site suggests Middle or Upper Palaeolithic date for artefact (McComb, 1989)	?	Scothern, 1992. McComb, 1989; Turk and Kavur, 1997
'Late Upper Palaeolithic'	Csaklya, Hungary	Whistle	?	Unspecified	Length: 60 mm. Central hole		?	Seewald, 1934; Scothern, 1992
'Upper Palaeolithic'	Dolni-Vestonice, Moravia	Whistle	Lion's tooth	Unspecified	Pierced with stone		?	Seewald, 1934; Scothern, 1992
'Upper Palaeolithic'	Dolni-Vestonice, Moravia	Four bone pipes	Red deer metacarpal	Unspecified	4 specimens, varying lengths, 1 with resin plug still in place. 'Block and duct principle within Pan pipes?' (Scothern, 1992)		?	Megaw, 1968; Scothern, 1992

(Continued)

Appendix Table 1. Continued

Cross-reference	Period	Origin	Described as	Material (if known)	Details of Age	Description	Status & Comments	Location (if known)	Main References
	'Late Upper Palaeolithic'	Gourdan, France	Bone pipe	?	Unspecified		Shows definite cut-marks. Not demonstrably a sound producer (Scothern, 1992, p. 121)	?	Seewald, 1934; Scothern, 1992
	'Upper Palaeolithic'	Horodnica, Poland	Flute fragment	?	Unspecified	Single hole in proximal end		?	Seewald, 1934; Scothern, 1992
	'Late Upper Palaeolithic'	Kesslerloch bei Thayngen, Switzerland	Flute fragment?	Swan bone	Unspecified		Questioned	Lost	Seewald, 1934; Scothern, 1992
	'Upper Palaeolithic'	Laugerie-Basse, France	Flute/whistle fragment?	?	Unspecified		Show definite cut-marks. Not demonstrably a sound producer (Scothern, 1992, p. 121)	Lost	Piette, 1907; Seewald, 1934; Scothern, 1992
	'Upper Palaeolithic'	Roque-Saint-Christophe, Périgord, France	Flute	Bird bone	Unspecified	4 holes on one surface, 2 holes on the opposite surface; extremities badly preserved		British Museum	Masset & Perlès, 1978; Brade, 1982; Fages and Mourer-Chauviré, 1983

Appendix Table 1 presents items that have been reputed to be possible flute-type sound-producers. It does not include objects that have, since original publication, and on the basis of further analysis, been subsequently deemed unlikely to be sound-producers (these are listed in Appendix Table 2). Note that not all of the objects listed here have been subjected to further analysis since original publication and may, on such future analysis, also be deemed unlikely. Where comments about their status are available, these are noted. Appendix Table 1 features 105 items.

Appendix Table 2. Inventory of Palaeolithic objects originally reputed to be pipes and flutes but since deemed unlikely

Number	Period	Origin	Described as	Material (if known)	Details of Age	Description	Status & Comments	Location (if known)	Main References
	Mousterian	Divje babe I, Slovenia	Flute	Femur of juvenile cave bear (*Ursus spelaeus*)	C14 date of layer 8, 43,100 ± 700 BP (Nelson, 1997)	Length of diaphysis: 113.6 mm. 2 complete holes on posterior side, diameters: 9.7 and 9.0 mm. Distance between the centres of the holes, 35 mm	Debated. Much contention regarding origins of the 2 complete holes (carnivore v. human agency), and original total number of holes. Despite complicated taphonomy carnivore origin remains most likely explanation	National Museum of Slovenia, Ljubljana	Turk (ed.), 1997; Nelson, 1997 (in Turk (ed.), 1997); Kunej and Turk, 2000; Morley, 2006; numerous other papers
	Middle Stone Age	Haua Fteah, Libya	Whistle	?	80,000–60,000 BP?	Fragment Length: 28 mm. 1 central perforation, possible remains of second and possible mouthpiece. Diameter of perforation: 34 mm. Distance	Dubious. Uncertainty about stratigraphy; may be disturbed, from succeeding layer, dated to 34,000 ± 2,800 (Scothern, 1992, p. 71)	Cambridge Museum of Archaeology and Anthropology	McBurney, 1967 Davidson, 1991; Scothern, 1992; Turk and Kavur, 1997; D'Errico and Villa, 1997

(*Continued*)

Appendix Table 2. Continued

Number	Period	Origin	Described as	Material (if known)	Details of Age	Description	Status & Comments	Location (if known)	Main References
						from 'mouthpiece' to perforation, 8 mm, to second 'perforation', 17.5 mm. (Scothern, 1992, quoting McBurney, 1967)	Chemically damaged surface. Hole may be due to carnivore tooth damage (d'Errico and Villa, 1997; Turk and Kavur, 1997)		
	Mousterian	Kent's Cavern, Britain	?	Hare bone	?	Three large holes, three small	Unlikely to be a sound-producer (Scothern, 1992)	?	Seewald, 1934; Megaw, 1960; Scothern, 1992
	Aurignacian	Bukovácer Höhle, Lokve, Croatia	Flute?	Cave bear (Ursus spelaeus) rib	Aurignacian II	Curved, 3 finger-holes on one side	More likely the product of natural processes than human agency (Scothern, 1992). Does not contain a cavity (Albrecht et al., 2001) so distinctly unlikely	?	Bayer, 1929; Megaw, 1960; Brade, 1975; Fages and Mourer-Chauviré, 1983; Collins, 1986; Scothern, 1992; Rottländer 1996; Albrecht et al., 1998; Holderman and Serangeli, 1998a

Aurignacian	Cro-Magnon, France	Flute/whistle	?	Aurignacian. 32–17 kya	Bone pipe/flute?	Questionable (Scothern, 1992)	?	Seewald, 1934; Scothern, 1992
Aurignacian	Istállóskö, Hungary	Multi-pitch whistle	Juvenile cave bear femur diaphysis (*Ursus spelaeus*)	Aurignacian. 32–17 kya. Layer in which it was found C14 dated to Upper Aurignacian; 31,540 ± 660 BP and 30,900 ± 600 (Allsworth-Jones, 1986, quoted in Turk and Kavur, 1997)	Fragmentary. Length: 107 mm. Three holes, diameters: 7 mm, 6 mm and 10–13 mm, the former on the posterior side, the latter pair on the anterior, and separated by 65 mm. All holes appear punched, although the first appears to be hollowed or chiselled too. Also possibly chewed by rodents. (Turk and Kavur, 1997) Reconstruction produces A, B flat, B, E	Animals contributed to the making of at least two of the three holes (Albrecht, Holderman, and Serangeli, 2001)	?	Horusitzky, 1955; Vértes, 1955; Brade, 1982; Soproni, 1985; Allsworth-Jones, 1986; Scothern, 1992; Turk and Kavur, 1997. Albrecht et al., 1998. Holderman and Serangeli, 1998a and 1998b Albrecht, Holderman and Serangeli, 2001
Aurignacian	Les Bernoux, France	Whistle/paint tube?	Bird	Aurignacian. 32–17 kya		Questionable (Scothern, 1992)	Musée du Périgord, Périgueux, France	Scothern, 1992

(Continued)

Appendix Table 2. Continued

Number	Period	Origin	Described as	Material (if known)	Details of Age	Description	Status & Comments	Location (if known)	Main References
	Aurignacian	Les Bernoux, France	Whistle?	Bird	Aurignacian. 32–17 kya		Questionable (Scothern, 1992)	Musée du Périgord, Périgueux, France	Scothern, 1992
	Aurignacian	Potočka Zijalka, Slovenia	Pierced mandible/flute?	Cave bear (*Ursus spelaeus*), sub-adult or adult (Turk and Kavur, 1997)	'Upper Aurignacian'. No C14 date available (Turk and Kavur, 1997)	Three holes, diameters 5–6 mm, 5 mm and 5 mm. Centre-centre distances 19 mm and 24 mm, punched or pierced. Several nearby signs of carnivore damage	More likely the product of natural processes than human agency (Scothern, 1992). At least eight other lower jaws from the site have similar holes (plus other bones); only the number of holes leads to attribution of flute (Brodar and Brodar, 1983; Albrecht et al., 2001)	?	Brodar and Bayer, 1928; Seewald, 1934; Brodar and Brodar, 1983. Scothern, 1992; Hahn and Munzel, 1995; Albrecht et al., 1998; Holderman and Serangeli, 1998a and b; Albrecht et al., 2001
	Aurignacian	Saint-Avil-S énieur, Vallée de la Couze, France	Pipe	Bird	Aurignacian	Length: 33 mm. Notched engraving down either side	More likely the product of natural processes than	?	? (Described in Scothern, 1992, p. 85; not in inventory)

Aurignacian	Spy, Belgium	2 whistles?	Bird diaphyses	Aurignacian, 32–17 kya	2 specimens Lengths: 39 and 58 mm	Questionable (Scothern, 1992)	Carpentier Collection, University of Liège?	Otte, 1979; Scothern, 1992
Gravettian/ Perigordian	Mammutova, Poland	Whistle?	Bird	Gravettian. 28–22 kya.	Engraved. 8 cm long.	Questionable (Scothern, 1992)	?	Otte, 1981; Scothern, 1992
Gravettian/ Perigordian?	Molodova, Ukraine	Flute	?	Gravettian? 28–22 kya?	Accidental finger holes?	Age and function questioned by Scothern, 1992	?	Cernys, 1955; 1956; Fages and Mourer-Chauviré, 1983; Scothern, 1992
Gravettian/ Perigordian	Pair-non-Pair, Gironde, France	Pierced epiphysis/ whistle?	Reindeer (*Rangifer tarandus*)	Gravettian. 28–22 kya	Length: 80 mm	Possible haft (Scothern, 1992)	?	Daleau, 1881; Scothern, 1992
Gravettian/ Perigordian	Pekarna, Moravia, Czech Republic	Whistle?	Chamois	Gravettian. 28–22 kya		Questionable (Scothern, 1992)	?	Absolon, 1936; Scothern, 1992
Gravettian/ Perigordian	Pekarna, Moravia, Czech Republic	Whistle?	Reindeer metatarsal (*Rangifer tarandus*)	Gravettian. 28–22 kya		Questionable (Scothern, 1992)	?	Absolon, 1936; Scothern, 1992.
Gravettian/ Perigordian	Pekarna, Moravia, Czech Republic	6 pipes/ whistles?	?	Gravettian. 28–22 kya	6 specimens, subjected to acoustic analysis	Questionable (Scothern, 1992)	?	Absolon, 1936; Scothern, 1992
Gravettian/ Perigordian	Pekarna, Moravia, Czech Republic	Whistle?	Ulna of Greylag goose, hollow	Gravettian. 28–22 kya		'unlikely to be musical' (Scothern, 1992)	?	Absolon, 1936; Scothern, 1992
Gravettian/ Perigordian	Wildschauer, Germany	Bone tube/ whistle?	?	Gravettian. 28–22 kya	Engraved. Length: 104 mm	Questionable (Scothern, 1992)	?	Otte, 1981; Scothern, 1992

(*Continued*)

Appendix Table 2. Continued

Number	Period	Origin	Described as	Material (if known)	Details of Age	Description	Status & Comments	Location (if known)	Main References
	Solutrean	Badegoule, France	Flute?	Reindeer radius (*Rangifer tarandus*)	Solutrean. 21–19 kya	'Comparable to Spy examples' Scothern, 1992 Hole on ventral surface, two intentionally bored holes on proximal and distal epiphyses	'Unlikely to be capable of producing sound' (Scothern, 1992, p. 115). Distal hole too close to the 'blow hole'; other is arbitrarily placed.	?	Cheynier, 1949; Brade, 1975; Brade, 1982; Scothern, 1992
	Solutrean	Badegoule, France	Flute?	Reindeer radius (*Rangifer tarandus*)	Solutrean. 21–19 kya	Fragmentary. One hole at distal end. Found in same level as above specimen	'Must be similarly rejected' (Scothern, 1992, p. 115)	?	Cheynier, n.d.; Scothern, 1992
	Solutrean	Badegoule, France	2 pipes	?	Solutrean. 21–19 kya		Questioned. 'May simply be a series of unrelated bone fragments' (Scothern, 1992, p. 115)	?	Da Veiga Ferreira and Cardosa, 1975; Scothern, 1992
	Solutrean	La Riera cave, Spain	Whistle?	Red deer tibia	Upper terminal Solutrean, C14	Artefact no. 76 Not in inventory	Questioned, though possibly originally able to	?	Scothern, 1992

Magdalenian	La Paloma, Asturias, Spain	?	Magdalenian? 18–12 kya?	Worked tibia of red deer, fragmentary, broken lengthways, only half of bone remains (Scothern, 1992, pp. 110–11)	function as small block and duct whistle (Scothern, 1992, p. 111)	Barandiaran, 1973; Scothern, 1992
Magdalenian		?		Perforated bone. Very narrow bore, and perforated through both surfaces. Uncertainty over date; may be as late as Neolithic	Unlikely, due to narrowness of bore and perforation through both sides of the bone (Scothern, 1992, p. 114)	?
Magdalenian	Liege Hohle, Enns Valley, Austria	Flute?	'Final Palaeolithic'	4 holes, 2 at either end	Questionable (Scothern, 1992)	Horusitzky, 1955; Scothern, 1992
Magdalenian	Mas-d'Azil, France	Pierced whistle/ haft?	Reindeer (*Rangifer tarandus*)	Magdalenian. 18–12 kya	Length: 45 mm	Musical function questioned. Possibly bone haft rather than sound-producer (Scothern, 1992, p. 121)	Pequart and Pequart, 1960; Scothern, 1992
Magdalenian	Molodova V, Ukraine	Flute, horn	Elk horn. Or	Magdalenian. C14 dated to	Length: 210 mm. 4 holes on	Lengthwise perforation only	?	Hausler, 1960; Megaw, 1960;

(Continued)

Appendix Table 2. Continued

Number	Period	Origin	Described as	Material (if known)	Details of Age	Description	Status & Comments	Location (if known)	Main References
				reindeer antler (Lucius, 1970) or moose antler (Megaw, 1968)	17,000 ± 1,400 BP (Hoffecker, 1988)	anterior surface, 2 on posterior, diameters: 5 mm × 2 mm, 6 mm × 3 mm, 2 mm × 2 mm and 2 mm × 4 mm on front, and 1.5-2 mm for the two on rear (Hahn annd Münzel, 1995; Turk and Kavur, 1997)	reaches as far as the fourth opening (Scothern, 1992, p. 120; Brade, 1982, p. 140); also, holes very small, so probably not functional (Megaw, 1968; Turk and Kavur, 1997), too randomly placed on posterior and too close together on anterior to influence sound (Scothern, 1992, p. 120). Very unlikely to have been intended as a sound-producer		Megaw, 1968; Lucius, 1970; Bibikov, 1978; Hoffecker, 1988; Scothern, 1992; Hahn and Münzel, 1995; Brade 1982
Magdalenian?		Molodova V, Ukraine	Flute?	?	'Late Magdalenian?' C14 dated to 11,900 ± 238 or 12,300 ± 140 BP	7 holes on one side and 2 on rear, hole diameter 2 mm	Age and functionality questioned by Scothern, 1992	?	Megaw, 1968; Scothern, 1992

'Upper Palaeolithic'	Bukovace Hohle, Croatia	Bone with bored hole	Bear	Unspecified	Hole diameter too small for functionality (Scothern, 1992) (quoted in Turk and Kavur, 1997, probably from Hoffecker, 1988)	?	Kormos, 1912; Brade, 1975; Scothern, 1992
'Upper Palaeolithic'	Mammuthohle bei Wiezchowie, Poland	Flute?	?	Unspecified	Single bored hole 'Similar to La Placard' (Scothern, 1992)	?	Seewald, 1934; Scothern, 1992
'Upper Palaeolithic'	Pekarna, Moravia, Czech Republic	2 Whistles?	Bird bone epiphyses	Unspecified	'No human agency can be conclusively demonstrated other than for the purposes of marrow extraction' (Scothern, 1992, p. 86) Questionable (Scothern, 1992) 2 specimens. 'Refuse from needle manufacture' (Scothern, 1992)	?	Absolon, 1936; Scothern, 1992
'Upper Palaeolithic'	Salzhofen, Toten Gebirge, Austria	Flute?	Bear upper thigh bone	Unspecified	Doubtful musical function (Scothern, 1992) 'Bored through dorsal surface for purpose of marrow extraction' (Scothern, 1992, p. 86) Doubtful musical function (Scothern, 1992)	?	Seewald, 1934; Mottl, 1950; Scothern, 1992

Appendix Table 2 presents items that have, at various times, been reputed to be possible flute-type sound-producers, but which have, since original publication, on the basis of further analysis, been subsequently deemed unlikely to be sound-producers. Appendix Table 2 features forty items.

Appendix Table 3. Inventory of Palaeolithic reputed phalangeal whistles

Number	Period	Origin	Type	Material	Details of Age	Description	Status & Comments	Location	Reference
W001-xM	Mousterian	La Quina, France	Phalangeal whistles?	Reindeer phalanges (*Rangifer tarandus*)	?	Numerous examples ('over twenty' from Mousterian layers, Scothern, 1992, p. 58) from Mousterian onwards. Pressure punctured, often with accompanying toothmarks	Doubtful—natural damage or carnivore activity more likely (Scothern, 1992, p. 58)	Some La Quina specimens in Museum of Archaeology and Anthropology, Cambridge	Martin, 1906, 1909; Seewald, 1934; Dauvois, 1989; Scothern, 1992
W002-xM	Mousterian	Prolom II, Crimea	Phalangeal whistles	*Saiga tartarica*	Mousterian, dated by lithic association	74 examples of pierced phalanges (41 from layer 2 represent 55.4% of total), Stepanchuk, 1993, p. 33		?	Stepanchuk, 1993
W003 A	Aurignacian	Aurignac, France	Phalangeal whistle	?	Aurignacian. 32–17 kya			?	Lartet and Christy, 1910; Dauvois, 1989; Scothern, 1992
W004-xA	Aurignacian	Bockstein Im Lonetal, Germany	Phalangeal whistle	?	Aurignacian. 32–17 kya	Several examples, late/Final Aurignacian and Magdalenian		?	Seewald, 1934; Scothern, 1992

W005 A	Aurignacian	Bourdeilles, Dordogne, France	Phalangeal whistle	?	Aurignacian. 32–17 kya		?	Seewald, 1934; Scothern, 1992
W006–007 A	Aurignacian	Castel-Merle, France	Phalangeal whistle	?	Aurignacian. 32–17 kya	2 examples, including one fragment	Musée du Périgord, Périgueux, France	Scothern, 1992
W008 A	Aurignacian	Gorge d'Enfer B, France	Phalangeal whistle	?	Aurignacian. 32–17 kya		?	Seewald, 1934; Dauvois, 1989; Scothern, 1992
W009 A	Aurignacian	Istállóskó, Hungary	Phalangeal whistle	Two reindeer phalanges (*Rangifer tarandus*)	Aurignacian II		?	Megaw, 1960; Scothern, 1992
W010–011 A	Aurignacian	Les Roches, France	2 phalangeal whistles	?	Aurignacian. 32–17 kya		?	Seewald, 1934; Scothern, 1992
W012 A	Aurignacian	Saint-Jean-de-Verges, French Pyrenees	Phalangeal whistle	Reindeer (*Rangifer tarandus*)	Aurignacian. 32–17 kya	Piercing appears man-made (cf. Dauvois, 1989, fig. 1, p. 6)	?	Dauvois, 1989
W013-x A	Aurignacian	Sirgenstein, Switzerland	Phalangeal whistles	Reindeer phalanges (*Rangifer tarandus*)	Late Aurignacian, Final Aurignacian, and Proto-Solutrean	Various specimens	?	Seewald, 1934; Scothern, 1992

(Continued)

Appendix Table 3. Continued

Number	Period	Origin	Type	Material	Details of Age	Description	Status & Comments	Location	Reference
W014 A	Aurignacian	Spy, Belgium	Perforated phalange/whistle?	?	Aurignacian. 32–17 kya	4.5 cm long	Questioned	Carpentier Collection, University of Liege	Otte, 1979; Scothern, 1992
W015–017 A	Aurignacian	Trou du Sureau, Montaigle, France	Phalangeal whistle	?	Mid-Aurignacian	Three examples		?	Seewald, 1934; Scothern, 1992
W018–019 G/P	Gravettian/Perigordian	Dolni Vestinice, Moravia, Czech Republic	Phalangeal whistle	?	Gravettian. 28–22 kya	Two examples		?	Scothern, 1992
W020 G/P	Gravettian/Perigordian	Pair-non-Pair, Gironde, France	Phalangeal whistle	?	Upper Perigordian/Gravettian. 28–22 kya			?	Daleau, 1881; Scothern, 1992
W021–026 G/P	Gravettian/Perigordian	Pavlov, Moravia, Czech Republic	6 phalangeal whistles	?	Gravettian. 28–22 kya	6 specimens, 5 of which produce a clear tone		?	Megaw, 1968; Scothern, 1992
W027 G/P	Gravettian/Perigordian	Pekarna, Moravia, Czech Republic	Phalangeal whistle	?	Gravettian. 28–22 kya	Pierced by 2 holes; produces C and G pitches	Fragmentary and 'likely to be the product of natural agencies' (Scothern, 1992, p. 61)	?	Absolon, 1936; Buchner, 1956; Harrison, 1978; Scothern, 1992
W028–029 G/P	Gravettian/Perigordian	Pekarna, Moravia,		?	Gravettian. 28–22 kya		Secondary damage such that difficult to	?	Absolon, 1936;

		Czech Republic	2 phalangeal whistles			identify agency responsible for piercing (Harrison, 1978)	Buchner, 1956; Harrison, 1978; Scothern, 1992
W030 G/P	Perigordian	Les Eyzies, France	Phalangeal whistle	Reindeer (*Rangifer tarandus*)	'from a "Perigordian" layer' (Harrison, 1978, p. 16). Listed by Scothern (1992) with W083 UP	'neatly made, circular hole in the distal end of anterior surface' (Harrison, 1978). This is unusual location for hole and not possible through natural damage	Musée Nationale de Prehistoire, Les Eyzies	Almost certainly an artefact (Harrison, 1978); Hole NOT natural damage	Harrison, 1978; Scothern, 1992
W031 G/P	Perigordian	Petit Puyrouseau, Dordogne, France	Phalangeal whistle	Reindeer (*Rangifer tarandus*)	From and 'Upper Perigordian' layer (Harrison, 1978)	Museum code: F534	Musée du Périgord, Périgueux	Damage identical to that of experimentally naturally pierced example	Harrison, 1978
W032 S	Solutrean	Le Mazarat, Dordogne, France	Phalangeal whistle	Reindeer (*Rangifer tarandus*)	'From a Solutrean level' (Harrison, 1978)	4.7 cm long. Small hole in lateral aspect of shaft (Harrison, 1978)	?	Almost certainly an artefact (Harrison, 1978); Hole NOT natural damage	Harrison, 1978; Scothern, 1992
W033–035 Ma	Magdalenian	Castel Merle, France	3 phalangeal whistles	?	Magdalenian. 18–12 kya	3 examples, perforated	Musée du Périgord, Périgueux	Highly worn, holes ill-defined, and irregular in shape (Scothern, 1992, p. 61)	Scothern, 1992

(*Continued*)

Appendix Table 3. Continued

Number	Period	Origin	Type	Material	Details of Age	Description	Status & Comments	Location	Reference
W036–037 Ma	Magdalenian	Castel Merle, France	2 phalangeal whistles	?	Magdalenian. 18–12 kya	2 examples, perforated	Have regular circular holes with V-shaped profile resulting from burin-type tool (p. 61)	Musée du Périgord, Périgueux	Scothern, 1992
W038–039 Ma	Magdalenian	Gourdan, Haute-Garonne, France	2 phalangeal whistles	?	Magdalenian. 18–12 kya			?	Seewald, 1934; Dauvois, 1989; Scothern, 1992
W040 Ma	Magdalenian	Goyet, Belgium	Phalangeal whistle	?	Magdalenian. 18–12 kya			?	Seewald, 1934; Scothern, 1992
W041 Ma	Magdalenian	Grotte de l'Homme, France	Phalangeal whistle	?	Magdalenian. 18–12 kya			?	Seewald, 1934; Scothern, 1992
W042 Ma	Magdalenian	Isturitz, French Pyrenees	Phalangeal whistle	Reindeer (*Rangifer tarandus*)	Magdalenian. 18–12 kya	2 specimens			Dauvois, 1989
W043 Ma	Magdalenian	Jankovich Cave, Hungary	Phalangeal whistle.	?	Magdalenian. 18–12 kya			?	Seewald, 1934; Scothern, 1992
W044 Ma	Magdalenian	La Madeleine, France	Phalangeal whistle	Reindeer (*Rangifer tarandus*)	Magdalenian. 18–12 kya	Large hole (38.5 sq. mm) in distal end of posterior surface	Feature 'regularly-shaped and definite blow-holes'	British Museum, London	Harrison, 1978;

					Audible with 100% reliability over 1.25 km (Harrison, 1978)	Scothern, 1992
W045 Ma	Magdalenian	La Madeleine, France	Phalangeal whistle	Reindeer (*Rangifer tarandus*)	Magdalenian. 18–12 kya	Small circular hole in mid-shaft position (Harrison, 1978)	(Scothern, 1992 p. 61). At least one of these examples is considered to be 'undoubtedly man-made' (though damaged after deposition) by Scothern (1992, p. 60). Almost certainly artefacts (Harrisson, 1978)	British Museum, London	Harrison, 1978; Scothern, 1992
W046–048 Ma	Magdalenian	La Madeleine, France	3 phalangeal whistles	Reindeer (*Rangifer tarandus*)	Magdalenian. 18–12 kya	3 specimens	Feature 'regularly-shaped and definite blow-holes' (Scothern, 1992 p. 61)	?	Harrison, 1978; Scothern, 1992
W049–051 Ma	Magdalenian	La Madeleine, France	3 phalangeal whistles	Reindeer (*Rangifer tarandus*)	Magdalenian. 18–12 kya	3 specimens (one of them is artefact 1938.34.271; Harrison, 1978)	Feature 'regularly-shaped and definite blow-holes' (Scothern, 1992 p. 61). Artefact 1938.34.271 has 'uneven, though obviously man-made' hole, with a tapered edge (Harrisson, 1978)	Pitt-Rivers Museum, Oxford	Harrison, 1978; Megaw, 1960; Scothern, 1992
W052 Ma	Magdalenian	Laugerie-Basse, France	Phalangeal whistle	?	Magdalenian. 18–12 kya	Pitt Rivers Mus. 1938.34.272 (Harrison, 1978)	'regularly-shaped and definite blow-holes' (Scothern, 1992, p. 61) 'Uneven but obviously man-made hole' with	Pitt-Rivers Museum, Oxford (Harrison, 1978)	Harrison, 1978; Dauvois, 1989; Scothern, 1992

(*Continued*)

Appendix Table 3. Continued

Number	Period	Origin	Type	Material	Details of Age	Description	Status & Comments	Location	Reference
W053 Ma	Magdalenian	Laugerie-Basse, France	Phalangeal whistle	?	Magdalenian. 18–12 kya	Pitt Rivers Mus. 1909.4.1 (Harrison, 1978)	tapered edge (Harrison, 1978, p. 14) 'regularly-shaped and definite blow-holes' (Scothern, 1992, p. 61). 'Extremely well-preserved', edges of hole cut at right-angles to bone surface (Harrison, 1978, p. 14)	Pitt-Rivers Museum, Oxford (Harrison, 1978)	Harrison, 1978; Dauvois, 1989; Scothern, 1992
W054–055 Ma	Magdalenian	Laugerie-Basse, France	2 phalangeal whistles	?	Magdalenian. 18–12 kya	2 specimens. 4.9 cm and 5.2 cm long	'regularly-shaped and definite blow-holes' (Scothern, 1992, p. 61)	Musée du Périgord, Périgueux	Dauvois, 1989; Scothern, 1992
W056 Ma	Magdalenian	Laugerie-Basse, France	Phalangeal whistle	?	Magdalenian. 18–12 kya		Smith (1911) suggested it was possible that the hole was made by a hyaena. Harrison (1978) suggests that the damage is depositional, as phalange unlikely to survive the attentions of hyaena	British Museum, London	Smith, 1911; Harrison, 1978; Dauvois, 1989; Scothern, 1992
W057 Ma	Magdalenian	Laugerie-Haute, France	Phalangeal whistle	?	Magdalenian. 18–12 kya	Phalanx fragment, 3.3 cm long	Clearly an intentionally produced hole (cf. Dauvois, 1989, fig. 2, p. 9)	Musée du Périgord, Périgueux	Dauvois, 1989; Scothern, 1992

W058 Ma	Magdalenian	Le Chaffaud, France	Phalangeal whistle	?	Magdalenian. 18–12 kya		?	Seewald, 1934; Scothern, 1992
W059-x Ma	Magdalenian	Le Placard, France	Phalangeal whistles	?	Magdalenian I, II, III	Several examples	?	Seewald, 1934; Dauvois, 1989; Scothern, 1992
W060 Ma	Magdalenian	Les Eyzies, France	Phalangeal whistle	?	Magdalenian. 18–12 kya		British Museum, London	Feature 'regularly-shaped and definite blow-holes' (Scothern, 1992, p. 61)	Scothern, 1992
W061 Ma	Magdalenian	Lussac-les-Châteaux, France	Phalangeal whistle	Reindeer (*Rangifer tarandus*)	Magdalenian. 18–12 kya		?	Dubious-looking piercing; appears natural (cf. Dauvois, 1989)	Dauvois, 1989
W062 Ma	Magdalenian	Mâcon, France	Phalangeal whistle	Horse and reindeer (*Rangifer tarandus*)	Magdalenian. 18–12 kya		?		Seewald, 1934; Scothern, 1992
W063 Ma	Magdalenian	Mas-d'Azil, France	Phalangeal whistle	Reindeer (*Rangifer tarandus*)	Magdalenian. 18–12 kya		?		Seewald, 1934; Dauvois, 1989; Scothern, 1992
W064–066 Ma	Magdalenian	Petersfels, Germany	3 phalangeal whistles	?	Magdalenian. 18–12 kya	3 specimens	?		Seewald, 1934 Scothern, 1992

(*Continued*)

Appendix Table 3. Continued

Number	Period	Origin	Type	Material	Details of Age	Description	Status & Comments	Location	Reference
W067-x Ma	Magdalenian	Schaffhausen, Switzerland	Phalangeal whistles and pipes	Various, including horse and fox	Magdalenian. 18–12 kya	Several specimens		?	Seewald, 1934; Scothern, 1992
W068-x Ma	Magdalenian	Schussenreid, Genrmany	Phalangeal whistles	Reindeer (*Rangifer tarandus*)	Magdalenian. 18–12 kya	Several specimens		?	Seewald, 1934; Scothern, 1992
W069-x Ma	Magdalenian	Hohler Stein bei Kallenhardt, Germany	Phalangeal whistles	?	Magdalenian. 18–12 kya, and Tardenoisian.	Several specimens		?	Seewald, 1934; Scothern, 1992
W070 UP	Late Upper Palaeolithic	Sun Hole, Somerset, England	Phalangeal whistle	?	15–8.3 kya	Damaged	Hole damage typical of piercing occurring after deposition (Harrison, 1978)	University of Bristol Spelaeological Society Collection	Harrison, 1978; Scothern, 1992
W071 UP	'Upper Palaeolithic'	Abri Blanchard, France	Phalangeal whistle	?	Unspecified			?	Harrison, 1978; Dauvois, 1989; Scothern, 1992
W072 UP	Upper Palaeolithic	Banwell Hole Cave, England	Phalangeal whistle	Reindeer (*Rangifer tarandus*)	'Pleistocene'	Produces tone with fundamental freq. 3586 ±18Hz Would produce 2nd and 3rd harmonics of 10,759 and 17,900	Piercing almost certainly the product of natural depositional process	University of Bristol Spelaeological Society Collection	Harrison, 1978

					Hz respectively (Harrison, 1978)		
W073 UP	'Upper Palaeolithic'	Bruniquel, Tarn-et-Garonne, France	Phalangeal whistle	?	Unspecified		?	Seewald, 1934; Scothern, 1992
W074 UP	'Upper Palaeolithic'	Bruniquel, Tarn-et-Garonne, France	Phalangeal whistle	?	Unspecified		?	Harrison, 1978; Dauvois, 1989; Scothern, 1992
W075–079 UP	'Upper Palaeolithic'	Castel Merle, France	5 phalangeal whistles	?	Unspecified	5 perforated specimens, one with 2 holes	Musée du Périgord, Périgueux	Scothern, 1992
W080 UP	Upper Palaeolithic	Croze de Tayac, Périgord, France	Phalangeal whistle	Not specified	Reindeer (*Rangifer tarandus*)		?	Dauvois, 1989
W081 UP	Upper Palaeolithic	Gargas, Pyrenees	Phalangeal whistle	Not specified	Reindeer (*Rangifer tarandus*)	Fragmentary	?	Dauvois, 1989
W082 UP	Upper Palaeolithic	Laugerie-Basse, France	Phalangeal whistle	Not specified	Reindeer (*Rangifer tarandus*)		?	Dauvois, 1989
W083 UP	'Upper Palaeolithic'	Les Eyzies, France	Phalangeal whistle	?	Unspecified		?	Harrison, 1978; Scothern, 1992
W084–085 UP	'Upper Palaeolithic'	Maisières Canal, Belgium	2 phalangeal whistles	?	Unspecified		I.R.Sc.N.	Otte, 1979; Scothern, 1992
W086–87 UP	Upper Palaeolithic	Mas-d'Azil, Pyrenees	2 phalangeal whistles	Not specified	Reindeer (*Rangifer tarandus*)	2 specimens	?	Dauvois, 1989

(Continued)

Appendix Table 3. Continued

Number	Period	Origin	Type	Material	Details of Age	Description	Status & Comments	Location	Reference
W088–092 UP	'Upper Palaeolithic'	Périgord, France	5 phalangeal whistles	?	Unspecified	5 specimens, few details given		?	Engel, 1874; Scothern, 1992
W093 UP	Upper Palaeolithic	Petersfels, Germany	Phalangeal whistle	Not specified	Reindeer (*Rangifer tarandus*)	Small circular hole in mid-shaft position (Harrison, 1978)	Almost certainly an artefact (Harrison, 1978)	?	Coles and Higgs, 1969; Harrison, 1978
W094 UP	'Upper Palaeolithic'	Petit Puyrousseau, France	Phalangeal whistle	?	Unspecified			?	Harrison, 1978; Scothern, 1992
W095 UP	'Upper Palaeolithic'	Salzhofen, Toten gebirge, Austria	Phalangeal whistle	?	Unspecified			?	Seewald, 1934; Scothern, 1992
W096 UP	Upper Palaeolithic	Solutré, France	Phalangeal whistle	Not specified	Reindeer (*Rangifer tarandus*)			?	Dauvois, 1989
W097 UP	Upper Palaeolithic	Trilobite, France	Phalangeal whistle	Not specified	Reindeer (*Rangifer tarandus*)			?	Dauvois, 1989

Appendix Table 3 presents items that have, in various reports, been reputed to be possible phalangeal whistle sound-producers; where comments about their likely status are available in the literature, these are noted. More than 186 examples are recorded (several entries are reported as 'multiple examples').

References

Absolon, C. (1936) Les Flûtes paléolithiques de l'Aurignacien et du Magdelénian de Moravie (analyse musicale et ethnologique comparative, avec démonstrations). *Congrès Préhistorique de France.* II.XI session, Toulouse and Foix: 770–84.

Addis, L. (1999) *Of Mind and Music.* Ithaca, NY: Cornell University Press.

Adolphs, R. (2003) Cognitive neuroscience of human social behaviour. *Nature Reviews, Neuroscience* 4, 165–78.

Ahuja, A. (2000) Why we are touched by the sound of music. *The Times (Science Supplement),* 23 Feb. 2000, 43.

Aiello, L. (1996) Terrestriality, bipedalism and the origin of language. *Proceedings of the British Academy* 88, 269–89.

Aiello, L., and Dunbar, R. (1993) Neocortex size, group size, and the evolution of language. *Current Anthropology* 34, 184–93.

Aitken, P. (1981) Cortical control of conditioned and spontaneous vocal behavior in rhesus-monkeys. *Brain and Language* 13, 171–84.

Aitkin, L., Merzenich, M., Irvine, D., Clarey, J., and Nelson, J. (1986) Frequency representation in auditory cortex of the common marmoset (*Callithrix jacchus jacchus*). *Journal of Comparative Neurology* 252, 175–85.

Albrecht, G., Holderman C.-S., Kerig, T., Lechterbeck, J., and Serangeli, J. (1998) 'Flöten' aus Bärenknochen—die frühesten Musikinstrumente? *Archäologisches Korrespondenzblatt* 28, 1–19.

Albrecht, G., Holderman, C.-S., and Serangeli, J. (2001) Towards an archaeological appraisal of specimen no. 652 from Middle-Palaeolithic level D/(layer 8) of the Divje babe I. *Archeoloski vestnik* 52, 11–15.

Alcaro, A., Huber, R., and Panksepp, J. (2007) Behavioural functions of the mesolimbic dopaminergic system: an affective neuroethological perspective. *Brain Research Reviews* 56, 283–321.

Alcock, K., Passingham, R., Watkins, K., and Vargha-Khadem, F. (2000) Pitch and timing abilities in inherited speech and language impairment. *Brain and Language* 75, 34–46.

Alebo, L. (1986) Manufacturing of drumskins and tendon strings for prehistoric musical instruments. In C. Lund (ed.) *The Second Conference Of the ICTM Study Group On Music Archaeology,* i. *General Studies.* Stockholm: Royal Swedish Academy of Music, 41–8.

Alemseged, Z., Spoor, F., Kimbel, W., Bobe, R., Geraads, D., Reed, D., and Wynn, J. (2006) A juvenile early hominin skeleton from Dikika, Ethiopia. *Nature* 443, 296–301.

Alexander, R., and Jayes, A. (1980) Fourier analysis of forces exerted in walking and running. *Journal of Biomechanics* 13, 383–90.

Allsworth-Jones, P. (1986) *The Szeletian and the Transition from Middle to Upper Palaeolithic in Central Europe.* Oxford: Oxford University Press.

Altenmüller, E. (2003) How many music centres are there in the brain? In I. Peretz and R. Zatorre (eds.) *The Cognitive Neuroscience of Music.* Oxford: Oxford University Press, 346–53.

Álvarez, R., and Siemens, L. (1988) The lithophonic use of large natural rocks in the prehistoric canary islands. In E. Hickman and D. Hughes (eds.) *The Archaeology Of Early Music Cultures; Third International Meeting of the ICTM Study Group On Music Archaeology.* Bonn: Verlag für systematische Musikwissenschaft, 1–10.

Andersson, B., and Ulvaeus, B. (1978). Thank you for the music. *Abba: The Album.* London: PolyGram Records.

Andrews, P., and Fernández-Jalvo, Y. (1997) Surface modifications of the Sima de los Huesos fossil humans. *Journal of Human Evolution* 33, 191–217.

Anonymous (2000) Music in our bones. *The Times,* 21 Feb.

Arensberg, B. (1989) Anatomy of Middle Palaeolithic populations in the Middle East. In P. Mellars and C. Stringer (eds.) *The Human Revolution: Behavioural and Biological Perspectives in the Origins of Modern Humans.* Edinburgh: Edinburgh University Press, 165–71.

Arensberg, P., Schepartz, L., Tillier, A., VanDerMeersch, B., and Rak, Y. (1990) A reappraisal of the anatomical basis for speech in middle Palaeolithic hominids. *American Journal of Physical Anthropology* 83, 137–46.

Armstrong, D., Stokoe, W., and Wilcox, S. (1994) Signs of the origins of syntax. *Current Anthropology* 35, 349–68.

Arsuaga, J. (1997) Sima de los Huesos (Sierra de Atapuerca, Spain): the site. *Journal of Human Evolution* 33, 109–27.

Arsuaga, J., Martinez, I., Lorenzo, C., Gracia, A., Muñoz, A., Alonso, O., and Gallego, J. (1999) The human cranial remains from Gran Dolina lower Pleistocene site (Sierra de Atapuerca, Spain). *Journal of Human Evolution* 37, 431–57.

Ayotte, J., Peretz, I., and Hyde, K. (2002) Congenital amusia: A group study of adults afflicted with a music-specific disorder. *Brain* 125, 238–51.

Baddeley, A., and Hitch, G. (1974) Working Memory. In G. Bower (ed.) *Recent Advances in Learning and Memory.* New York: Academic Press, 47–90.

Baghemil, B. (1988) The morphology and phonology of Katajjait (Inuit Throat-Games). *Canadian Journal of Linguistics* 33, 1–58.

Bahn, P. (1983) Late Pleistocene economies in the French Pyrenees. In G. Bailey (ed.) *Hunter-Gatherer Economy In Prehistory: A European Perspective.* Cambridge: Cambridge University Press, 167–85.

Bahn, P., and Vertut, J. (1997) *Journey Through the Ice Age.* London: Weidenfeld & Nicolson.

Bahuchet, S. (1999) Aka Pygmies. In R. Lee and R. Daly (eds.) *The Cambridge Encyclopedia of Hunters and Gatherers.* Cambridge: Cambridge University Press, 190–4.

Balkwill, L. and Thompson, W. (1999) A cross-cultural investigation of the perception of emotion in music: psycho-physical and cultural cues. *Music Perception* 17, 43–64.

Barac, V. (1999) From primitive to pop: foraging and post-foraging hunter-gatherer music. In R. Lee and R. Daly (eds.) *The Cambridge Encyclopedia of Hunters and Gatherers.* Cambridge: Cambridge University Press, 434–50.

Barandiaran, I. (1973) *Arte Mueble del Paleolítico Cantábrico.* Universidad de Zaragoza Monografías Arqueológicas XIV. Saragossa: Universidad de Zaragoza, 215.

Barkow, J., Cosmides, L., and Tooby, J. (eds.) (1992) *The Adapted Mind.* Oxford: Oxford University Press.
Barrett, L., Dunbar, R., and Lycett, J. (2001) *Human Evolutionary Psychology.* Basingstoke: Palgrave Macmillan.
Bastiani, G., and Turk, I. (1997) Appendix: results from the experimental manufacture of a bone flute with stone tools. In I. Turk (ed.) *Mousterian 'Bone Flute'.* Ljubljana: Znanstvenoraziskovalni Center Sazu, 176–8.
Bayer, J. (1929) Die Olschewa Kultur. *Eiszeit und Urgeschichte* 6, 83–100.
Beaudry, N. (1978) Towards transcription and analysis of Inuit throat-games: macrostructure. *Ethnomusicology* 22, 261–73.
Belin, P., Fecteau, S., and Bédard, C. (2004) Thinking the voice: neural correlates of voice perception. *Trends in Cognitive Sciences* 8, 129–35.
Benade, A. (1990) *Fundamentals of Musical Acoustics.* New York: Dover.
Benson, D. (1985) Language in the left hemisphere. In D. Benson and E. Zaidel (eds.) *The Dual Brain.* New York: Guildford Press, 193–203.
Bergeson, T. (2002) Perspectives on Music and Music Listening in Infancy. PhD thesis, University of Toronto.
Bergounioux, F., and Glory, A. (1952) *Les Premiers Hommes: Précis d'anthropologie préhistorique.* Paris: Didier, 289–95, fig. 151.
Bermejo, M., and Omedes, A. (1999) Preliminary vocal repertoire and vocal communication of wild bonobos (*Pan paniscus*) at Lilungo (Democratic Republic of Congo). *Folio Primatologica* 70, 328–57.
Bermúdez de Castro, J., and Martinón-Torres, M. (2004) Paleodemography of the Atapuerca-Sima de Los Huesos hominin sample: a revision and new approaches to the paleodemography of the European Middle Pleistocene population. *Journal of Anthropological Research* 60, 5–26.
Bertoncini, J., Morais, J., Bijeljac-Babic, R., McAdams, S., Peretz, I., and Mehler, J. (1989) Dichotic perception and laterality in neonates. *Brain and Language* 37, 591–605.
Besson, M., and Schön, D. (2003) Comparisons between language and music. In I. Peretz and R. Zatorre (eds.) *The Cognitive Neuroscience of Music.* Oxford: Oxford University Press, 269–93.
Best, C., Hoffman, H., and Glanville, B. (1982) Development of infant ear asymmetries for speech and music. *Perceptual Psychophysics* 31, 75–85.
Bettinger, R. (1991) *Hunter-Gatherers: Archaeological and Evolutionary Theory.* London: Plenum Press.
Bhudhara Das (2006) Sacred sound. In C. Mills (ed.) *The School of Braja: Indian Devotional Music.* Martinez CA: Gaudiya Vedanta Publications, 16–37.
Bibikov, S. (1978) A stone age orchestra. In D. Hunter and P. Whitten (eds.) *Readings in Physical Anthropology and Archaeology.* London: Harper & Row, 134–48.
Bispham, J. (2006) Rhythm in music: What is it? Who has it? And why? *Music Perception* 24, 125–34.
Blacking, J. (1969/1995) The value of music in human experience. *The 1969 Yearbook of the International Folk Music Council.* Republished as Blacking (1995), chapter 1.
Blacking, J. (1995) *Music, Culture and Experience,* ed. P. Bohlman and B. Nettl. Chicago: University of Chicago Press.

Blake, E. (2011) 'Stone Tools' as Portable Sound-Producing Objects in Upper Palaeolithic Contexts: The Application of an Experimental Study. PhD thesis, University of Cambridge.

Blood, A., and Zatorre, R. (2001) Intensely pleasurable responses to music correlate with activity in brain regions implicated in reward and emotion. *Proceedings of the National Academy of Sciences of the USA*. 98, 11818–23.

Blount, B. (1994) Comment on Armstrong, Stokoe, and Wilcox. *Current Anthropology* 35, 358–9.

Bogen, J. (1985) The dual brain: some historical and methodological aspects. In D. Benson and E. Zaidel (eds.) *The Dual Brain*. New York: Guildford Press, 27–43.

Bohlman, P. (2002) *World Music*. Oxford: Oxford University Press.

Boivin, N. (2004) Rock art and rock music: petroglyphs of the South Indian Neolithic. *Antiquity* 78, 38–53.

Boivin, N., Brumm, A., Lewis, H., Robinson, D., and Korisettar, R. (2007) Sensual, material, and technological understanding: exploring prehistoric soundscapes in south India. *Journal of the Royal Anthropological Institute* 13, 267–94.

Borchgrevink, H. (1980) Cerebral lateralisation of speech and singing after intracarotid Amytal injection. In M. Taylor Sarno and O. Hooks (eds.) *Aphasia: Assessment and Treatment*. Stockholm: Almkvist & Wiksell, 186–91.

Borchgrevink, H. (1982) Prosody and musical rhythm are controlled by the speech hemisphere. In M. Clynes (ed.) *Music, Mind and Brain*. New York: Plenum Press, 151–7.

Borchgrevink, H. (1991) Prosody, musical rhythm, tone pitch and response initiation during Amytal hemisphere anaesthesia. In J. Sundberg, L. Nord, and R. Carlson (eds.), *Music, Language, Speech and Brain*. Basingstoke: MacMillan Press, 327–43.

Borg, E., and Counter, S. (1989) The middle-ear muscles. *Scientific American* 261 (Aug.), 63–8.

Bourdier, F. (1967) *Préhistoire de France*. Paris: Flammarion editions, 412, fig. 151.

Bowles, S. and Gintis, H. (1998) The moral economy of community: structured populations and the evolution of pro-social norms. *Evolution and Human Behaviour* 19, 3–25.

Boyd, R., and Richerson, P. (2005) *The Origin and Evolution of Cultures*. Oxford: Oxford University Press.

Brade, C. (1975) Die mittelalterlichen kernspaltflöten Mittel- und Nord Europas: Ein Beitrag zur Überlieferung prähistorischer und zur Typologie mittelalterlicher Kernspaltflöten. *GottingerSchriften zür vor und Frügeschichte* 14. Neumunster: Wacholz.

Brade, C. (1982) The prehistoric flute: did it exist? *Galpin Society Journal* 35, 138–50.

Bramble, D., and Lieberman, D. (2004) Endurance running and the evolution of Homo. *Nature* 432, 345–52.

Brandt, A., Gebrian, M., and Slevc, L. (2012) Music and early language acquisition. *Frontiers in Psychology: Auditory Cognitive Neuroscience* 3 (Article 327), 1–17.

Bråten, S. (2007) *On Being Moved: From Mirror Neurons to Empathy*. Amsterdam: John Benjamins.

Bråten, S., and Trevarthen, C. (2007) Prologue: from infant intersubjectivity and participant movements to simulations and conversations in cultural common sense. In Bråten (2007), 21–34.
Breen, M. (1994) I have a dreamtime: aboriginal music and black rights in Australia. In S. Broughton, M. Ellingham, D. Muddyman, and R. Trillo (eds.) *World Music: The Rough Guides*. London: Rough Guides Limited, 655–62.
Brodar, S., and Bayer, J. (1928) Die Potocka Zijalka, eine Hochstation der Aurignacschwankung in den Ostalpen. *Praehistorica* 1, 3–13.
Brodar, S., and Brodar, M. (1983) *Potocka zijalka: Visokoalpska postaja aurignacienskih lovcev: Potocka zijalka, eine hochalpine Aurignacjägerstation*. Ljubljana: Slovenska akademija znanosti in umetnosti.
Broughton, S., Ellingham, M., Muddyman, D., and Trillo, R. (1994) *World Music: The Rough Guide*. London: Rough Guides Limited.
Brown, S. (2000) The 'musilanguage' model of music evolution. In N. Wallin, B. Merker, and S. Brown (eds.) *The Origins of Music*. Cambridge, MA: MIT Press, 271–300.
Brown, S., Martinez, M., and Parsons, L. (2006a) Music and language side by side in the brain: a PET study of the generation of melodies and sentences. *European Journal of Neuroscience* 23, 2791–2803.
Brown, S., Martinez, M., and Parsons, L. (2006b) The neural basis of human dance. *Cerebral Cortex* 16, 1157–67.
Brown, S., Merker, B., and Wallin, N. (2000) An introduction to evolutionary musicology. In N. Wallin, B. Merker, and S. Brown (eds.) *The Origins of Music*. Cambridge, Mass.: MIT Press, 3–24.
Brugge, J. (1985) Patterns of organization in auditory cortex. *Journal of the Acoustical Society of America* 78, 353–9.
Brust, J. (2003) Music and the neurologist: a historical perspective. In I. Peretz and R. Zatorre (eds.) *The Cognitive Neuroscience of Music*. Oxford: Oxford University Press, 181–91.
Buccino, G., Binkofski, F., Fink, G., Fadiga, L., Fogassi, L., Gallese, V., Seitz, R., Zilles, K., Rizzolatti, G., and Freund, H.-J. (2001) Action observation activates premotor and parietal areas in a somatotopic manner: an fMRI study. *European Journal of Neuroscience* 13, 400–4.
Buchanan, T., Lutz, K., Mirzazade, S., Specht, K., Shah, N., Zilles, K., and Jancke, L. (2000) Recognition of emotional prosody and verbal components of spoken language: an fMRI study. *Cognitive Brain Research* 9, 227–38.
Buchner, A. (1956) *Musical Instruments Through The Ages*, trans. I. Irwin. London: Spring Books.
Budil, I. (1994) A functional reconstruction of the supralaryngeal vocal tract of the fossil hominid from Petralona. In J. Wind, A. Jonker, R. Allot, and L. Rolfe (eds.) *Studies In Language Origins 3*. Amsterdam: Benjamins, 1–19.
Buisson, D. (1990) Les Flûtes paléolithiques d'Isturitz. *Bulletin de la Société Préhistorique Française* 87, 420–33.
Burghardt, G. (2005) *The Genesis of Animal Play*. Cambridge, MA: MIT Press.
Burling, R. (1993) Primate calls, human language, and nonverbal communication. *Current Anthropology* 34, 25–53.

Burnham, D., Kitamura, C., and Vollmer-Conna, U. (2002) What's new, pussycat? On talking to babies and animals. *Science* 296, 1435.

Burns, E. (1999) Intervals, scales and timing. In D. Deutsch (ed.) *The Psychology of Music* (2nd edn., San Diego CA: Academic Press, 215–64.

Bundo, D. (2002) Mode of participation in singing and dancing performances among the Baka. In *Proceedings of the 9th International Conference On Hunting and Gathering Societies.* Available online at http://www.abdn.ac.uk/chags9/Bundo.htm.

Butcher C., and Goldin-Meadow, S. (2000) Gesture and the transition from one- to two-word speech: when hand and mouth come together. In D. McNeill (ed.) *Language and Gesture.* Cambridge: Cambridge University Press, 235–58.

Butler, D. (1989) Describing the perception of tonality in music: a critique of the tonal heirarchy theory and proposal for a theory of intervallic rivalry. *Music Perception* 6, 219–42.

Byrne, R. (1995) *The Thinking Ape: Evolutionary Origins of Intelligence.* Oxford: Oxford University Press.

Caldwell, D. (2009) Palaeolithic whistles or figurines? A preliminary survey of prehistoric phalangeal figurines. *Rock Art Research* 26, 65–82.

Callaghan, C. (1994) Comment on Armstrong, Stokoe & Wilcox. *Current Anthropology* 35, 359–60.

Calvin, W. (1996) *How Brains Think.* London: Weidenfeld & Nicolson.

Campbell, M., and Greated, C. (1987) *The Musician's Guide to Acoustics.* London: Dent & Sons.

Cantalupo, C., and Hopkins, W. (2001) Asymmetric Broca's area in great apes. *Nature* 414, 505.

Capitan, L., and Peyrony, D. (1912) Station préhistorique de La Ferrassie, commune de Savignac-Deu-Bugue (Dordogne). *Revue Anthropologique* 22, 76–99.

Carlson, N. (1994) *The Physiology of Behavior.* Boston: Allyn & Bacon.

Carter, R. (1998) *Mapping the Mind.* London: Weidenfeld & Nicolson.

Cartwright, J., and Davies, R. (2000) *Evolution and Human Behaviour: Darwinian Perspectives on Human Nature.* Basingstoke: Palgrave Macmillan.

Cassoli, P., and Tagliacozzo, A. (1997) Butchering and cooking of birds in the Palaeolithic site of Grotta Romanelli (Italy). *International Journal of Osteoarchaeology* 7, 303–20.

Cernys, A. (1955) Une flûte paléolitique. *Kraktie soobscentija Instituta Istorii materialnoi kyltyry, Moskva* 59, 129–30.

Cernys, A. (1956) Nouvelles découvertes dans le Fuilles de la Station de Molodova 5 sur le Dneistr. *Kraktie soobscentija Instituta Istorii materialnoi kyltyry, Moskva* 63, 150–2, fig. 70.

Chase, P. (1989) How different was Middle Palaeolithic subsistence? A zooarchaeological perspective on the Middle to Upper Palaeolithic transition. In P. Mellars and C. Stringer (eds.) *The Human Revolution: Behavioural and Biological Perspectives in the Origins of Modern Humans.* Edinburgh: Edinburgh University Press, 321–37.

Chase, P. (1990) Sifflets du paléolithique moyen(?): Les Implications d'un coprolite de coyote naturel. *Bulletin de la Société Préhistorique Française* 87, 165–7.

Chase, P., and Nowell, A. (1998) Taphonomy of a suggested Middle Palaeolithic bone flute from Slovenia. *Current Anthropology* 39, 549–53.

Chen, W., Kato, T., Zhu, X., Adriany, G., and Ugurbil, K. (1996) Functional activation mapping of human brain during music imagery processing. *NeuroImage* 3, S205–6.
Cheynier, A. (1949) *Badegoule, station Solutreéne et Protomagdalénienne*. Archives de l'Institute de Paléontologie Humaine 23. Paris: Institute de Paléontologie Humaine.
Cheynier, A. (1963) *La Caverne de Pair-non-Pair, Gironde. Fouilees de Francois Daleau*. Bordeaux: Societé Archéologique de Bordeaux, fig. 22.
Cheynier, A. (n.d.) *Comment vivait l'homme des cavernes à l'âge du renne?* Paris: Arnoux, 280, fig. 58; table 1.
Clayton, M., Sager, R., and Will, U. (2005) In time with the music: the concept of entrainment and its significance for ethnomusicology. *European Meetings in Ethnomusicology* 11, 3–142.
Clegg, M. (2012) The evolution of the human vocal tract: specialized for speech? In N. Bannan (ed.) *Music, Language and Human Evolution*. Oxford: Oxford University Press, 58–80.
Clegg, M., and Aiello, L. (2000) Paying the price for speech? An analysis of mortality statistics for choking on food. *American Journal of Physical Anthropology* 111 suppl. 30, 126.
Clynes, M. (1977) *Sentics: the Touch of Emotions*. New York: Souvenir Press.
Clynes, M. (1982) *Music, Mind and Brain: The Neuropsychology of Music*. New York: Plenum Publishing Corporation.
Coles, J., and Higgs, E. (1969) *The Archaeology of Early Man*. London: Faber & Faber.
Collins, D. (1975) Early man. In D. Collins (ed.) *Origins of Europe*. London: Allen & Unwin, 19–125.
Collins, D. (1986) *Palaeolithic Europe: A Theoretical and Systematic Study*. Tiverton: Clayhanger.
Conard, N., and Malina, M., (2006) Schmuck und vielleicht auch Musik am Vogelherd bei Niederstotzingen-Stetten ob Lontal, Kreis Heidenheim. *Archäologische Ausgrabungen in Baden-Württemberg 2005*, 21–5.
Conard, N., Malina, M., and Münzel, S. (2009) New flutes document the earliest musical tradition in southwestern Germany. *Nature* 460, 737–40.
Conard, N., Malina, M., Münzel, S., and Seeberger, F. (2004) Eine Mammutelfenbeinflöte aus dem Aurignacien des Geissenklösterle: Neue Belege für eine Musikalische Tradition im Frühen Jungpaläolithikum auf der Schwäbischen Alb. *Archäologisches Korrespondenzblatt* 34, 447–62.
Coolidge, F., and Wynn, T. (2007) The working memory account of Neandertal cognition—how phonological storage capacity may be related to recursion and the pragmatics of modern speech. *Journal of Human Evolution* 52, 707–10.
Coolidge, F., and Wynn, T. (2009) *The Rise of Homo sapiens: The Evolution of Modern Thinking*. Oxford: Wiley-Blackwell.
Corballis, M. (1992) On the evolution of language and generativity. *Cognition* 44, 197–226.
Cousté, R., and Krtolitza, Y. (1961) La Flûte paléolithique de l'Abri Lespaux, à Saint-Quentin-de-Baron (Gironde). *Bulletin de la Société Préhistorique Français* 58, 28–30.
Cousté, R., and Krtolitza, Y. (1965) L'Abri Lespaux (Cmn de Saint-Quentin-de-Baron) et le question du Périgordien en Gironde: Note preliminaire. *Revue Historique et Archéologique du Libournais* 116, 47–54.

Crelin, E. (1987) *The Human Vocal Tract: Anatomy, Function, Development and Evolution*. New York: Vantage.
Cross, I. (1999a) Is music the most important thing we ever did? Music, development and evolution. In Suk Won Yi (ed.) *Music, Mind and Science*. Seoul: Seoul National University Press, 10–39.
Cross, I. (1999b) *Lithoacoustics—Music in Stone*. Unpublished preliminary study reported on web page, http://www.mus.cam.ac.uk/~ic108/lithoacoustics/
Cross, I. (2001) Music, mind and evolution. *Psychology of Music* 29, 95–102.
Cross, I. (2003a) Music and Emotion: lecture notes, available online at http://www.mus.cam.ac.uk/~ic108/IBMandS/emotion/musicemotion.html
Cross, I. (2003b) Music and biocultural evolution. In M. Clayton and T. Herbert (eds.) *The Cultural Study of Music: A Critical Introduction*. London: Routledge, 19–30.
Cross, I. (2003c) Music and evolution: consequences and causes. *Contemporary Music Review* 22, 79–89.
Cross, I. (n.d.) *The 'Lithoacoustics' Project: Musical behaviours and the Archaeological Record; An Experimental Study*. Cambridge: Faculty of Music.
Cross, I., and Morley, I. (2009) The evolution of music: theories, definitions and the nature of the evidence. In S. Malloch and C. Trevarthen (eds.) *Communicative Musicality*. Oxford: Oxford University Press, 61–81.
Cross, I., Zubrow, E., and Cowan, F. (2002) Musical behaviours and the archaeological record: a preliminary study. In J. Mathieu (ed.) *Experimental Archaeology. British Archaeological Reports International Series* 1035, 25–34.
Curnoe, D., and Tobias, P. (2006) Description, new reconstruction, comparative anatomy, and classification of the Sterkfontein Stw 53 cranium, with discussions about the taxonomy of other southern African early Homo remains. *Journal of Human Evolution* 50, 36–77.
Da Veiga Ferreira, O., and Cardosa, J. (1975) Flauta, chamariz ou negaza de caca de osso encontrada no Castro de Liceia (Barcarena). *Boletim Cultural da Junta Distrital de Lisboa*, III. series, no. 81, 3–9.
Daleau, F. (1881) La Grotte Pair-non-Pair, Commune de Marcamps (Gironde). *Congrès de l'Association Française pour l'Avancement des Sciences*, 10th Session, Algiers, 755.
Daleau, F. (1884–5) *Excursions: 6 Juillet 1884–Mai 1885*. MS of Musée d'Aquitaine, Bordeaux, 99–100.
Daleau, F. (1885) Presentation d'os travaillés a l'époque paleólithique. *Congrès de l'Association Française pour l'Avancement des Sciences*, 14th Session, Grenoble, 161.
Daleau, F. (1896) Les Gravures sur rocher de la Caverne de Pair-non-Pair. *Société Archéologique de Bordeaux* 21, 235–50, plates VI and XI.
Dams, L. (1984) Preliminary Findings at the 'Organ Sanctuary' in the Cave of Nerja, Malaga, Spain. *Oxford Journal Of Archaeology* 3, 1–14.
Dams, L. (1985) Palaeolithic Lithophones: Descriptions and Comparisons. *Oxford Journal of Archaeology* 4, 31–46.
Daniel, H. (1990) The vestibular system and language evolution. In J. Wind, E. Pulleybank, E. de Groler, and G. Bichakjian (eds.) *Studies in Language Origins*, i, Amsterdam: Benjamins, 257–71.
Darwin, C. (1871) *The Descent of Man and Selection in Relation to Sex*, London: John Murray.

Darwin, C. (1872/1998) *The Expression of Emotions in Man and Animals*. London: Harper Collins.
Dauvois, M. (1989) Son et musique paléolithiques. *Les Dossiers d'Archéologie* 142, 2–11.
Dauvois, M. (1999) Mesures acoustiques at témoins sonores osseux paléolithiques. In *Préhistoire d'os, recuil d'études sur l'industrie osseuse préhistorique offert à Mme Henriette Camps-Febrer*. Aix-en-Provence: Publications de l'Université de Provence.
Davidson, I. (1991) The archaeology of language origins—a review. *Antiquity* 65, 39–48.
Davies, S. (2001) Philosophical perspectives on music's expressiveness. In P. Juslin and J. Sloboda (eds.) *Music and Emotion: Theory and Research*. Oxford: Oxford University Press, 23–44.
Davis, P., Zhang, S., Winkworth, A., and Bandler, R. (1996) Neural control of vocalisation: respiratory and emotional influences. *Journal of Voice* 10, 23–38.
Davison, G., and Neale, J. (1994) *Abnormal Psychology* (6th edn.). New York: John Wiley & Sons.
d'Errico, F. (1991) Carnivore Traces or Mousterian Skiffle? *Rock Art Research* 8, 61–3.
d'Errico F., Henshilwood, C., Lawson, G., Vanhaeren, M., Tillier, A.-M., Soressi, M., Bresson, F., Maureille, B., Nowell, A., Lakarra, J., Backwell, L., and Julien, M. (2003) Archaeological evidence for the emergence of language, symbolism and music—an alternative multidisciplinary perspective. *Journal of World Prehistory* 17, 1–70.
d'Errico F., Henshilwood, C., and Nilssen, P. (2001) An engraved bone fragment from c.70,000 year-old Middle Stone Age levels at Blombos Cave, South Africa: Implications for the origin of symbolism and language. *Antiquity* 75, 309–18.
d'Errico, F., and Lawson, G. (2006) The sound paradox: how to assess the acoustic significance of archaeological evidence? In G. Lawson and S. Scarre (eds.) *Archaeoacoustics*. Cambridge: McDonald Institute Monographs, 41–57.
d'Errico, F., and Villa, P. (1997) Holes and grooves: the contribution of microscopy and taphonomy to the problem of art origins. *Journal of Human Evolution* 33, 1–31.
d'Errico, F., Villa, P., Pinto Llona, A., and Ruiz Idarraga, R. (1998) A Middle Palaeolithic origin of music? Using cave-bear bone accumulations to assess the Divje babe I bone 'flute'. *Antiquity* 72, 65–79.
de Bayle des Hermens, R. (1974) Un radius de cygne sauvage utilisé et décoré dans le Magdalénien final du Rond-du-Barry. *L'Anthropologie* 78, 49–52.
de Beaune, S., Coolidge, F., and Wynn, T. (2009) *Cognitive Archaeology and Human Evolution*. Cambridge: Cambridge University Press.
Deacon, T. (1997) *The Symbolic Species*. London: Allen Lane.
Dean, C, Leakey, M., Reid, D., Schrenk, F., Schwartzk, G., Stringer, C., and Walker, A. (2001) Growth processes in teeth distinguish modern humans from Homo erectus and earlier hominins. *Nature* 414, 628–31.
Decety, J., and Chaminade, T. (2003) Neural correlates of feeling sympathy. *Neuropsychologia* 41, 127–38.
DeGelder, B., and Vroomen, J. (2000) The perception of emotions by ear and by eye. *Cognition and Emotion* 14, 289–311.
DeGusta, D., Gilbert, W., and Turner, S. (1999) Hypoglossal canal size and hominid speech. *Proceedings of the National Academy of Sciences of the USA* 96, 1800–4.

Delporte, H. (1980–1) La Collection Saint-Perier et le paléolithique supérieur d'Isturitz: Une acquisition prestigieuse. *Antiquités Nationales, Saint-Germain-en-Laye* 12–13, 20–6.

Dennell, R. (1997) The world's oldest spears. *Nature* 385, 767–8.

Dennett, D. (1998) *Brainchildren: Essays on Designing Minds*. London: Penguin.

Devereux, P. (2006) Ears and years: aspects of acoustics and intentionality in antiquity. In C. Scarre and G. Lawson (eds.) *Archaeoacoustics*. Cambridge: McDonald Institute Monographs, 23–30.

Dissanayake, E. (2000) Antecedents of the modern arts in early mother–infant interaction. In N. Wallin, B. Merker, and S. Brown (eds.) *The Origins of Music*. Cambridge MA: MIT Press, 389–410.

Dissanayake, E. (2009) Bodies swayed to music: the temporal arts as integral to ceremonial ritual. In S. Malloch and C. Trevarthen (eds.) *Communicative Musicality*. Oxford: Oxford University Press, 533–44.

Donald, M. (1991) *Origins of the Modern Mind: Three Stages in the Evolution of Culture and Cognition*. Cambridge MA: Harvard University Press.

Donald, M. (2001) *A Mind So Rare: The Evolution of Human Consciousness*. London: Norton.

Dowling, W., and Harwood, D. (1986) *Music Cognition*. London: Academic Press.

Drake, C., and Bertrand, D. (2003) The quest for universals in temporal processing in music. In I. Peretz and R. Zatorre (eds.) *The Cognitive Neuroscience of Music*. Oxford: Oxford University Press, 21–31.

Duchin, L. (1990) The evolution of articulate speech: comparative anatomy of *Pan* and *Homo*. *Journal of Human Evolution* 19, 687–97.

Duffau, H., Capelle, L., Denvil, D., Gatignol, P., Sichez, N., Lopes, M., Sichez, J.-P., and Van Effenterre, R. (2003) The role of dominant premotor cortex in language: a study using intraoperative functional mapping in awake patients. *Neuroimage* 20, 1903–14.

Dunbar, R. (1998) Theory of mind and the evolution of language. In J. Hurford, M. Studdert-Kennedy, and C. Knight (eds.) *Approaches to the Evolution of Language*. Cambridge: Cambridge University Press, 92–110.

Dunbar, R., Duncan, N., and Marriott, A. (1997) Human conversational behaviour. *Human Nature* 8, 231–46.

Dunbar, R., Knight, C., and Power, C. (eds.) (1999) *The Evolution of Culture*. Edinburgh: Edinburgh University Press.

Einwögerer, T., and Käfer, B. (1998) Eine jungpaläolithische Knochenflöte aus der Station Grubgraben bei Kammern, Niederösterreich. *Archäologisches Korrespondenzblatt* 28, 21–30.

Ekman, P. (1980) *The Face of Man: Expressions of Universal Motions in a New Guinea Village*. New York: Garland SPTM Press.

Ekman, P., and Friesen, W. (1971) Constants across cultures in the face and emotion. *Journal of Personality and Social Psychology* 17, 124–9.

Elowson, A., Snowdon, C., and Lazaro-Perea, C. (1998a) 'Babbling' and social context in infant monkeys: parallels to human infants. *Trends in Cognitive Sciences* 2, 31–7.

Elowson, A., Snowdon, C., and Lazaro-Perea, C. (1998b) Infant 'babbling' in a nonhuman primate: complex vocal sequences with repeated call-types. *Behaviour* 135, 643–64.

Emler, N. (1992) The truth about gossip, *Social Psychology Newsletter* 27, 23–37.
Enloe J., David, F., and Baryshnikov, G. (2000) Hyenas and hunters: zooarchaeological investigations at Prolom II cave, Crimea. *International Journal of Osteoarchaeology* 10, 310–24.
Engel, C. (1874) *A Descriptive Catalogue of Musical Instruments in the South Kensington Museum*. 2nd edn., London: G. E. Eyre & W. Spottiswoode.
Epp, H. (1988) Way of the migrant herds: dual dispersion strategy amongst bison. *Plains Anthropologist* 33, 95–111.
Esparza San-Juan, X. (1990) El Paleolitico superior de Isturitz en la Baja Navarra (Francia). PhD thesis, Departamento de Prehistoria e Historia, Madrid, 3 vols.
Fages, G., and Mourer-Chauviré, C. (1983) La Flûte en os d'oiseau de la grotte sépulcrale de Veyreau (Aveyron) et inventaire des flutes préhistoriques d'Europe. In F. Poplin (ed.) *La Faune et l'Homme Préhistorique: Dix études en hommage a Jean Bouchud. Mémoires de la Société Préhistorique Française* 16, 95–103.
Fagg, M. (1997) *Rock Music*, Pitt-Rivers Museum Occasional Papers No. 14. Oxford: Pitt-Rivers Museum.
Falk, D. (2000) Hominid brain evolution and the origin of music. In N. Wallin, B. Merker, and S. Brown (eds.) *The Origins of Music*. Cambridge MA: MIT Press, 197–216.
Falk, D. (2004a) *Braindance: New Discoveries about Human Origins and Brain Evolution* (rev. and expanded edn.). Gainesville: University Press of Florida.
Falk, D. (2004b) Prelinguistic evolution in early hominins: whence motherese? *Behavioural and Brain Sciences* 27, 491–541.
Feld, S. (1982) *Sound and Sentiment: Birds, Weeping, Poetics and Song in Kaluli Expression*. Philadelphia: University of Pennsylvania Press.
Fernald, A. (1989a) Emotion and meaning in mothers' speech to infants, paper presented to Society for Research in Child Development, Kansas City, April 1989.
Fernald, A. (1989b) Intonation and communicative intent in mothers' speech to infants: is melody the message? *Child Development* 60, 1497–1510.
Fernald, A. (1992a) Meaningful melodies in mothers' speech to infants. In H. Papousek, U. Jurgens, and M. Papousek (eds.) *Nonverbal Vocal Communication*. Cambridge: Cambridge University Press, 262–82.
Fernald, A. (1992b) Human maternal vocalisations to infants as biologically relevant signals: an evolutionary perspective. In J. Barkow, L. Cosmides, and J. Tooby (eds.) *The Adapted Mind: Evolutionary Psychology and the Generation of Culture*. Oxford: Oxford University Press, 391–428.
Fernald, A. (1993) Approval and disapproval—infant responsiveness to vocal affect in familiar and unfamiliar languages. *Child Development* 64, 657–74.
Fernald, A. (1994) Human maternal vocalisations to infants as biologically relevant signals: an evolutionary perspective. In P. Bloom (ed.) *Language Acquisition: Core Readings*. Cambridge MA: MIT Press, 51–94.
Fernald, A., and Simon, T. (1984) Expanded intonation contour in mothers' speech to newborns. *Developmental Psychology* 20, 104–13.
Fernald, A., Taeschner, T., Dunn, J., Papousek, M., Boysson-Bardies, B., and Fukui, I. (1989) A cross-language study of prosodic modifications in mothers' and fathers' speech to pre-verbal infants. *Journal of Child Language* 16, 477–501.

Feyereisen, P. (1997) The competition between gesture and speech production in dual-task paradigms. *Journal of Memory and Language* 36, 13–33.

Fidelholtz, J. (1991) On dating the origin of the modern form of language. In J. Wind, W. Von Raffler-Engel, and A. Jonker (eds.) *Studies In Language Origins*, ii. Amsterdam: Benjamins, 99–113.

Field, T., Woodson, R., Greenberg, R., and Cohen, D. (1982) Discrimination and imitation of facial expression in neonates. *Science* 218, 179–81.

Fink, R. (1997) Neanderthal Flute: Oldest Musical Instrument's 4 Notes Matches 4 of Do, Re, Mi Scale, http://www.webster.sk.ca/greenwich/fl-compl.htm (updated 1998, 1999, 2000).

Fink, R. (2000) Chewed Or Chipped? Who Made The Neanderthal Flute?, http://www.webster.sk.ca/greenwich/chewchip.htm.

Fink, R. (2002) The Neanderthal flute and origins of the scale: fang or flint? A response. In E. Hickman, A. D. Kilmer, and R. Eichman (eds.) *Studien zur Musikarchäologie III*. Rahden: Verlag Marie Leidorf, 83–7.

Finlayson, C., Brown, K., Blasco, R., Rosell, J., Negro, J., Bortolotti, G., Finlayson, G., Marco, A., Pacheco, F., Vidal, J., Carrion, J., Fam, D. and Llanes, J. (2012). Birds of a feather: Neanderthal exploitation of raptors and corvids. *PloS ONE* 7, e45927 (9 pages).

Fitch, W. (1999) Acoustic size exaggeration of size in birds via tracheal elongation: comparative and theoretical analyses. *Journal of Zoology* 248, 31–48.

Fitch, W. (2000a) The phonetic potential of nonhuman vocal tracts: comparative cineradiographic observations of vocalising animals. *Phonetica* 57, 205–18.

Fitch, W. (2000b) The evolution of speech: a comparative review. *Trends in Cognitive Science* 4, 258–67.

Fitch, W. (2009a) Fossil cues to the evolution of speech. In R. Botha and C. Knight (eds.) *The Cradle of Language*. Oxford: Oxford University Press, 108–30.

Fitch, W. (2009b) Biology of music: another one bites the dust. *Current Biology* 19, R403–4.

Foley, R. (1992) Studying Human Evolution by Analogy. In S. Jones, R. Martin, and D. Pilbeam (eds.) *The Cambridge Encyclopedia of Human Evolution*. Cambridge: Cambridge University Press, 335–40.

Foley, R., and Lee, P. (1991) Ecology and energetics of encephalisation in hominid evolution. *Philosophical Transactions of the Royal Society of London, Series B—Biological Sciences* 334, 223–32.

Foucher, P., and Normand, C. (2004) Étude de l'industrie lithique des niveaux solutréens de la grotte d'Isturitz (Isturitz/Saint-Martin-d'Arberoue, Pyrénées-atlantiques). *Antiquités Nationales* 36, 69–103.

Franz E., Zelaznik, H., and Smith, A. (1992) Evidence of common timing processes in the control of manual, orofacial, and speech movements. *Journal of Motor Behavior* 24, 281–7.

Frayer, D., and Nicolay, C. (2000) Fossil evidence for the origin of speech sounds. In N. Wallin, B. Merker, and S. Brown (eds.) *The Origins of Music*. Cambridge MA: MIT Press, 217–34.

Freeman, W. (1995) *Societies of Brains: A Study in the Neurobiology of Love and Hate*. Mahwah: Erlbaum.

Freeman, W. (2000) A neurobiological role for music in social bonding. In N. Wallin, B. Merker, and S. Brown (eds.) *The Origins of Music*. Cambridge MA: MIT Press, 411–24.

Fujii, T., Fukatsu, R., Watabe, S., Ohnuma, A., Teramura, T., Kimura, I., Saso, S., and Kogure, K. (1990) Auditory sound agnosia without aphasia following right temporal lobe lesion. *Cortex* 26, 263–8.

Gabrielsson, A., and P. Juslin (1996) Emotional expression in music performance: between the performer's intention and the listener's experience. *Psychology of Music* 24, 68–91.

Gallese, V. (2001) The 'Shared Manifold' hypothesis: from mirror neurons to empathy. In E. Thompson (ed.) *Between Ourselves: Second-Person Issues in the Study of Consciousness*. Thorverton: Imprint Academic, 33–50.

Gallese, V., Fadiga, L., Fogassi, L., and Rizzolatti, G. (1996) Action recognition in the premotor cortex. *Brain* 119, 593–609.

Gallese, V., Keysers, C., and Rizolatti, G. (2004) A unifying view of the basis of social cognition. *Trends in Cognitive Sciences* 8, 396–403.

Gamble, C. (1983) Culture and society in the Upper Palaeolithic of Europe. In G. Bailey (ed.) *Hunter-Gatherer Economy In Prehistory: A European Perspective*. Cambridge: Cambridge University Press, 201–11.

Gamble, C. (1999) *The Palaeolithic Societies of Europe*. Cambridge: Cambridge University Press.

Gardner, P. (1991) Forager Pursuit of Individual Autonomy. *Current Anthropology* 32, 543–72.

Garfinkel, Y. (2003) *Dancing at the Dawn of Agriculture*. Austin: University of Texas Press.

Geissman, T. (2000) Gibbon songs and human music in an evolutionary perspective. In N. Wallin, B. Merker, and S. Brown (eds.) *The Origins of Music*. Cambridge MA: MIT Press, 103–23.

Geissman, T., and Orgeldinger, M. (1998) Duet or divorce! *Folia Primatologica*, 69, 283.

Geschwind, N., Quadfasel, F., and Segarra, J. (1965) Isolation of the speech area. *Neuropsychologia* 6, 327–40.

Glory, A. (1964) La Grotte de Roucador. *Bulletin de la Société Préhistorique Française* 61, pp. clxvi–clxix.

Glory, A. (1965) Nouvelles découvertes de dessins rupestres sur le causse de Gramat. *Bulletin de la Société Préhistorique Française* 62, 528–36.

Glory, A., Vaultier, M., and Farinha Dos Santos, M. (1965) La Grotte ornée d'Escoural (Portugal). *Bulletin de la Société Préhistorique Française* 62, 110–17.

Goldbeck, C. (2001) Der Steinzeitflöte. http://www.quarks.de/musik/002.htm.

Goldhahn, J. (2002) Roaring rocks: an audio-visual perspective on hunter-gatherer engravings in Northern Sweden and Scandinavia. *Norwegian Archaeological Review* 35, 29–61.

Goldstein, A. (1980) Thrills in response to music and other stimuli. *Physiological Psychology* 3, 126–9.

González Morales, M. (1986) Inventory of the bone and antler industry from the 1976–1979 excavations at La Riera cave. In L. Straus and G. Clark (eds.) *La Riera Cave: Stone Age Hunter-Gatherer Adaptations in Northern Spain*. Arizona State University Anthropological Research Papers No. 36. Tempe: Arizona State University Press, 385–419.

Goodall, J. (1986) *The Chimpanzees of Gombe: Patterns of Behavior.* Cambridge MA: Harvard University Press.

Gosselin, N., Samson, S., Adolphs, R., Noulhiane, M., Roy, M., Hasboun, D., Baulac, M., and Peretz, I. (2006) Emotional responses to unpleasant music correlates with damage to the parahippocampal cortex. *Brain* 129, 2585-92.

Gotfredsen, A. (1997) Sea bird exploitation on coastal Inuit sites, West and Southeast Greenland. *International Journal of Osteoarchaeology* 7, 271-86.

Gould, R. (1969) *Yiwara: Foragers of the Australian Western Desert.* New York: Scribner's.

Goutas, N. (2004) Caractérisation et évolution du Gravettien en France par l'approche techno-économique des industries en matières dures animals (étude de six gisements du Sud-ouest). PhD thesis, l'Université de Paris I, Panthéon-Sorbonne.

Gowlett, J. (1984) Mental abilities of early man: a look at some hard evidence. In R. Foley (ed.) *Hominid Evolution and Community Ecology: Prehistoric Human Adaptation in Biological Perspective.* London: Academic Press, 167-92.

Grahn, J., and Brett, M. (2007) Rhythm perception in motor areas of the brain. *Journal of Cognitive Neuroscience* 19, 893-906.

Grahn, J., and Rowe, J. (2009) Feeling the beat: premotor and striatal interactions in musicians and non-musicians during beat processing. *Journal of Neuroscience* 29, 7540-8.

Grauer, V. (2006) Echoes of our forgotten ancestors [main paper and critical replies]. *Worlds of Music* 48, 5-134.

Grauer, V. (2007) New perspectives on the Kalahari debate: a tale of two 'genomes'. *Before Farming* 2007/2, article 4, 1-14.

Gray, P., Krause, B., Atema, J., Payne, R., Krumhansl, C., and Baptista, L. (2001) The music of nature and the nature of music. *Science* 291, 52-4.

Greiser, D., and Kuhl, P. (1988) Maternal speech to infants in a tonal language: support for universal prosodic features in motherese. *Developmental Psychology* 24, 14-20.

Gros-Louis, J. (2002) Contexts and behavioural correlates of trill vocalisations in wild white-faced Capuchin monkeys (*Cebus capucinus*). *American Journal of Primatology* 57, 189-202.

Gundlach, R. (1935) Factors determining the characterization of musical phrases. *American Journal of Psychology* 47, 624-43.

Haddon, A. (1898) *The Study of Man.* New York: John Murray.

Hadingham, E. (1980) *Secrets of the Ice Age: The World of the Cave Artists.* London: Heinemann.

Hagan, E., and Bryant, G. (2003) Music and dance as a coalitional signaling system. *Human Nature,* 14, 21-51.

Hahn, J., and Münzel, S. (1995) Knochenflöten aus den Aurignacien des Geissenklösterle bei Blaubeuren, Alb-Donau-Kreis. *Fundberichte aus Baden-Württemberg* 20, 1-12.

Harding, J. (1973) The bull-roarer in history and in antiquity. *African Music* 5, 40-2.

Hardy, B., and Moncel, M.-H. (2011) Neanderthal use of fish, mammals, birds, starchy plants and wood 125-250,000 years ago. PLOS ONE 6, e23768. doi:10.1371/journal.pone.0023768, http://www.plosone.org/article/info%3Adoi%2F10.1371%2Fjournal.pone.0023768.

Harrison, R. (1978) A pierced reindeer phalanx from Banwell Bone Cave and some experimental work on phalanges. *Proceedings of the University of Bristol Spelaeological Society* 15, 7–22.

Harrold, F. (1988) The Chatelperronian and the early Aurignacian in France. In J. Hoffecker and C. Wolf (eds.) *The Early Upper Palaeolithic. BAR International Series* 437. Oxford: Archaeopress, 157–91.

Hatfield, E., Cacioppo, J., and Rapson, R. (1994) *Emotional Contagion*. Cambridge: Cambridge University Press.

Hauser, M., and McDermott, J. (2003) The evolution of the music faculty: A comparative perspective. *Nature Neuroscience* 6, 663–8.

Hausler, A. (1960) Neue funde Steinzeitlicher Musikinstrumente in Ost-Europe. *Acta Musicologia* 32, 151–5.

Hedges, K. (1993) Places to see and places to hear: rock art and features of the sacred landscape. In J. Steinbring, A. Watchman, P. Faulstich, and P. Taçon (eds.) *Time and Space: Dating and Spatial Considerations in Rock Art Research*. Occasional AURA Publication No. 8. Melbourne: Australian Rock Art Research Association, 121–7.

Heim, J. (1985) L'Apport de l'ontogenèse à la phylogénie des Néanderthaliens. *Second International Congress on Human Palaeontology, Turin*, 184.

Henschen-Nyman, O. (1988) Cup-marked standing stones in Sweden. In E. Hickman and D. Hughes (eds.) *The Archaeology of Early Music Cultures: Third International Meeting of the ICTM Study Group on Music Archaeology*. Bonn: Verlag für systematische Musikwissenschaft GmbH, 11–16.

Henshilwood C., Sealy, J., Yates, R., Cruz-Uribe, K., Goldberg, P., Grine, F., Klein, R., Poggenpoel, C., Van Niekerk, K., and Watts, I. (2001) Blombos Cave, Southern Cape, South Africa: preliminary report on the 1992–1999 excavations of the Middle Stone Age levels. *Journal of Archaeological Science* 28, 421–48.

Hersh, K. (2010) *Paradoxical Undressing*. London: Atlantic Books.

Hewes, G. (1973) Primate communication and the gestural origin of language. *Current Anthropology* 14, 5–24.

Hewes, G. (1992) Primate communication and the gestural origin of language. *Current Anthropology* 33, 65–84.

Hewes, G. (1994) Comment on Armstrong, Stokoe & Wilcox. *Current Anthropology* 35, 360–1.

Higham, T., Basell, L., Jacobi, R., Wood, R, Bronk Ramsey, C., and Conard, N. (2012) Testing models for the beginnings of the Aurignacian and the advent of figurative art and music: The radiocarbon chronology of Geißenklösterle. *Journal of Human Evolution* 62, 664–76.

Hitchcock, R. (1999) Introduction: Africa. In R. Lee and R. Daly (eds.) *The Cambridge Encyclopedia of Hunters and Gatherers*. Cambridge: Cambridge University Press, 175–84.

Hoffecker, J. (1988) Early Upper Palaeolithic sites of the European USSR. In J. Hoffecker and C. Wolf, *The Early Upper Palaeolithic. BAR International Series* 437. Oxford: Archaeopress, 237–72.

Holderman C.-S., and Serangeli, J. (1998a) Flöten an Höhlenbärenknochen: Spekulationen oder Beweise? *Mitteilungsblatt der Gesellschaft für Urgeschichte* 6, 7–19.

Holderman C.-S., and Serangeli, J. (1998b) Einige Bemerkungen zur Flöte von Divje babe I (Slowenien) und deren Vergleichsfunde aus dem Östereichischen Raum und angenzenden Gebeiten. *Archäologie Österreichs* 9, 31–8.

Holloway, R. (1981) Volumetric and asymmetry determinations on recent hominid endocasts: Spy I and II, Djebel Ihroud I, and the Salè *Homo erectus* specimens, with some notes on Neanderthal brain size. *American Journal of Physical Anthropology* 55, 385–93.

Holloway, R. (1983) Human palaeontological evidence relevant to language behaviour. *Human Neurobiology* 2, 105–14.

Horusitzky, Z. (1955) Eine Knochenflöte aus der Höhle von Istállóskö. *Acta Archaeologica Academiae Scientiarum Hungaricae* 5, 133–45.

Hull, D. (1980) Individuality and selection. *Annual Review of Ecology and Systematics* 11, 311–32.

Hülle, W. (1977) *Die Ilsenhöhle unter Burg Ranis/Thüringen. Eine Paläolitische Jägerstation.* Stuttgart: Gustav Ficher Verlag.

Huron, D. (1999) An instinct for music: is music an evolutionary adaptation? *The 1999 Ernest Bloch Lectures: Lecture 2, University of California, Berkeley Department of Music.* Available online at http://csml.som.ohio-state.edu/Music220/Bloch.lectures/2.Origins.html.

Huron, D. (2001) Is music an evolutionary adaptation? *Annals of the New York Academy of Sciences* 930, 43–61.

Huyge, D. (1990) Mousterian skiffle? Note on a Middle Palaeolithic engraved bone from Schulen, Belgium. *Rock Art Research* 7, 125–32.

Huyge, D. (1991) The 'Venus' of Laussel in the light of ethnomusicology. *Archeologie in Vlaanderen* 1, 11–18.

Ichikawa, M. (1999) The Mbuti of Northern Congo. In R. Lee and R. Daly (eds.) *The Cambridge Encyclopedia of Hunters and Gatherers.* Cambridge: Cambridge University Press, 210–14.

Imig, T., Ruggero, M., Kitzes, L., Javel, E., and Brugge, J. (1977) Organisation of auditory cortex in the owl monkey. *Journal of Comparative Neurology* 171, 111–28.

Ingold, T. and Gibson, K. (1993) *Tools, Language and Cognition in Human Evolution.* Cambridge: Cambridge University Press.

Ito, S., and Hinoki, M. (1991) Human and animal semicircular canal function during circular walking. *Acta Oto-Laryngologica*, supplemental 481, 339–42.

Iverson, J. and Goldin-Meadow, S. (1998) Why people gesture as they speak. *Nature* 396, 228.

Jablonski, N., and Aiello, L. (eds.) (1998) *The Origin and Diversification of Language.* Wattis Symposium Series in Archaeology: Memoirs of the California Academy of Sciences 24. San Francisco: California Academy of Sciences.

Jacobs, A. (1972) *New Dictionary of Music* (2nd edn.). Harmonsdworth: Penguin Books.

Jeannerod, M. (2004) Vision and action cues contribute to self-other distinction. *Nature Neuroscience* 7, 422–3.

Jelinek, J. (1990) *Art in the Mirror of Ages: The Beginnings of Artistic Activities.* Brno: Moravian Museum, Anthropos Institute.

Joffe, T. (1997) Social pressures have selected for an extended juvenile period in primates. *Journal of Human Evolution* 32, 593–605.
Johanson, D., and Edgar, B. (2006) *From Lucy to Language* (2nd edn.). New York: Simon & Schuster.
Johnston, T. (1989) Song categories and musical style of the Yupik Eskimo. *Anthropos* 84, 423–31.
Jones, M., and Boltz, M. (1989) Dynamic attending and responses to time. *Psychological Review* 96, 459–91.
Jones, T. (1983) Australia. In D. Arnold (ed.) *The New Oxford Companion To Music*, i. Oxford: Oxford University Press, 117–19.
Jürgens, U. (1976) Projections from the cortical larynx area in the squirrel monkey. *Experimental Brain Research* 25, 401–11.
Jürgens, U. (1992) On the neurobiology of vocal communication. In H. Papousek, U. Jürgens, and M. Papousek (eds.) *Nonverbal Vocal Communication*. Cambridge: Cambridge University Press, 31–42.
Jürgens, U. (1998) Neuronal control of mammalian vocalisation, with special reference to the squirrel monkey. *Naturwissenschaften* 85, 376–88.
Jürgens, U. (2002) Neural pathways underlying vocal control. *Neuroscience and Biobehavioural Reviews* 26, 235–58.
Jürgens, U., and Alipour, M. (2002) A comparative study of the cortico-hypoglossal connections in primates, using biotin dextranamine. *Neuroscience Letters* 328, 245–8.
Jürgens, U., and Von Cramon, D. (1982) On the role of the anterior cingulate cortex in phonation—a case-report. *Brain and Language* 15, 234–48.
Jürgens, U., and Zwirner, P. (1996) The role of the periaqueductal grey in limbic and neocortical vocal fold control. *Neuroreport* 7, 2921–3.
Jungers, W., Pokempner, A., Kay, R., and Cartmill, M. (2003) Hypoglossal canal size in living hominoids and the evolution of human speech. *Human Biology* 75, 473–84.
Juslin, P., and Sloboda, J. (2001) *Music and Emotion: Theory and Research*. Oxford: Oxford University Press.
Käfer, B., and Einwögerer, T. (2002) Die jungpaläolithische Knochenflöte aus der Station Grubgraben bei Kammern, Nieder-österreich, und ihre Spielweise im Nachbau mit Silexwerkzeugen. In E. Hickman, A. D. Kilmer, and R. Eichman (eds.) *Studien zur Musikarchäologie III*. Rahden: Verlag Marie Leidorf, 91–105.
Kappas, A., Hess, U., and Scherer, K. (1991) Voice and emotion. In R. Feldman and B. Rime (eds.) *Fundamentals of Nonverbal Behavior: Studies in Emotion and Social Interaction*. Cambridge: Cambridge University Press, 200–38.
Karow, C., Marquardt, T., and Marshall, R. (2001) Affective processing in left and right hemisphere brain-damaged subjects with and without subcortical involvement. *Aphasiology* 15, 715–29.
Kay, R., Cartmill, M., and Balow M. (1998) The hypoglossal canal and the origin of human vocal behaviour. *Proceedings of the National Academy of Science of the USA*, 5417–19.
Kehoe, A. (1999) Blackfoot and other hunters of the North American Plains. In R. Lee and R. Daly (eds.) *The Cambridge Encyclopedia of Hunters and Gatherers*. Cambridge: Cambridge University Press, 36–40.

Keil, C. (1995) The theory of Participatory Discrepancies: A progress report. *Ethnomusicology* 39, 1–19.

Kimura, D. (1993) *Neuromotor Mechanisms in Human Communication.* Oxford: Oxford University Press.

Kirschner, S., and Tomasello, M. (2009) Joint drumming: Social context facilitates synchronization in preschool children. *Journal of Experimental Child Psychology* 102, 299–314.

Kirzinger, A., and Jürgens, U. (1982) Cortical lesion effects and vocalization in the squirrel-monkey. *Brain Research* 233, 299–315.

Kisliuk, M. (1991) Confronting the Quintessential: Singing, Dancing and Everyday Life Among the Biaka Pygmies (Central African Republic). PhD thesis, New York University.

Kitamura, C., Thanavisthuth, C., Burnham, D., and Luksaneeyanawin, S. (2002) Universality and specificity in infant-directed speech: pitch modifications as a function of infant age and sex in a tonal and non-tonal language. *Infant Behaviour and Development* 24, 372–92.

Kivy, P. (1989) *Sound Sentiment.* Philadelphia: Temple University Press.

Klein, R. (2009) *The Human Career: Human Biological and Cultural Origins* (3rd edn.). Chicago: Chicago University Press.

Kochetkova, V. (1978) *Paleoneurology.* Winston & Sons, Washington.

Koelsch, S., Grossmann, T., Gunter, T., Hahne, A., and Friederici, A. (2003) Children processing music: Electric brain responses reveal musical competence and gender differences. *Journal of Cognitive Neuroscience* 15, 683–93.

Koelsch, S., and Siebel, W. (2005) Toward a neural basis for music perception. *Trends in Cognitive Sciences* 9, 578–84.

Koelsch, S., Fritz, T., Schulze, K., Alsop, D., and Schlaug, G. (2005) Adults and children processing music: An fMRI study. *Neuroimage* 25, 1068–76.

Kogan, N. (1997) Reflections on aesthetics and evolution. *Critical Review* 11, 193–210.

Kordos, L., and Begun, D. (1997) A new reconstruction of RUD 77, a partial cranium of *Dryopithecus brancoi* from Rudabanya, Hungary. *American Journal of Physical Anthropology* 103, 277–94.

Kormos, T. (1912) Die ersten spuren des Urmenschen im Karst-Gebirge. *Földtani Közlony* 42, 98–101.

Krause, J., Lalueza-Fox, C., Orlando, L., Enard, W., Green, R., Burbano, H., Hublin, J., Hänni, C., Fortea, J., de la Rasilla, M., Bertranpetit, J., Rosas, A., and Pääbo, S. (2007) The derived FOXP2 variant of modern humans was shared with Neandertals. *Current Biology* 17, 1908–12.

Kraut, R., and Johnston, R. (1979) Social and emotional messages of smiling: an ethological approach. *Journal of Personality and Social Psychology* 37, 1539–53.

Krumhansl, C. (1997) An exploratory study of musical emotion and psychophysiology. Canadian Journal of Experimental Psychology 51, 336–52.

Krumhansl, C., and Schenck, D. (1997) Can dance reflect the structural and expressive qualities of music? A perceptual experiment on Balanchine's choreography of Mozart's Divertimento No. 15. *Musicae Scientiae* 1, 63–85.

Kudo, H., and Dunbar, R. (2001) Neocortex size and social network size in primates. *Animal Behaviour* 62, 711–22.

Kuhl, P. (1988) Auditory perception and the evolution of speech. *Human Evolution* 3, 19–43.
Kuhn, S., and Stiner, M. (1998) The earliest Aurignacian of Riparo Mochi (Liguria, Italy). *Current Anthropology* 39, suppl. 3, 175–89.
Kunej, D., and Turk, I. (2000) New perspectives on the beginnings of music: archaeological and musicological analysis of a Middle Palaeolithic bone 'flute'. In N. Wallin, B. Merker, and S. Brown (eds.) *The Origins of Music*. Cambridge MA: MIT Press, 235–68.
Kurtén, B. (1976) *The Cave Bear Story: Life and Death of a Vanished Animal*. New York: Columbia University Press.
Lai, C., Fisher, S., Hurst, J., Vargha-Khadem, F., and Monaco, A. (2001) A forkhead-domain gene is mutated in a severe speech and language disorder. *Nature* 413, 519–23.
Laitman, J. (1984) The anatomy of human speech. *Natural History* 93, 20–7.
Laitman, J. and Heimbuch, R. (1982) The basicranium of Plio-Pleistocene hominids as an indicator of their upper respiratory systems. *American Journal of Physical Anthropology* 59, 323–44.
Laitman, J., Heimbuch, R., and Crelin, E. (1979) The basicranium of fossil hominids as an indicator of their upper respiratory systems. *American Journal of Physical Anthropology* 51, 13–34.
Laitman, J., and Reidenberg, J. (1988) Advances in understanding the relationship between the skull base and larynx with comments on the origins of speech. *Human Evolution* 3, 99–109.
Laland, K., and Brown, G. (2011) *Sense and Nonsense: Evolutionary Perspectives on Human Behaviour* (2nd edn.). Oxford: Oxford University Press.
Laland, K. Odling-Smee, J., and Feldman, M. (2000) Niche construction, biological evolution and cultural change. *Behavioral and Brain Sciences*. 23, 131–75.
Laland, K. Odling-Smee, J., and Myles, S. (2010) How culture shaped the human genome: bringing genetics and the human sciences together. *Nature Reviews Genetics* 11, 137–48.
Lang, A. (1884) *Custom and Myth*. London: Harper & Bros.
Langlais, M. (2007) Dynamiques culturelles des sociétés magdaléniennes dans leurs cadres environnementaux: Enquête sur 7 000 ans d'évolution de leurs industries lithiques entre Rhône et Èbre. PhD thesis, l'Université de Toulouse II et Universitat de Barcelona.
Lartet, E., and Christy, H. (1910) *Relique Aquitanicae; Being a Contribution to the Archaeology and Palaeontology of Perigord and the Adjoining Provinces of Southern France*. London: Williams & Norgate.
Lavy, M. (2001) Emotion and the Experience of Listening to Music: A Framework for Empirical Research. PhD thesis, University of Cambridge.
Lawson, G., and d'Errico, F. (2002) Microscopic, experimental and theoretical re-assessment of Upper Palaeolithic bird-bone pipes from Isturitz, France: ergonomics of design, systems of notation and the origins of musical traditions. In E. Hickman, A. Kilmer, and R. Eichman (eds.) *Studien zur Musikarchäologie III*. Rahden: Verlag Marie Leidorf, 119–42.
Lawson, G., Scarre, C., Cross, I., and Hills, C. (1998) Mounds, megaliths, music and mind: some acoustical properties and purposes of archaeological spaces. *Archaeological Review from Cambridge* 15, 111–34.

Layton, R. (2006) Habitus and narratives of rock art. In J. Keyser, G. Poetschat, and M. Taylor (eds.) *Talking with the Past: The Ethnography of Rock Art.* Portland: Oregon Archaeological Society, 73–99.

Láznicková-Gonysěvová, M. (2002) Art mobilier magdalénien en matières dures animales de Moravie (République tchèque): Aspects technologique et stylistique (Portable art on hard animal tissues from Moravia. Technological and stylistical aspects). *L'Anthropologie* 106, 525–64.

Le Gonidec, M.-B., Garcia, L., and Caussé, R. (1996) Au sujet d'un flûte paléolithique: En souvenir du Dominique Buisson. *Antiquités Nationales* 28, 149–52.

Leakey, R. (1994) *The Origin of Humankind.* London: Weidenfeld & Nicolson.

Lee, R., and Daly, R. (1999) *The Cambridge Encyclopedia of Hunters and Gatherers.* Cambridge: Cambridge University Press.

Lefevre, C. (1997) Sea bird following in southern Patagonia: a contribution BQ understanding the nomadic round of the Canoeros Indians. *International Journal of Osteoarchaeology* 7, 260–70.

Leocata, F. (2001) Osservazioni sui 'flauti' paleolitici. *Rivista di scienze preistoriche* 51, 177–200.

Levenson, R., Ekman, P., and Friesen, V. (1990) Voluntary facial action generates emotion-specific autonomic nervous system activity. *Psychophysiology* 27, 363–84.

Lévi-Strauss, C. (1970) *The Raw and the Cooked*, trans. J. and D. Weightman. London: Cape.

Levitin, D., and Bellugi, U. (1997) Musical abilities in individuals with Williams Syndrome. Paper presented at the 1997 Society for Music Perception and Cognition. Massachusetts Institute of Technology, Cambridge MA.

Lewis, J. (2009) As well as words: Congo pygmy hunting, mimicry and play. In R. Botha and C. Knight (eds.) *The Cradle of Language.* Oxford: Oxford University Press, 236–56.

Lewis-Williams, D. (2002) *The Mind in the Cave.* London: Thames & Hudson.

Lewkowicz, D. J. (1998) Infants' response to the audible and visible properties of the human face: II. Discrimination of differences between singing and adult-directed speech. *Developmental Psychobiology* 32, 261–74.

Liberman, A., Cooper, F., Shankweiler, D., and Studdert-Kennedy, M. (1967) Perception of the speech code. *Psychological Review* 74, 431–61.

Librado, F. (1977) *The Eye of the Flute: Chumash Traditional History and Ritual*, ed. T. Hudson). Santa Barbara CA: Santa Barbara Museum of Natural History.

Lieberman, P. (1984) *The Biology and Evolution of Language.* Cambridge MA: Harvard University Press.

Lieberman, P. (1989) The origins of some aspects of human language and cognition. In P. Mellars and C. Stringer (eds.) *The Human Revolution: Behavioural and Biological Perspectives in the Origins of Modern Humans.* Edinburgh: Edinburgh University Press, 391–414.

Lieberman, P. (1991) *Uniquely Human: The Evolution of Speech, Thought and Selfless Behaviour.* Cambridge MA: Harvard University Press.

Lieberman, P. (1992) Human speech and language. In S. Jones, R. Martin, and D. Pilbeam (eds.) *The Cambridge Encyclopedia of Human Evolution.* Cambridge: Cambridge University Press, 134–7.

Lieberman, P. (1994) Comment on Armstrong, Stokoe & Wilcox. *Current Anthropology* 35, 362–3.
Lieberman, P., and Crelin, E. (1971) On the speech of Neanderthal man. *Linguistic Enquiry* 11, 203–22.
Lieberman, P., Crelin, E., and Klatt, D. (1972) Phonetic ability and related anatomy of the newborn and adult human, Neanderthal man and the chimpanzee. *American Anthropology* 74, 287–307.
Lieberman, P., Klatt, D., and Wilson, W. (1969) Vocal tract limitations on the vowel repertoires of rhesus monkeys and other nonhuman primates. *Science* 164, 1185–7.
Liégois-Chauvel, C., Giraud, K., Badier, J.-M., Marquis, P., and Chauvel, P. (2003) Intracerebral evoked potentials in pitch perception reveal a functional asymmetry of human auditory cortex. In I. Peretz and R. Zatorre (eds.) *The Cognitive Neuroscience of Music*. Oxford: Oxford University Press, 152–67.
Livingstone, F. (1973) Did The Australopithecines Sing? *Current Anthropology* 14, 25–9.
Llinas, R. (2001) *I of the Vortex, From Neuroscience to Self*. Cambridge, MA: MIT Press.
Locke, D. (1996) Africa: Ewe, Mande, Dagbamba, Shona and BaAka. In J. Titon (ed.) *Worlds of Music: An Introduction to the Music of the World's People* (3rd edn.). New York: Schirmer, 83–144.
Locke, J. (1998) Social sound-making as a precursor to modern language. In J. Hurford, M. Studdert-Kennedy, and C. Knight (eds.) *Approaches to the Evolution of Language*. Cambridge: Cambridge University Press, 190–201.
Locke, J. (2000) Movement patterns in spoken language. *Science* 288, 449–51.
Locke, J., Bekken, K., McMinnLarson, L., and Wein, D. (1995) Emergent control of manual and vocal motor activity in relation to the development of speech. *Brain and Language* 51, 498–508.
Lucius, E. (1970) Das Problem der Chronologie Jungpaläolitischer Stationen im Bereiche der Europäischen USSR. *Mittel Prähistorische Kommunikations*, 13–14.
McAllester, D. (1996) North America/Native America. In J. Titon (ed.) *Worlds of Music: An Introduction to the Music of the World's People* 3rd edn. New York: Schirmer, 33–82.
McBrearty, S., and Brooks, A. (2000) The revolution that wasn't: a new interpretation of the origin of modern human behaviour. *Journal of Human Evolution* 39, 453–563.
McBurney, C. (1967) *The Haua Fteah (Cyrenaica) and the Stone Age of the South East Mediterranean*. Cambridge: Cambridge University Press.
McClave, E. (1994) Gestural beats: the rhythm hypothesis. *Journal of Psycholinguistic Research* 23, 45–66.
McComb, P. (1989) *Upper Palaeolithic Osseous Artefacts from Britain and Belgium*. BAR International Series 481. Oxford: Archaeopress.
MacDonald, K. (1998) The avifauna of the Haua Fteah (Libya). *Archaeozoologia* 9, 83–101.
McFarland, R., and Kennison, R. (1989) Asymmetry in the relationshipbetween finger temperature changes and emotional state in males. *Biofeedback and Self Regulation* 14, 281–90.

MacLarnon, A. (1993) The vertebral canal. In A. Walker and R. Leakey (eds.) *The Nariokotome Homo erectus Skeleton*. Cambridge MA: Harvard University Press, 359–90.

MacLarnon, A., and Hewitt, G. (1999) The evolution of human speech: the role of enhanced breathing control. *American Journal of Physical Anthropology* 109, 341–63.

McNeill, D. (1992) *Hand and Mind: What Gestures Reveal About Thought*. Chicago: University of Chicago Press.

McNeill, D. (2000) *Language and Gesture*. Cambridge: Cambridge University Press.

McNeill, W. (1995) *Keeping Together in Time*. Cambridge MA: Harvard University Press.

McPherron, S., Alemseged, Z., Marean, C., Wynn, J., Reed, D., Geraads, D., Bobe, R., and Béarat, H. (2010) Evidence for stone-tool-assisted consumption of animal tissues before 3.39 million years ago at Dikika, Ethiopia. *Nature* 466, 857–60.

Magriples, U., and Laitman, J. (1987) Developmental change in the position of the fetal human larynx. *American Journal of Physical Anthropology* 72, 463–72.

Maioli, W. (1991) *Le origini: Il suono e la musica*. Milan: Jaca Book.

Malloch, S., and Trevarthen, C. (2009a) Musicality: communicating the vitality of human life. In Malloch and Trevarthen (2009b), 1–12.

Malloch, S., and Trevarthen, C. (2009b) *Communicative Musicality: Exploring the Basis of Human Companionship*. Oxford: Oxford University Press.

Malvesin-Fabre, G. (1946) La Stratigraphie de Pair-non-Pair. *Procès Verbaux de la Société Linnéene de Bordeaux* 93 (1943–6), 175–86.

Mang, E. (2000) Intermediate vocalizations: an investigation of the boundary between speech and songs in young children's vocalisations. *Bulletin of the Council for Research in Music Education* 147, 116–21.

Marin, O., and Perry, D. (1999) Neurological aspects of music perception and performance. In D. Deutsch (ed.) *The Psychology of Music* (2nd edn.). New York: Academic Press, 653–724.

Marshack, A. (1970) Notations dans les gravures du paléolithique supérieur: Nouvelles méthodes d'analyse. *Publications de l'Institut de Préhistoire de l'Université de Bordeaux No 8*. Bordeaux: Université de Bordeaux.

Marshack, A. (1972) *The Roots of Civilisation*. London: Weidenfeld & Nicolson.

Marshack, A. (1990) Early hominid symbol and evolution of the human capacity. In P. Mellars (ed.) *The Emergence of Modern Humans: An Archaeological Perspective*. Edinburgh: Edinburgh University Press, 457–98.

Martin, D. (1999) Cape York Peoples, North Queensland, Australia. In R. Lee and R. Daly (eds.) *The Cambridge Encyclopedia of Hunters and Gatherers*. Cambridge: Cambridge University Press, 335–8.

Martin, H. (1906) Presentation d'ossements de renne portant des lesions d'origine humaine et animale. *Bulletin de la Société Préhistorique Française* 3, 385–8.

Martin, H. (1909) *Recherches sur l'évolution du Mousterien dans le gisement de La Quina (Charante)*. Paris: Schleicher Frères.

Martínez, I., Arsuaga, J., Quam, R., Carretero, J., Gracia, A., and Rodriguez, L. (2008) Human hyoid bones from the middle Pleistocene site of the Sima de los Huesos (Sierra de Atapuerca, Spain). *Journal of Human Evolution* 54, 118–24.

Martínez, I., Rosa, M., Arsuaga, J., Jarabo, P., Quam, R., Lorenzo, C., Gracia, A., Carretero, J., Bermúdez de Castro, J., and Carbonell, E. (2004) Auditory capacities in middle Pleistocene humans from the Sierra de Atapuerca in Spain. *Proceedings of the National Academy of Sciences of the U.S.A.* 101, 9976–81.

Masataka, N. (2000) The role of modality and input in the earliest stages of language acquisition: studies of Japanese sign language. In C. Chamberlain, J. Morford, and R. Mayberry (eds.) *Language Acquisition by Eye*. Mahwah: Erlbaum, 3–24.

Masset, C., and Perlès, C. (1978) *Travail et société au Paléolithique. Préhistoire 1.* Documentation Photographique n° 6037. Paris: La Documentation Française.

Matravers, D. (1998) *Art and Emotion*. Oxford: Oxford University Press.

Mattson, D., Knight, R., and Blanchard, B. (1992) Cannibalism and predation on black bears by grizzly bears in the Yellowstone Ecosystem, 1975–1990. *Journal of Mammalogy* 73, 422–5.

Mayberry, R., and Jaques, J. (2000) Gesture production during stuttered speech: insights into the nature of gesture-speech integration. In D. McNeill (ed.) *Language and Gesture*. Cambridge: Cambridge University Press, 199–214.

Mazzucchi, A., Marchini, C., Budai, R., and Parma, M. (1982) A case of receptive amusia with prominent timbre perception defect. *Journal of Neurology, Neurosurgery, and Psychiatry* 45, 644–7.

Megaw, J. (1960) Penny Whistles and Prehistory. *Antiquity* 34, 6–13.

Megaw, J. (1968) The earliest musical instruments in Europe. *Archaeologia* 21, 124–32.

Mellars, P. (1989) Major issues in the emergence of modern humans. *Current Anthropology* 30, 349–85.

Mellars, P. (1994) The Upper Palaeolithic revolution. In B. Cunliffe (ed.) *Prehistoric Europe: An Illustrated History*. Oxford: Oxford University Press.

Mellars, P. (2000) The emergence of modern cognitive abilities, paper presented at the One-day conference on *The Speciation of Modern Homo sapiens*, at the British Academy, 28 March.

Mellars, P., Boyle, K., Bar-Yosef, O., and Stringer, C. (2007) *Rethinking the Human Revolution*. Cambridge: McDonald Institute Monographs.

Mellars, P., and Gibson, K. (1996) *Modelling the Early Human Mind*. Cambridge: McDonald Institute Monographs.

Mellars, P., and Stringer, C. (1989) *The Human Revolution: Behavioural and Biological Perspectives in the Origins of Modern Humans*. Edinburgh: Edinburgh University Press.

Menon, V., and Levitin, D. (2005) The rewards of music listening: response and physiological connectivity of the mesolimbic system. *Neuroimage* 26, 175–84.

Menon, V., Levitin, D., Smith, B., Lembke, A., Krasnow, B., Glazer, D., Glover, G., and McAdams, S., 2002. Neural correlates of timbre change in harmonic sounds. *Neuroimage* 17, 1742–54.

Menuhin, Y., and Davis, C. (1979) *The Music of Man*. London: Methuen.

Merker, B. (1999) Synchronous chorusing and the origins of music. *Musicae Scientiae*, Special Issue, 59–73.

Merker, B. (2000) Synchronous chorusing and human origins. In N. Wallin, B. Merker, and S. Brown (eds.) *The Origins of Music*. Cambridge, MA: MIT Press, 315–27.

Messinger D., and Fogel A. (1998) Give and take: the development of conventional infant gestures. *Merrill-Palmer Quarterly-Journal of Developmental Psychology* 44, 566–90.

Meyer, L. (1956) *Emotion and Meaning in Music*. Chicago: Chicago University Press.

Miller, G. (2000a) Evolution of human music through sexual selection. In N. Wallin, B. Merker, and S. Brown (eds.) *The Origins of Music*. Cambridge MA: MIT Press, 329–60.

Miller, G. (2000b) *The Mating Mind: How Sexual Choice Shaped the Evolution of Human Nature*. London: Heinemann.

Minetti, A., and Alexander, R. (1997) A theory of metabolic costs for bipedal gaits. *Journal of Theoretical Biology* 186, 467–76.

Mitchell, R., and Gallaher, M. (2001) Embodying music: matching music and dance in memory. *Music Perception* 19, 65–85.

Mithen, S. (1996) *The Prehistory of the Mind*. London: Phoenix.

Mithen, S. (1998) Was there a creative explosion? In S. Mithen (ed.) *Creativity in Human Evolution and Prehistory*. London: Routledge, 193–210.

Mithen, S. (2005) *The Singing Neanderthals: The Origins of Music, Language, Mind and Body*. London: Weidenfeld & Nicolson.

Mithen, S. (2006) Reply (Review Feature on *The Singing Neanderthals: The Origins of Music, Language, Mind and Body* by Steven Mithen). *Cambridge Archaeological Journal* 16, 97–112.

Mohr, J., Pessin, M., Finkelstein, S., Funkenstein, H., Duncan, G., and Davis, K. (1978) Broca's aphasia: pathological and clinical. *Neurology* 28, 311–24.

Molinari, M., Leggio, M., De Martin, M., Cerasa, A., and Thaut, M. (2003) Neurobiology of rhythmic motor entrainment. *Annals of the New York Academy of Sciences* 999, 313–21.

Molnar-Szakacs, I., and Overy, K. (2006) Music and mirror neurons: from motion to 'e'motion. *Social Cognitive and Affective Neuroscience* 1, 235–41.

Mondragon-Ceballos, R. (2002) Machiavellian intelligence in primates and the evolution of social brain. *Salud Mental* 25, 29–39.

Montagu, J. (2007) *Origins and Development of Musical Instruments*. Lanham: Scarecrow Press.

Morel, A., Garraghty, P., and Kaas, J. (1993) Tonotopic organization, architectonic fields, and connections of auditory cortex in macaque monkeys. *Journal of Comparative Neurology* 335, 437–59.

Morley, I. (2000) The Origins and Evolution of the Human Capacity for Music. MA Thesis, University of Reading, Reading, Berkshire.

Morley, I. (2002a) Evolution of the physiological and neurological capacities for music. *Cambridge Archaeological Journal* 12, 195–216.

Morley, I. (2002b) A cross-continental chorus: commonalities in the music of hunters and gatherers. In *Proceedings of the 9th International Conference On Hunting and Gathering Societies*. Available online at http://www.abdn.ac.uk/chags9/1morley.htm

Morley, I. (2003) The Evolutionary Origins and Archaeology of Music. An Investigation into the Prehistory of Human Musical Capacities and Behaviours. PhD Thesis, University of Cambridge. Published online at http://www.dar.cam.ac.uk/dcrr.

Morley, I. (2006) Mousterian musicianship? The case of the Middle Palaeolithic Divje babe I bone. *Oxford Journal of Archaeology* 25, 317–33.
Morley, I. (2007) Time, cycles and ritual behaviour. In C. Malone and D. Barrowclough (eds.) *Cult in Context: Reconsidering Ritual in Archaeology*. Oxford: Oxbow Books, 205–9.
Morley, I. (2009) Music and ritual: parallels, practice, and the Palaeolithic. In C. Renfrew and I. Morley (eds.) *Becoming Human: Innovation in Prehistoric Material and Spiritual Culture*. Cambridge: Cambridge University Press, 159–75.
Morley, I. (*in press*) Rocks, rhombes and racleurs—beyond piped music in the Palaeolithic. In P. Pettitt and B. Gravina (eds.) *Title tbc*. Oxford: Oxford University Press.
Mortillet, A. (1906) Le Placard (Charente) et les diverses industries qu'elle a livrées. *Congrès préhistorique de France*, Session de Vannes, 241–65.
Morton, E. (1977) On the occurrence and significance of motivation-structural rules in some bird and mammal sounds. *American Naturalist* 111, 855–69.
Morton, E. (1994) Sound symbolism and its role in non-human vertebrate communication. In L. Hinton, J. Nichols, and J. Ohala (eds.) *Sound Symbolism*. Cambridge: Cambridge University Press, 348–65.
Morton, J. (1999) The Arrernte of Central Australia. In R. Lee and R. Daly (eds.) *The Cambridge Encyclopedia of Hunters and Gatherers*. Cambridge: Cambridge University Press, 329–38.
Mottl, M. (1950) Die Paläolithischen Funde aus der Salzhofenhohle im Toten Gebirge. *Archaeologia Austriaca* 5, 18–22.
Moyle, R. (1990) *Polynesian Sound-producing Instruments*. Princes Risborough: Shire Publications.
Mulvaney, J. (1999) The chain of connection: the material evidence. In N. Peterson (ed.) *Tribes and Boundaries In Australia*. Canberra: Australian Institute of Aboriginal Studies, 72–94.
Münzel, S., Seeberger, F., and Hein, W. (2002) The Geißenklösterle flute—discovery, experiments, reconstruction. In E. Hickman, A. Kilmer, and R. Eichman (eds.) *Studien zur Musikarchäologie III*. Rahden: Verlag Marie Leidorf, 107–18.
Myers, F. (1999) Pintupi-speaking Aboriginals of the Western Desert. In R. Lee and R. Daly (eds.) *The Cambridge Encyclopedia of Hunters and Gatherers*. Cambridge: Cambridge University Press, 348–57.
Nattiez, J. (1983) Some aspects of Inuit vocal games. *Ethnomusicology* 27, 457–75.
Nelson, D. (1997) Radiocarbon dating of bone and charcoal from Divje babe I cave. In I. Turk (ed.) *Mousterian 'Bone Flute' and other finds from Divje babe I cave site in Slovenia*. Ljubljana: Institut za Archaeologijo, Znanstvenoraziskovalni Center Sazu, 51–64.
Nelson, E. (1899/1983) *The Eskimo About Bering Strait*. Washington: Smithsonian Institution.
Nettiez, J. (1983) Some aspects of Inuit vocal games. *Ethnomusicology* 27, 457–75.
Nettl, B. (1956) *Music in Primitive Culture*. Cambridge MA: Harvard University Press.
Nettl, B. (2000) An ethnomusicologist contemplates universals in musical sounds and musical cultures. In N. Wallin, B. Merker, and S. Brown (eds.) *The Origins of Music*. Cambridge MA: MIT Press, 463–72.

Nettl, B. (1989) *Blackfoot Musical Thought: Comparative Perspectives*. Kent OH: Kent State University Press.

Nettl, B. (1992) North American Indian music. In B. Nettl, C. Capwell, P. Bohlman, I. Wong, and T. Turino (eds.) *Excursions in World Music*. Englewood Cliffs NJ: Prentice Hall, 260–77.

Nicolson, N. (1977) A comparison of early behavioral development in wild and captive chimpanzees. In S. Chavalier-Skolnikoff and F. Poirier (eds.) *Primate Bio-Social Development*. New York: Garland, 529–62.

Nishitani, N., Schürmann, M., Amunts, K., and Hari, R. (2005) Broca's region: from action to language. *Physiology* 20, 60–9.

Nobe, S. (1996) Cognitive Rhythms, Gestures, and Acoustic Aspects of Speech. PhD thesis, University of Chicago. Cited in Mayberry and Jaques (2000).

Noffsinger, D. (1985) Dichotic-listening techniques in the study of hemispheric asymmetries. In D. Benson and E. Zaidel (eds.) *The Dual Brain*. New York: Guildford Press, 127–41.

Normand, C., de Beaune, S., Costamagno, S., Diot, M.-F., Henry-Gambier, D., Goutas, N., Laroulandie, V., Lenoble, A., O'Farrell, M., Rendu, W., Rios Garaizar, J., Schwab, C., Tarriño Vinagre, A., Texier, J.-P., and White, R. (2007) Nouvelles données sur la séquence aurignacienne de la grotte d'Isturitz (communes d'Isturitz et de Saint-Martin-d'Arberoue; Pyrénées-Atlantiques). In J. Evin (ed.), '... *Aux conceptions d'aujourd'hui*', vol. 3 of the Proceedings of the XXVIe Congrès préhistorique de France—Avignon, 21–5 Sept. 2004, 277–93.

Nyklicek, I., Thayer, J., and Van Doornen, L. (1997) Cardiorespiritory differentiation of musically-induced emotions. *Journal of Psychophysiology* 11, 304–21.

Ojemann, G., Ojemann, J., Lettich, E., and Berger, M. (1989) Cortical language localization in left, dominant hemisphere: An electrical stimulation mapping investigation in 117 patients. *Journal of Neurosurgery* 71, 316–26.

Okanoya, K., and Merker, B. (2007) Neural substrates for string-context mutual segmentation: a path to human language. In C. Lyon, C. Nehaniv, and A. Cangelosi (eds.) *Emergence of Communication and Language*. London: Springer, 421–34.

Otte, M. (1979) *Le Paléolithique supérieur ancien en Belgique*. Monographies d'Archéologie Nationale 5. Brussels: Musées Royaux d'Art et d'Histoire.

Otte, M. (1981) *Le Gravettien en Europe Centrale*. Bruges: De Tempel (Dissertationes Archaeologicae Gandenses 20).

Otte, M. (2000) On the suggested bone flute from Slovenia. *Current Anthropology* 41, 271–2.

Ouzman, S. (1998) Towards a mindscape of landscape: rock-art as expression of world-understanding. In C. Chippindale and P. Taçon (eds.) *The Archaeology of Rock-Art*. Cambridge: Cambridge University Press, 30–41.

Ouzman, S. (2001) Seeing is deceiving: rock art and the non-visual. *World Archaeology* 33, 237–56.

Palastanga, N., Field, D., and Soames, R. (2002) *Anatomy and Human Movement*. London: Butterworth Heinemann.

Panksepp, J. (1995) The emotional sources of 'chills' induced by music. *Music Perception* 13, 171–207.

Panksepp, J. (1998) The periconscious substrates of consciousness: affective states and the evolutionary origins of the SELF. *Journal of Consciousness Studies* 5, 566–82.
Panksepp, J. (2005) Affective consciousness: core emotional feelings in animals and humans. *Consciousness and Cognition* 14, 19–69.
Panksepp, J., and Bernatsky, G. (2002) Emotional sounds and the brain: the neuroaffective foundations of musical appreciation. *Behavioural Processes* 60, 133–55.
Panksepp, J., and Trevarthen, C. (2009) The neuroscience of emotion in music. In Malloch and Trevarthen (2009b), 105–46.
Papaeliou, C., and Trevarthen, C. (1998) The infancy of music. *Musical Praxis* 1, 19–33.
Papousek, H. (1996a) Musicality in infant research: biological and cultural origins of early musicality. In I. Deliège and J. Sloboda (eds.) *Musical Beginnings: Origins and Development of Musical Competence.* Oxford: Oxford University Press, 37–55.
Papousek, M. (1996b) Intuitive parenting: a hidden source of musical stimulation in infancy. In I. Deliège and J. Sloboda (eds.) *Musical Beginnings: Origins and Development of Musical Competence.* Oxford: Oxford University Press, 88–112.
Papousek, M., Papousek, H., and Symmes, D. (1991) The meanings of melodies in motherese in tone and stress languages. *Infant Behavior and Development*, 14, 415–40.
Parsons, L. (2003) Exploring the functional neuroanatomy of music performance, perception, and comprehension. In I. Peretz and R. Zatorre (eds.) *The Cognitive Neuroscience of Music.* Oxford: Oxford University Press, 247–68.
Passemard, E. (1923) Une flûte aurignacienne d'Isturitz. *Congrès de l'Association Française pour l'Avancement des Sciences* 46th Session, Montpellier, 474–6.
Passemard, E. (1944) La Caverne d'Isturitz en Pays Basque. *Préhistoire* 9, 7–95.
Patel, A. (2003) Language, music, syntax and the brain. *Nature Neuroscience* 6, 674–81.
Patel, A. (2008) *Music, Language and the Brain.* Oxford: Oxford University Press.
Patel, A., Iverson, J., Bregman, M., and Schultz, I. (2009) Experimental evidence for synchronization to a musical beat in a nonhuman animal. *Current Biology* 19, 827–30.
Patel, A., Peretz, I., Tramo, M., and Labrecque, R. (1998) Processing prosodic and musical pattern: a neuropsychological investigation. *Brain and Language* 61, 123–44.
Patel, A., Wong, M., Foxton, J., Lochy, A, and Peretz, I. (2008) Speech intonation perception deficits in musical tone deafness (congenital amusia). *Music Perception* 25, 357–68.
Peek, P. (1994) The sounds of silence: cross-world communication and the auditory arts in African societies. *American Ethnologist* 21, 474–94.
Pequart, M., and Pequart, S. (1960) *Grotte du Mas d'Azil (Ariège).* Paris: Annales de Paléontologie.
Peretz, I. (1993) Auditory atonalia for melodies. *Cognitive Neuropsychology* 10, 21–56.
Peretz, I. (2001) Listen to the brain: a biological perspective on musical emotions. In P. Juslin and J. Sloboda (eds.) *Music and Emotion.* Oxford: Oxford University Press, 105–34.
Peretz, I. (2003) Brain specialization for music: new evidence from congenital amusia. In I. Peretz and R. Zatorre (eds.) *The Cognitive Neuroscience of Music.* Oxford: Oxford University Press, 192–203.
Peretz, I., and Gagnon, L. (1999) Dissociation between recognition and emotional judgements for melodies. *Neurocase* 5, 21–30.

Peretz, I., and Kolinsky, R. (1993) Boundaries of seperability between melody and rhythm in music discrimination: a neurological perspective. *Quarterly Journal of Experimental Psychology* 46A, 301–25.

Peretz, I., Kolinsky, R., Tramo, M., Labrecque, R., Hublet, C., DeMeurisse, G., and Belleville, S. (1994) Functional dissociations following bilateral lesions of auditory-cortex. *Brain* 117, 1283–1301.

Peretz, I., and Morais, J. (1980) Modes of processing melodies and ear asymmetry in non-musicians. *Neuropsychologia* 18, 477–89.

Peretz, I., and Morais, J. (1983) Task determinants of ear differences in melody processing. *Brain and Cognition* 2, 288–92.

Peretz, I., and Zatorre, R. (2003) *The Cognitive Neuroscience of Music*. Oxford: Oxford University Press.

Peterson, N. (1999) Introduction: Australia. In R. Lee and R. Daly (eds.) *The Cambridge Encyclopedia of Hunters and Gatherers*. Cambridge: Cambridge University Press, 317–23.

Peterson, S., Fox, P., Posner, M., Mintum, M., and Raichle, M. (1988) Positron Emission Tomographic studies of the cortical anatomy of single-word processing. *Nature* 331, 585–9.

Pétillon, J.-M. (2004) Lecture critique de la stratigraphie magdalénienne de la Grande Salle d'Isturitz (Pyrénées-Atlantiques). *Antiquités Nationales* 36, 105–31.

Petitto, L., and Marentette, P. (1991) Babbling in the manual mode—evidence for the ontogeny of language. *Science* 251, 1493–6.

Petrides, M., Cadoret, G., and Mackey, S. (2005) Orofacial somatomotor responses in the macaque monkey homologue of Broca's area. *Nature* 435, 1235–8.

Petsche, H., Rappelsberger, P., Filz, O., and Gruber, G. (1991) EEG studies in the perception of simple and complex rhythms. In J. Sundberg and L. Nord (eds.) *Music, Language, Speech and Brain*. Basingstoke: MacMillan, 318–26.

Piette, E. (1874) La Flûte composée a l'âge du Renne. *Comptes Rendus de l'Académie des Sciences* 79, 1277.

Piette, E. (1875) Les Vestiges de la période neolithique comparés à ceux des âges antérieurs. *Congrès de l'Association Française pour l'Avancement des Sciences*, Congrès de Nantes, 20.

Piette, E. (1907) *L'Art pendant l'âge du Renne*. Paris: Masson et Cie Éditeurs, plate 1, figs. 6–7.

Pinker, S. (1994) *The Language Instinct: How The Mind Creates Language*. New York: William Morrow.

Pinker, S. (1997) *How The Mind Works*. London: Allen Lane.

Platel, H., Price, C., Wise, J., Lambert, J., Frackowiak, R., Lechevalier, B., and Eustache, F. (1997) The structural components of music perception. *Brain* 120, 229–43.

Ploog, D. (2002) Is the neural basis of vocalisation different in non-human primates and *Homo sapiens*? *Proceedings of the British Academy* 106, 121–34.

Poeppel, D., and Hickok, G. (2004) Towards a new functional anatomy of language. *Cognition* 92, 1–12.

The PoinTIS Spinal Cord Occupational Therapy site of the SCI Manuals for Providers, http://calder.med.miami.edu/providers/OCCUPATIONAL/over2.html.

Popescu, M., Otsuka, A., and Ioannides, A. (2004) Dynamics of brain activity in motor and frontal cortical areas during music listening: a magnetoencephalographic study. *Neuroimage* 21, 1622–38.

Provasi, J., and Bobin-Begue, A. (2003) Spontaneous motor tempo and rhythmical synchronisation in 2½- and 4-year-old children. *International Journal of Behavioral Development* 27: 220–31.

Rebuschat, P., Rohrmeier, P., Hawkins, J., and Cross, I. (2011) *Language and Music as Cognitive Systems*. Oxford: Oxford University Press.

Recanzone, G., Schreiner, C., and Merzenich, M. (1993) Plasticity in the frequency representation of primary auditory cortex following discrimination training in adult owl monkeys. *Journal of Neuroscience* 13, 87–103.

Reddy, V. (2003) On being the object of attention: implications for self–other consciousness. *Trends in Cognitive Sciences* 7, 397–402.

Renfrew, C., and Bahn, P. (2000) *Archaeology: Theories, Methods and Practice* (3rd edn.). London: Thames & Hudson.

Renfrew, C., and Morley, I. (2009) *Becoming Human: Innovation in Prehistoric Material and Spiritual Culture*. Cambridge: Cambridge University Press.

Renfrew, C., and Scarre, C. (1998) *Cognition and Material Culture: The Archaeology of Symbolic Storage*. Cambridge: McDonald Institute Monographs.

Repp, B. (1991) Some Cognitive and Perceptual Aspects of Speech and Music. In J. Sundberg, L. Nord, and R. Carlson (eds.) *Music, Language, Speech and Brain*. Basingstoke: MacMillan Press, 257–68.

Reznikoff, I. (2008) Sound resonance in prehistoric times: A study of Paleolithic painted caves and rocks. Paper presented to *Acoustics Paris '08*. 24 June–4 July.

Reznikoff, I., and Dauvois, M. (1988) La Dimension sonore des grottes ornées. *Bulletin de la Société Préhistorique Française* 85, 238–46.

Richards, K. (2010) *Life*. London: Weidenfeld & Nicolson.

Richerson, P., and Boyd, R. (2005) *Not By Genes Alone: How Culture Transformed Human Evolution*. Chicago: Chicago University Press.

Riches, D. (1984) Hunters, herders and potlatchers: towards a sociological theory of prestige. *Man* 19, 234–51.

Riches, D. (1995) Hunter-gatherer structural transformations. *Journal of the Royal Anthropological Institute* 1, 679–701.

Richman, B. (2000) How music fixed "nonsense" into significant formulas: On rhythm, repetition, and meaning. In N. Wallin, B. Merker, and S. Brown (eds.) *The Origins of Music*. Cambrige MA: MIT Press, 301–14.

Richter D., Waiblinger J., Rink W., and Wagner G. (2000) Thermoluminescence, electron spin resonance and C-14-dating of the Late Middle and Early Upper Palaeolithic site of Geissenklosterle Cave in southern Germany. *Journal of Archaeological Science* 27, 71–89.

Rizzolatti, G., and Arbib, M. (1998) Language within our grasp. *Trends in Neurosciences* 21, 188–94.

Rizzolatti, G., and Craighero, L. (2004) The mirror-neuron system. *Annual Review of Neuroscience* 27, 169–92.

Rizzolatti, G., Fadiga, L., Gallese, V., and Fogassi, L. (1996) Premotor cortex and the recognition of motor actions. *Cognitive Brain Research* 3, 131–41.

Rizzolatti, G., Fogassi, L., and Gallese, V. (2006) Mirrors in the mind. *Scientific American* 295, 30–7.

Rodríguez, L. Cabo, L., and Egocheaga, J. (2003) Breve nota sobre el hioides Neandertalense de Sidrón (Piloña, Asturias). In M. Aluja, A. Malgosa, and R. Nogués (eds.) *Antropología y Diversidad*, i. Barcelona: Edicions Bellaterra, 484–93.

Roederer, J. (1984) The search for a survival value of music. *Music Perception* 1, 350–6.

Ross, C., and Ravosa, M. (1993) Basicranial flexion, relative brain size, and facial kyphosis in nonhuman primates. *American Journal of Physical Anthropology* 91, 305–24.

Rottländer, R. (1996) Frühe Flöten und die Ausbildung der musikalischen Hörgewohnheiten des paläolithischen Menschen. In I. Campen, J. Hahn, and M. Uerpmann (eds.) *Spuren der Jagd—Die Jagd nach Spuren. Festschrift für Hansjürgen Müller-Beck.* Tübingen: Tubinger Monographien zur Urgeschichte 11, 35–40.

Rousseau, J.-J. (1781/1993) *Essai sur l'origine des langues.* Paris: Flammarion.

Roussot, A. (1970) Flûtes et sifflets paléolithiques en Gironde. *Révue Historique de Bordeaux et du Département de la Gironde*, 19, 5–12.

Roussot, A., and Ferrier, J. (1970) Le Grotte de Fontarneaud (Gironde). *Bulletin de la Société Préhistorique Française* 68, 505–20.

Roux, V., and Bril, B. (2005) *Stone Knapping: The Necessary Conditions for a Uniquely Hominin Behaviour.* Cambridge: McDonald Institute Monographs.

Sacks, O. (1996) *An Anthropologist on Mars.* New York: Random House.

Sacks, O. (2008) *Musicophilia: Tales of Music and the Brain.* London: Vintage.

Sadie, S. (2001) *The New Grove Dictionary of Music and Musicians* (2nd edn.). London: Grove.

Saint-Périer, R. de (1947) Les Derniers Objets Magdaleniens d'Isturitz. *L'Anthropologie* 51, 393–415.

Saint-Périer, R. de, and Saint-Périer, S. de (1952) La Grotte d'Isturitz III. Les Solutréens, Les Aurignaciens et Les Moustériens. *Archives de l'Institut de Paléontologie Humaine* 25.

Sakai, K., Hikosaka, O., Miyauchi, S., Takino, R., Tamada, T., Iwata, N., and Nielsen, M. (1999) Neural representation of a rhythm depends on its interval ratio. *Journal of Neuroscience* 19, 10074–81.

Samson, S., and Ehrlé, N. (2003) Cerebral substrates for music temporal processes. In I. Peretz and R. Zatorre (eds.) *The Cognitive Neuroscience of Music.* Oxford: Oxford University Press, 204–16.

Scarre, C. (1989) Painting By Resonance. *Nature* 338, 382.

Scarre, C., and Lawson, G. (2006) *Archaeoacoustics.* Cambridge: McDonald Institute Monographs.

Schachner, A., Brady, T., Pepperberg, I., and Hauser, M. (2009) Spontaneous motor entrainment to music in multiple vocal mimicking species. *Current Biology* 19, 831–6.

Schacter, S., and Singer, J. (1962) Cognitive, social, and physiological determinants of emotional states. *Psychological Review* 69, 379–99.

Schaeffner, A. (1936) *L'Origine des instruments de musique, Introduction ethnologique à l'histoire de la musique instrumentale.* Paris: Mouton.

Schaller, G. (1963) *The Mountain Gorilla.* Chicago: University of Chicago Press.

Schaller, G., Qitao, T., Johnson, K., Xiaoming, W., Heming, S., and Jinchu, H. (1989) The feeding ecology of giant pandas and Asiatic black bears in the Tangjiahe reserve, China. In J. Gittleman (ed.) *Carnivore Behaviour, Ecology and Evolution*. Ithaca NY: Cornell University Press, 212–41.

Schellenberg, E., and Trehub, S. (1994) Frequency ratios and the perception of tone patterns. *Psychonomic Bulletin and Review* 1, 191–201.

Schellenberg, E., and Trehub, S. (1996a) Natural musical intervals: evidence from infant listeners. *Psychological Science* 7, 272–7.

Schellenberg, E., and Trehub, S. (1996b) Children's discrimination of melodic intervals. *Developmental Psychology* 32, 1039–50.

Scherer, K. (1985) Vocal affect signalling: a comparative approach. *Advances in the Study of Behaviour* 15, 189–244.

Scherer, K. (1986) Vocal affect expression: a review and model for future research. *Psychological Bulletin* 99, 143–65.

Scherer, K. (1991) Emotion expression in speech and music. In J. Sundberg, L. Nord, and R. Carlson (eds.) *Music, Language, Speech and Brain*. Basingstoke: MacMillan Press, 146–56.

Schmidt, K., and Cohn, J. (2001) Human facial expressions as adaptations: evolutionary questions in facial expression research. *Yearbook of Physical Anthropology* 44, 3–24.

Schön, D., Magne, C., and Besson, M. (2004) The music of speech: Music training facilitates pitch processing in both music and language. *Psychophysiology* 41, 341–9.

Schrive, C. (1984) *Past and Present in Hunter-Gatherer Studies*. London: Academic Press.

Schulkin, J. (2013) *Reflections on the Musical Mind: An Evolutionary Perspective*. Princeton: Princeton University Press.

Schulz, G., Varga, M, Jeffries, K., Ludlow, C., and Braun, A. (2005) Functional neuroanatomy of human vocalization: An $H_2^{15}O$ PET Study. *Cerebral Cortex* 15, 1835–47.

Schweiger, A. (1985) Harmony of the spheres and the hemispheres: the arts and hemispheric specialisation. In D. F. Benson and E. Zaidel (eds.) *The Dual Brain*. New York: Guildford Press, 359–73.

Schweiger, A., and Maltzman, I. (1985) Behavioural and electrodermal measures of lateralisation for music perception in musicians and nonmusicians. *Biological Psychology* 20, 129–45.

Scothern, P. (1987) The Middle-Upper Palaeolithic Transition and the Evolution of Musical Behaviour. M.Mus. dissertation, University of London.

Scothern, P. (1992) The Music-Archaeology of the Palaeolithic within its Cultural Setting. PhD thesis, University of Cambridge.

Seewald, O. (1934) *Beitrage zur Kenntnis der Steinzeitlichen Muzikinstrumente Europas*. Vienna: Schroll.

Sergent, J., Zuck, E., Terriah, S., and MacDonald, B. (1992) Distributed neural network underlying musical sight-reading and keyboard performance. *Science* 257, 106–9.

Serjeantson, D. (1997) Subsistence and symbol: the interpretation of bird remains in archaeology. *International Journal of Osteoarchaeology* 7, 255–9.

Seyfarth, R., and Cheyney, D. (1992) Meaning and mind in monkeys. *Scientific American* 267, 78–84.

Seyfarth, R., and Cheyney, D. (1997) Behavioural mechanisms underlying vocal communication in nonhuman primates. *Animal Learning and Behaviour* 25, 249–67.

Shanahan, D. (2007) *Language, Feeling and the Brain*. London: Transaction Publishers.

Shanon, B. (1982) Lateralisation effects in music decision tasks. *Neuropsychologia* 18, 21–31.

Shenfield, T., Trehub, S., and Nakata, T. (2002) Salivary cortisol responses to maternal speech and singing. Presented at the International Conference on Infant Studies, Toronto, Ontario.

Shennan, S. (2002) *Genes, Memes and Human History*. London: Thames & Hudson.

Shepard, R. (1982) Geometrical approximations to the structure of musical pitch. *Psychological Review* 89, 305–33.

Simmons, T., and Nadel, D. (1998) The Avifauna of the early Epipalaeolithic site of Ohalo II (19,400 years BP), Israel: species diversity, habitat and seasonality. *International Journal of Osteoarchaeology* 8, 79–96.

Slater, P. (2000) Birdsong repertoires: their origin and use. In N. Wallin, B. Merker, and S. Brown (eds.) *The Origins of Music*. Cambridge MA: MIT Press, 49–63.

Sloboda, J. (1985) *The Musical Mind: The Cognitive Psychology of Music*. Oxford: Oxford University Press.

Sloboda, J. (1991) Music structure and emotional response: some empirical findings. *Psychology of Music* 19, 110–20.

Sloboda, J. (1998) Does music mean anything? *Musicae Scientiae* 2, 21–31.

Sloboda, J. (2001) The psychology of music: affect. In L. Macy (ed.) *The New Grove Dictionary of Music Online*. http://www.grovemusic.com (accessed 2 Sept. 2011).

Sloboda, J., Hermelin, B., and O'Connor, N. (1985) On the anatomy of the ritard: a study of timing in music. *Journal of the Acoustical Society of America* 97, 53–67.

Sloboda, J., and Juslin, P. (2001) Psychological perspectives on music and emotion. In P. Juslin and J. Sloboda (eds.) *Music and Emotion: Theory and Research*. Oxford: Oxford University Press, 71–104.

Smith, R. (1911) *Guide to the Antiquities of the Stone Age*. London: British Museum.

Snow, D. (2000) The emotional basis of linguistic and nonlinguistic intonation: implications for hemispheric specialization. *Developmental Neuropsychology* 17, 1–28.

Soffer, O. (1985) *The Upper Palaeolithic of the Central Russian Plain*. New York: Academic Press.

Sollas, W. (1924) *Ancient Hunters and Their Modern Representatives*. 3rd edn., London: MacMillan.

Solow, B. (1966) The pattern of craniofacial associations, *Acta Oolontologica Scandinavia* 24, suppl. 46, 125.

Soproni, I. (1985) The reconstruction of the Istállóskö flute. *Folio Archaeologia* 36, 33.

Sperber, D. (1996) *Explaining Culture*. Oxford: Blackwell.

Spoor, F. (1996) The ancestral morphology of the hominid bony labyrinth: the evidence from *Dryopithicus*. *American Journal of Physical Anthropology* 22 (suppl.), 219.

Spoor, F., Wood, B., and Zonneveld, F. (1994) Implications of hominid labyrinthine morphology for evolution of human bipedal locomotion. *Nature* 369, 645–8.
Spoor, F., and Zonneveld, F. (1998) Comparative review of the human bony labyrinth. *Yearbook of Physical Anthropology* 41, 211–51.
Steinbeis, N., and Koelsch, S. (2008) Shared neural resources between music and language indicate semantic processing of musical tension-resolution patterns. *Cerebral Cortex* 18, 1169–78.
Stepanchuk, V. (1993) Prolom II, a Middle Palaeolithic cave site in the Eastern Crimea with non-utilitarian bone artefacts. *Proceedings of the Prehistoric Society* 59, 17–37.
Stern, D. (1977) *The First Relationship: Infant and Mother*. Cambridge MA: Harvard University Press.
Stern, D., Speiker, S., Barnett, R., and MacKain, K. (1983) The prosody of maternal speech: infant age and context related changes. *Journal of Child Language* 10, 1–15.
Stiner, M., Achyuthan, H., Arsebük, G., Howell, F., Josephson, S., Juell, K., Pigati, J., and Quade, J. (1998) Reconstructing cave bear paleoecology from skeletons: a cross-disciplinary study of middle Pleistocene bears from Yarimburgaz Cave, Turkey. *Paleobiology* 24, 74–98.
Stiner M., Munro, N., and Surovell, T. (2000) The tortoise and the hare—Small-game use, the broad-spectrum revolution, and Paleolithic demography. *Current Anthropology* 41, 39–73.
Stockman, D. (1986) On the early history of drums and drumming in Europe and the mediterranean. In C. Lund (ed.) *The Second Conference Of the ICTM Study Group On Music Archaeology*, i. *General Studies*. Stockholm: Royal Swedish Academy of Music, 11–28.
Stokoe, W. (2000) Gesture to sign (language). In D. McNeill (ed.) *Language and Gesture*. Cambridge: Cambridge Unversity Press, 388–99.
Storr, A. (1992) *Music and the Mind*. London: Harper Collins.
Stringer, C., and Andrews, P. (2005) *The Complete World of Human Evolution*. London: Thames & Hudson.
Stringer, C., and Gamble, C. (1993) *In Search of the Neanderthals*. Thames & Hudson, London.
Svoboda J., Czudec, T., Havlicek, P., Lozek, V., Macoun, J., Prichystal, A., Svobodová, H., and Vlcek, E. (1994) *Paleolit Moravy a Slezska*. Brno: Dolnovestonicke studie 1.
Swenson, J., Dahle, B., and Sandegren, F. (2001) Intraspecific predation in Scandinavian brown bears older than cubs-of-the-year. *Ursus* 12, 81–92.
Sykes J. (1983) *Concise Oxford Dictionary* (7th edn.). Oxford: Oxford University Press.
Tartter, V. (1980) Happy talk: perceptual and acoustic effects of smiling on speech. *Perception and Psychophysics* 27, 24–7.
Tartter, V., and Braun, D. (1994) Hearing smiles and frowns in normal and whisper registers. *Journal of the Acoustical Society of America* 96, 2101–7.
Taylor, C. (1991) *The Native Americans*. Twickenham: Salamander Books (Tiger Books).
Thaut, M. (2005) *Rhythm, Music and the Brain*. Abingdon: Routledge.
Thaut, M., McIntosh, G., and Rice, R. (1997) Rhythmic facilitation of gait training in hemiparetic stroke rehabilitation. *Journal of Neurological Sciences* 151, 207–12.
Thieme, H. (1997) Lower Palaeolithic hunting spears from Germany. *Nature* 385, 807–10.

Thieme, H. (2005) The Lower Palaeolithic art of hunting: The Case of Schöningen 13 11-4, Lower Saxony, Germany. In C. Gamble and M. Porr (eds.) *The Hominid Individual in Context: Archaeological Investigations of Lower and Middle Palaeolithic Landscapes, Locales and Artefacts.* Abingdon: Routledge, 115-32.

Thompson, E. (2001) *Between Ourselves: Second-Person Issues in the Study of Consciousness.* Thorverton: Imprint Academic.

Toner, P. (2007) The gestation of cross-cultural music research and the birth of ethnomusicology. *Humanities Research* 14, 85-110.

Tonkinson, R. (1999) The Ngarrindjeri of Southeastern Australia. In R. Lee and R. Daly (eds.) *The Cambridge Encyclopedia of Hunters and Gatherers.* Cambridge: Cambridge University Press, 343-7.

Tooby, J., and Cosmides, L. (2001) Does beauty build adapted minds? Towards an evolutionary theory of aesthetics, fiction and the arts. *SubStance* 30, 6-27.

Trainor, L., Austin, C., and Desjardins, R. (2000) Is infant-directed speech prosody a result of the vocal expression of emotion? *Psychological Science* 11, 188-95.

Trainor, L., Clark, E., Huntley, A., and Adams, B. (1997) The acoustic basis of infant preferences for infant-directed singing. *Infant Behavior and Development* 60, 383-96.

Trainor, L., and Heinmiller, B. (1998) The development of evaluative responses to music: Infants prefer to listen to consonance over dissonance. *Infant Behavior and Development* 21, 77-88.

Trainor, L., and Schmidt, L. (2003) Processing emotions induced by music. In I. Peretz and R. Zatorre (eds.) *The Cognitive Neuroscience of Music.* Oxford: Oxford University Press, 310-24.

Tramo, M., Cariani, P., Delgutte, B., and Braida, L. (2003) Neurobiology of harmony perception. In I. Peretz and R. Zatorre (eds.) *The Cognitive Neuroscience of Music.* Oxford: Oxford University Press, 127-51.

Trehub, S. (2003) Musical predispositions in infancy: an update. In I. Peretz and R. Zatorre (eds.) *The Cognitive Neuroscience of Music.* Oxford: Oxford University Press, 3-20.

Trehub, S., Schellenberg, E., and Kamenetsky, S. (1999) Infants' and adults' perception of scale structure. *Journal of Experimental Psychology: Human Perception and Performance* 25, 965-75.

Trehub, S., Trainor, L., and Unyk, A. (1993) Music and speech processing in the first year of life. In H. Reese (ed.) *Advances in Child Development and Behaviour* 24. New York: Academic Press, 1-35.

Trevarthen, C. (1997) Foetal and neonatal psychology: Intrinsic motives and learning behaviour. In F. Cockburn (ed.) *Advances in Perinatal Medicine.* New York: Parthenon, 282-91.

Trevarthen, C. (1999) Musicality and the intrinsic motive pulse: evidence from human psychobiology and infant communication. *Musicae Scientiae*, special issue, 155-215.

Trevarthen, C. (2004) How infants learn how to mean. In M. Tokoro and L. Steels (eds.) *A Learning Zone of One's Own.* Amsterdam: IOS Press, 37-69.

Trevarthen, C., and Aitken, K. (1994) Brain development, infant communication, and empathy disorders: intrinsic factors in child mental health. *Development and Psychopathology* 6, 599–635.

Turino, T. (1992) The music of Sub-Saharan Africa. In B. Nettl, C. Capwell, P. Bohlman, I. Wong, and T. Turino (eds.) *Excursions in World Music*. Englewood Cliffs NJ: Prentice Hall, 165–95.

Turk, I. (1997) *Mousterian 'Bone Flute' and other finds from Divje babe I cave site in Slovenia*. Ljubljana: Institut za Archaeologijo, Znanstvenoraziskovalni Center Sazu.

Turk, I., Blackwell, B., Turk, J., and Pflaum, M. (2006) Results of computer tomography of the oldest suspected flute from Divje babé I (Slovenia) and its chronological position within global palaeoclimatic and palaeoenvironmental change during Last Glacial. *L'Anthropologie* 110, 293–317.

Turk, I., Dirjec, J., Bastiani, G., Pflaum, M., Lauko, T., Cimerman, F., Kosel, F., Grum, J., and Cevc, P. (2001) Nove analize 'piscali' iz Divjih bab I (Slovenija). *Arheoloski vestnik* 52, 25–79.

Turk, I., Dirjec, J., and Kavur, B. (1997) Description and explanation of the origin of the suspected bone flute. In I. Turk (ed.) *Mousterian 'Bone Flute'*. Ljubljana: Institut za Archaeologijo, Znanstvenoraziskovalni Center Sazu, 157–78.

Turk, I., and Kavur, B. (1997) Palaeolithic bone flutes—comparable material. In I. Turk (ed.) *Mousterian 'Bone Flute' and other finds from Divje babe I cave site in Slovenia*. Ljubljana: Institut za Archaeologijo, Znanstvenoraziskovalni Center Sazu, 179–84.

Turk, I., Pflaum, M., and Pekarovič, D. (2005) Rezultati računališke tomografije najstarejše domnevne piščali iz Divjih bab I (Slovenija): prispevek k teoriji luknjanja kosti [Results of computer tomography of the oldest suspected flute from Divje babe I (Slovenia): contribution to the theory of making holes in bones]. *Arheološki vestnik* 56, 9–36.

Turnbull, C. (1962) *The Forest People*. New York: Simon & Schuster.

Turner, R., and Ioannides, A. (2009) Brain, music and musicality: inferences from neuroimaging. In S. Malloch and C. Trevarthen (eds.) *Communicative Musicality: Exploring the Basis of Human Companionship*. Oxford: Oxford University Press, 147–81.

Turq, A., Normand, C., and Valladas, H. (1999) Saint-Martin d'Arberoue; grotte d'Isturitz. *Bilan scientifique 1998*. Bordeaux: Direction Régionale des Affaires Culturelles, Service Régional de l'Archéologie.

Tyrberg, T. (1998) *Pleistocene Birds of the Palearctic: A Catalogue*. Publications of the Nuttall Ornithological Club, No. 27. Cambridge MA: Nuttall Ornithological Club.

Tyrberg, T. (2008) *Supplement to Pleistocene Birds of the Palearctic: A Catalogue*. http://web.telia.com/~u11502098/pleistocene.pdf.

Van Lancker, D., and Canter, G. (1982) Impairment of voice and face recognition in patients with hemispheric damage. *Brain and Cognition* 1, 185–95.

Vanderhorst, V., Teresawa, E., and Ralston, H. (2001) Monosynaptic projections from the nucleus retroambiguus region to laryngeal motoneurons in the rhesus monkey. *Neuroscience* 107, 117–25.

Vaneechoutte, M., and Skoyles, J. (1998) The memetic origin of language: modern humans as musical primates, *Journal of Memetics—Evolutionary Models of*

Information Transmission 2, 129–69. Also available online at http://www.mmu.ac.uk/jom-emit/1998/vol2/vameechoutte_m&skoyles_jr.html.

VanWijngaarden-Bakker, L. (1997) The selection of bird bones for artefact production at Dutch Neolithic sites. *International Journal of Osteoarchaeology* 7, 339–45.

Vekua, A., Lordkipanidze, D., Rightmire, G., Agusti, J., Ferring, R., Maisuradze, G., Mouskhelishvili, A., Nioradze, M., de Leon, M., Tappen, M., Tvalchrelidze, M., and Zollikofer, C. (2002) A new skull of early homo from Dmanisi, Georgia. *Science* 297, 85–9.

Vértes, L. (1955) Neue Ausgraben und Palaeolitische Funde in der Höhle von Istállóskö. *Acta Academiae Scientiarum Hungarae* 5, 111.

Victor, P., and Robert-Lamblin, J. (1989) *La Civilisation du Phoque*. Paris: Armand Colin-Raymond Chabaud.

Vincent, A. (1988) L'Os comme artefact au Paléolithique moyen: Principes d'étude et premiers résultats. In L. Binford and J. Rigaud (eds.) *L'Homme de Néanderthal*, iv. *La Technique*. Liège: Études et Recherches Archéologiques de l'Université de Liège 31, 185–96.

Vircoulon, J., and Deffarge, R. (1977) La Station de Garrigue. *Revue de l'Histoire et Archéologie Libournais* 165, 96–8.

Wagner, H. (1989) The peripheral physiological differentiation of emotions. In H. Wagner and A. Manstead (eds.) *Handbook of Psychophysiology: Emotion and Social Behavior*. Chichester: Wiley & Sons, 77–98.

Walker, A. (1993) Perspective on the Nariokotome discovery. In A. Walker and R. Leakey, *The Nariokotome* Homo erectus *Skeleton*. Cambridge MA: Harvard University Press, 411–32.

Wallin, N. (1991) *Biomusicology: Neurophysiological, Neuropsychological and Evolutionary Perspectives on the Origins and Purpose of Music*. Stuyvesant NY: Pendragon Press.

Wallin, N., Merker, B., and Brown, S. (2000) *The Origins of Music*. Cambridge MA: MIT Press.

Waterman, C. (1991) Uneven development of African ethnomusicology. In B. Nettl and P. Bohlman (eds.) *Comparative Musicology and Anthropology of Music*. Chicago: University of Chicago Press, 169–84.

Watt, D., and Pincus, D. (2004) Neural Substrates of Consciousness: Implications for Clinical Psychiatry. In J. Panksepp (ed.) *Textbook of Biological Psychiatry*. Oxford: Wiley-Blackwell, 75–110.

Watt, R., and Ash, R. (1998) A Psychological Investigation Of Meaning In Music. *Musicae Scientiae* 2, 33–54.

Weissengruber, G., Forstenpointner, G., Peters, G., Kübber-Heiss, A., and Fitch, W. (2002) Hyoid apparatus and pharynx in the lion (*Panthera leo*), jaguar (*Panthera onca*), tiger (*Panthera tigris*), cheetah (*Acinonyx jubatus*), and domestic cat (*Felis silvestris f. catus*). *Journal of Anatomy (London)* 201, 195–209.

Werker, J., Pegg, J., and McLeod, P. (1994) A cross-language investigation of infant preference for infant-directed communication. *Infant Behaviour and Development* 17, 323–33.

Wharton, T. (2009) *Pragmatics and Non-Verbal Communication*. Cambridge: Cambridge University Press.

White, H. (1978) *Tropics of Discourse: Essays in Cultural Criticism*. London: Johns Hopkins University Press.
Wild, B., Erb, M., and Bartels, M. (2001) Are emotions contagious? Evoked emotions while viewing emotionally expressive faces: quality, quantity, time course and gender differences. *Psychiatry Research* 102, 109–24.
Williams, L. (1980) *The Dancing Chimpanzee*, 2nd edn. London: Allison & Busby.
Williamson, R., and Farrer, C. (1992) *Earth and Sky: Visions of the Cosmos in Native American Folklore*. Albuquerque: University of New Mexico Press.
Wilson, D. (1997) Incorporating group-selection into the adaptationist program: a case study involving human decision-making. In J. Simpson and D. Kendrick, (eds.) *Evolutionary Social Psychology*. Hillsdale NJ: Lawrence Erlbaum, 345–86.
Wilson, D., and Sober, E. (1994) Reintroducing group selection to the behavioural sciences. *Behavioral and Brain Sciences* 17, 585–654.
Wind, J. (1990) The evolutionary history of the human speech organs. In J. Wind, E. Pulleybank, E. de Groler, and G. Bichakjian (eds.) *Studies in Language Origins*, i. Amsterdam: Benjamins, 173–97.
Winston, C. (ed.) (1918) *Winston's Encyclopedia*. Philadelphia: The John C. Winston Company.
Wray, A. (1998) Protolanguage as a holistic system for social interaction. *Language and Communication* 18, 47–67.
Wray, A. (2000) Holistic utterances in protolanguage: the link from primates to humans. In C. Knight, M. Studdert-Kennedy, and J. Hurford: *The Evolutionary Emergence of Language: Social Function and the Origins of Linguistic Form*. Cambridge: Cambridge University Press, 285–302.
Wray, A. (2006) Joining the dots: the evolutionary picture of language and music (Review Feature on The Singing Neanderthals by Steven Mithen). *Cambridge Archaeological Journal* 16, 97–112.
Wynn, T. (1993) Two developments in the mind of early Homo. *Journal of Anthropological Archaeology* 12, 299–322.
Wynn, T., and Coolidge, F. (2003) The role of working memory in the evolution of managed foraging. *Before Farming* 2003/2, 1–16.
Wynn, T., and Coolidge, F. (2004) The expert Neandertal mind. *Journal of Human Evolution* 46, 467–87.
Wynn, T., and Coolidge, F. (2005) Working memory, its executive functions, and the emergence of modern thinking. *Cambridge Archaeological Journal* 15, 5–26.
Wynn, T., and Coolidge, F. (2006) The effect of enhanced working memory on language. *Journal of Human Evolution* 50, 230–1.
Wynn, T., and Coolidge, F. (2007) Did a small but significant enhancement in working-memory capacity power the evolution of modern thinking? In P. Mellars, K. Boyle, O. Bar-Yosef, and C. Stringer (eds.) *Rethinking the Human Revolution*. Cambridge: McDonald Institute Monographs, 79–90.
Wynn, T., and Coolidge, F. (2010) Beyond symbolism and language: an introduction to Supplement 1, Working Memory. *Current Anthropology* 51, suppl. 1, S5–16.
Wynn, T., Coolidge, F., and Bright, M. (2009) Hohlenstein-Stadel and the evolution of human conceptual thought. *Cambridge Archaeological Journal* 19, 73–83.

Zatorre, R. (1984) Musical perception and cerebral function: a critical review. *Musical Perception* 2, 196–221.

Zatorre, R. (2003) Neural specializations for tonal processing. In I. Peretz and R. Zatorre (eds.) *The Cognitive Neuroscience of Music*. Oxford: Oxford University Press, 231–47.

Zatorre, R., Evans, A., and Meyer, E. (1994) Neural mechanisms underlying melodic perception and memory for pitch. *Journal of Neuroscience* 14, 1908–19.

Zatorre, R., Evans, A., Meyer, E., and Gjedde, A. (1992) Lateralization of phonetic and pitch processing in speech perception. *Science* 256, 846–9.

Zatorre, R., Halpern, A., Perry, D., Meyer, E., and Evans, A. (1996) Hearing in the mind's ear: a PET investigation of musical imagery and perception. *Journal of Cognitive Neurocience* 8, 29–46.

Zentner, M., and Eorola, T. (2010) Rhythmic engagement with music in infancy. *Proceedings of the National Academy of Sciences of the USA* 107, 5768–73.

Zentner, M., and Kagan, J. (1998) Infants' perception of consonance and dissonance in music. Infant Behavior and Development 21, 483–92.

Zervos, C. (1959) *L'Art de l'époque du Renne en France*. Paris: Éditions Cahiers d'Art.

Zubrow, E., Cross, I., and Cowan, F. (2001) Musical behaviour and the archaeology of the mind. *Archaeologia Polona* 39, 111–26.

Index

Italic figures refer to items referenced in artwork and tables.

A.L.333–45 (hominin specimen) 147
A.L.333–105 (hominin specimen) 147
A.L.333–114 (hominin specimen) 147
Aartswoud (Netherlands) 95
Abbé Glory 117, 119, 127
Aborigines (Australian) 14, 15, 22–4, 224, 282
  bullroarers 105
abri Fontalès (France) 108
Abri Lafaye Bruniquel (France) 110, 111, 112
abri Morin (France) 108
abstract:
  signs 117–19, 125
  symbolism 301, 309
Ach Valley (Germany) 38, 42–51, 90, 94, 124, 128, 324
  pipes 43–6, *46*, *48*, *50*, 88
acoustic(s):
  earliest evidence for 3
  energy of vocal sound 133
  properties at Isturitz 53–4
  properties of caves 104, 115–17, 124, 126–7
  properties of flint 120
  properties of music and speech 170, 216
action-observation-network 249, 252
Addis, L. 263
adolescence, vocal physiology 135
Aegypius monachus, see black vulture
Aegypius sp., see vultures
aerophones 17, 99, 100–9
  free 105–9
Afalou 5 (hominin specimen) 140
affect (emotion) 167, 206, 213, 258
  affect and body language/posture/corporeal gesture 215, 234, 237, 240, 251, 313
  affective content processing 188, 189, 197, 205, 208, 212, 213, 250, 252, 273, 306, 313, 321
  affective function 183
  affective relationships 210
  affective response 205, 213, 226, 235
  affective state 168, 177, 206, 226, 244, 261, 263, 265
  affective vocal performance/production 185, 197, 207, 209, 210, 211, 212, 250, 273, 313
  affective vocalization and communication 209, 215, 219, 220, 222, 223, 225, 226, 227, 237, 240, 251, 319
  affective vocalization in primates 222, 237
  extrinsic affect 257

facial affect 208, 215, 218, 226, 312, 313
intrinsic affect 257
vocal affect 208, 226, 312, 313
see also emotion
Africa(n):
  bevelled flutes 49
  bullroarers 105
  early modern humans 91
  great apes 132, 143, 144
  Middle Stone Age 36
  new package of behaviours 96
  North, lack of bird bone 92
  Pygmies 14, 15, 19–22, 224, 282
  see also Aka, Mbuti
  ringing stones 120
aggression and pitch 207
agriculturalists, early *vs* hunter-gatherers 123
agriculture, spread of early 122
Aiello, L. 141, 221, 222, 225
Aka 13, 19–22
  songs 29
Alaska 24–9
  juggling-game songs 27
  Yupik people 14, 24–9, 30, 31, 224, 282
Alcock, K. 193, 230
alders 25
Alipour, M. 148
altriciality 293
Amazonian bullroarers 108
American Sign Language 233, 234
amplitude 150, 170
  modulation 218
AMS dating, cave art 119
Amud 1 (hominin specimen) 145
amusia 182, 184, 186, 187
amygdala 221, 265, 270
Anatolia, dancing representations 122
anatomically modern humans 3, 33, 38, 90–1, 125, 153, 177, 302
  inner ear 172
  see also early modern humans, *Homo sapiens*
ancestor spirits 23
animal(s):
  emotive expression 218
  -headed figures 89
  representations 53
  scarers 105

428 Index

animal(s): (cont.)
  vocal anatomy 142, 143, 144
  vocalization 135
  see also under names of individual species
Anser albifrons, see white-fronted goose
Anser fabalis, see bean goose
Anser sp., see geese
antelope 16, 20
anterior cingulate cortex 165, *165*, 166, *180*
anthropomorphs 102
antler:
  bullroarers 105, 107–8, *107*
  decorated 53
  mallets 15
  rasp 110
  rattles 99
apes 297
  African 144
  basicranium 141
  cognition 298
  great 132, 143–4, 146, 149, 151, 163, 173
  hypoglossal canal 147
  vocal anatomy 133
aphasia 182, 184–5
appoggiaturas 269
Aquila sp., see eagles
'archaic *Homo sapiens*' 299
Arctic Canada, whistles 105
Arcy-sur-Cure (France) 116
Ardipithecus kadabba 4
Ardipithecus ramidus 4
Arensberg, P. 140, 145
Ariège (France) 89, 115
Armstrong, D. 223, 233, 235, 236
Arrernte 23
art 96
  early evidence for 36
  Palaeolithic 53, 89–90, 324
  and symbolism 294
articulatory invariants 170
Artificial Memory Systems 87, 111
Ash, R. 225, 262
Asia, South-East, ringing stones 120
Asperger syndrome 265–7
astronomical phenomena 114
Asturias (Spain) 145
AT-1500 (hominin specimen) 145
AT-2000 (hominin specimen) 145
Atapuerca (Spain) 145
Atapuerca 4 and 5 (hominin specimens) 177
Atema, J. 39
auditory:
  abilities 131
  agnosia 189
  cortex *180*, 191, 192, 197, 198
  meatus, see ear canal

  nerve 271
  rhythm 248
  significance 117
  stimuli 7
  system 169–74
Aurignacian:
  avian faunal remains 92, 125
  pipe-like artefacts 88
  pipes 35, 42–6, *46*, 48–9, *48*, *50*, 51, 55, *85*
  split bone points 110
  whistles *101*
aurora borealis 28
Australia:
  Aborigines 14, 15, 22–4, 224, 282, 309
  bullroarers 105, 108
  ringing stones 120
australopithecines 140, 149, 151, 156–7, 173, 175, 319
  A. afarensis 4, 144, 145, 147, 222
  A. africanus 4, 138, 147, 222
  A. anamensis 4
  A. boisei 147
  A. garhi 4
  A. sediba 4
  labyrinths 172
  vocal anatomy 146
Austria:
  avian faunal remains 92
  sound-producers 89
autism 265–7, 273
automatic systems 269
autonomic nervous system 238, 264
avian fauna 91
  and environmental stress 94–6
  for subsistence 125
  and technological limitations 93
avocalias 184
awls 96
axes, stone 22
Ayotte, J. 186
Azerbaijan, avian faunal remains 91

BaAka, see Aka
babbling 203, 204, 231, 250, 251
baboons 221
Baddeley, A. 301
Badegoule (France) 108
Baghemil, B. 29
Baka 19
balance 172, 173
Balkans, dancing representations 122
Balow, M. 146
BaMbuti, see Mbuti
bark artefacts 19
bark pipes 51

Barkow, J. 280
Bartels, M. 239
basal ganglia *180*, 192, 238
basicranial flexion 135–44, *137*, 142, 145, 152, 176
basicranium 134, 140, 141, 174
Basque region (France), *see* Isturitz (France)
bas-relief engraving, of rasp 110, 112
bean goose 94
bear(s) 128
   central and eastern Europe 90
   head representations 119
bearded lammergeier/vulture 56, *86*, 87
beat perception 245
beaters 114, 115
   representations 121, 122
Beaudry, N. 28
beetles 20
Belgium 110
   flutes 88
Belin, P. 208
Benson, D. F. 189
Bering Strait 24
Bernatsky, G. 272
berries 16, 25
berry-picking songs 26, 27
Bersac (France) 108
Bertrand, D. 192
Besson, M. 8, 189
Bibikov, S. 115, 127
bilma 24
biodegradable materials, *see* organic materials
biological:
   reactions to music 268–73, 274, 282
   rhythms 243, 244
biological selection, *see* selection
bipedalism 141, 156, 158, 175, 249, 319, 320
   and the auditory system 172
   evolution of 132, 133
   and language 174
   and rhythm processing 193
bird bone:
   artefacts, western Europe 90
   engraved 94
   instruments 112
   pipes 35, 42, 44–5, 55–75, 77–81, 85, *85*, 87, 323
   vs mammoth ivory, use for pipes 50
birds:
   see also avian faunal data; individual species
   carrion 87
   identification songs 27
   as a subsistence resource 94
   vocal tract 155
bison 16
   horn representation 112, 113, *113*

representations 118, 121, 122
spirit 89
Bispham, J. 244
Blackfoot 13, 15–19
   songs 29
Blacking, J. 13, 224
blades:
   flint 96
   as sound-producers 120
Blake, E. 120
Blombos cave (South Africa) 96, 97
Blood, A. 269, 270
Blount, B. 234
blueberries 27
boars, representations 121
Bocksteinschmeide (Germany) 106
body:
   language 234, 251, 252, 298, 313, 321
   movements 237–42, 253
   see also dance
Bogen, J. 187, 188
Bois Roche (France) 106
Boivin, N. 115, 120
Bolivia, ringing stones 120
bonding 286, 301, 304–5
   in primates 220, 224, 284, 286
   inter-personal 281
   'muscular bonding' 240, 241
   role of music 26, 224, 280, 281, 283, 284, 319
   role of oxytocin 270, 281
   social 220
bone:
   artefacts 15, 97
   bullroarers 105
   decorated 53
   mammoth 100, 110, 114, 127
   points 96, 110
   rasp 109
   rattles 99
   tools, Upper Palaeolithic 42
   use of, for instrument manufacture 90–6
   -working 125
   see also pipes
bonobo chimpanzees 132, 173, 208, 220
boot-strapping 150, 276
Borchgrevink, H. 180
borer holes 38
Boule, M. 138
bovids, engravings 108
bovine horn 114
Bowles, S. 290
bows 20
   representations 121, 122
brain *162*, *165*, *180*
   anatomy *180*
   areas, reaction to music 270

brain (cont.)
   capacity for language 139
   function, interpretation from
      physiology 162
   and hearing 161–76
   hemispheres 180–1, 184, 189, 198, 231, 311, 319–20
   pathologies 184
   scanning 181–3
   size increase 159, 176, 211, see also encephalization
   specialization for music 196
   stem 167, 238
   -worms 247, 248
Brandt, S. A. 7, 199
Branta bernicla, see brent goose
Branta sp., see geese
breathing, control of 142, 148–53
brent goose 94
Breuil, H. 57, 69, 70, 71, 72
Broca's:
   aphasia 169, 185
   area 162, *162*, 163–5, 167, 175, *180*, 198, 231, 242
Brodmann's areas 163, *180*, 184
Broken Hill (Zambia) 138, 140
Bronze Age:
   cup-marks 120
   rasps 110
Brooks, A. S. 97
brown bears 40, 41
Brown, S. 184, 215, 216, 218, 219
Brust, J. 181, 186, 187
Bryant, G. 283, 284, 285
Buchanan, T. 189
Budil, I. 140
Buisson, D. 52, 55–6, 59–60, 66, 70, 76, 78, 80, 81, 84–5
Bukovač (Serbia) 88
bulbs 16
Bulgaria, avian faunal remains 91, 92
bullroarers 17, 90, 99, 100, 105–9
   reconstructions 108, 109
burin holes 38
burins 103
bustards 94
Butcher, C. 231

C14 dating, see radiocarbon dating
Cacatua galerita, see sulphur-crested cockatoo
calendrical:
   notation, 111, 113
   ritual 123
Callicebus 286
Callitrichidae family 203

Canada:
   Arctic, whistles 105
   Inuit 14, 24–9
   juggling-game songs 27
   throat-games 28
Canary Islands, ringing stones 119
cane 125
   flutes 21
   instruments 31
Cantonese infant-directed speech 205
Cape York 23
capuchin monkeys 202
carcass cutting 26
caribou 28
Carlson, N. 239
carnivore activity 102, 106, 110, 221
   perforations 37–40
   vs deliberate perforation 41
Cartmill, M. 146
Cassoli, P. 93
cat's-cradle 27
Çatalhöyük (Turkey) 121
caterpillars 20
cats:
   feral 22
   vocal capabilities 143
caves:
   art 53
   see also rock art
   decorated, acoustic properties 124, 126–7
   faunal remains 106
   and lithophones 100, 115–21
   use of bullroarers in 108
cave bears 40, 41
   femur pipe 38, 39
cave lions 40
Cebuella pygmaea, see pygmy marmosets
Cebus capucinus, see white-faced capuchin monkeys
cerebellum 165, *180*, 192, 267
ceremonies:
   Aboriginal 22
   and music 17, 20, 23, 29
   see also ritual
cetaceans 221
Chase, P. 102
Chauvet cave (France) 119
Cheney, D. L. 220
children:
   songs for 27
   vocal physiology 135
   see also infant(s)
chime bars 120
chimpanzees 20, 132, 143, 147, 169, 173, 241, 245, 298
   brain anatomy 167
   grooming 221

hyoid 144
rib cages *149*
vocal anatomy 136, *137*
vocalizations 218
  see also bonobo chimpanzees
China, chime bars 120
choking, mortality from 142
chorusing activities 241, 277
Circaetus sp., *see* eagles
circumcision 20
clapping 21, 24, 31
Clayton, M. 248
Clegg, M. 141
cloudberries 27
clubs 93
Clynes, M. 264
coalition signalling 283–5, 305
cochlea 173
cockatoos 245
cocoons, as rattles *18*
cognition:
  abilities 36, 125, 204
  capacity 97, 177
  and culture 297–301
  development 219; and music 291–4
  domains, combination of 195
  evolution 305; and music 294–303
  flexibility 293, 294
  modularity 294–6
  musical 2
  revolution 96
comedy in music 26
communal gatherings 87, 126
communality 122
  of music 268, 314, 315
communication:
  early human 280
  long-distance 104
  in music 255–74
  vocal signalling 199
comparative musicology 13
composers 261, 265
consonance 6, 7, 11, 166, 271, 272, 308
containers, bone 97
contour *see* pitch contour
Coolidge, F. 297, 301–3
cooperative behaviour 241
coprolites 102
corporeal movement 237–42
cortical systems 188, 269
cortico-motoneuronal connection 148
coughing 149
Cougnac (France) 117, 118
courtship songs 285–6
coyotes 102
cranes 95

cranial:
  capacity 151
  vault 162
creative explosion 294
  models of 91, 97
  see also cultural revolution,
    Upper Palaeolithic revolution
Crelin, E. 138
Cro-Magnon 1 (hominin specimen) 140
Cross, I. 8, 120, 260, 291, 292, 293, 295, 296
cultural: revolution 96–8
  see also creative explosion, Upper
    Palaeolithic revolution
cultural selection *see* selection
culture and cognition 297–301
Cumbrian slate 120
cup marks 119, 120
cut-marks 94, 164
cyclical motor control 230, 231, 237, 251
Cygnus cygnus, *see* whooper swans
Cygnus olor, *see* mute swans
Cygnus sp., *see* swans
Czech Republic 106
  rasps 110, 111, *111*

Dame a la corne, *see* Venus of Laussel
Dams, L. 117, 118, 119, 126, 127
dance 6, 8, 11, 17, 240, 241, 308
  Aka 20
  as a coalition signalling system 283–5
  communal 29
  coordination in 243
  as an indicator of fitness 286
  in later prehistory 121–4
  movements and the brain 184
  and music 237, 309, 316, 317
  representations 89, 120, 121, 122
  songs 25–6
  steps 218
  and vocalization 251
Daniel, H. 172, 174
Danube basin, dancing representations 122
darts, poisoned 19, 20
Darwin, C. 1, 3, 285
Darwinian theory 278
Dauvois, M. 102, 103, 105, 108, 112, 117, 119
Davidson, I. 37
Davies, S. 262, 263
Davis, P. J. 167
declarative memory *see* memory
deer:
  hooves 16, 17
  laryngeal lowering 156
  red 135
  representations 118, 121, 122

defecation 149
DeGusta, D. 147, 148
dentition 143
desert bushmen 15
developmental:
  psychology 153
  studies 2
  see also children, infant(s)
didjeridu 24, 31
digestive action 106
digging-sticks 23
DIK-1-1 (hominin specimen) 144
Dikika (Ethiopia) 144
disfluency 229
Dissanayake, E. 153, 211, 212, 281
dissonance 6, 7, 11, 271, 272, 308
distance-calls 241
Divje babe I (Slovenia) 38, 39, 93, 128
Djebel Ihroud I (Morocco) 164
Dmanisi (Georgia) 299
dogs 135
  vocal tract length 154
dolerite boulders 120
Dolní-Vestonice (Czech Republic) 88
dolphins 221
Donald, M. 218, 264, 297–300, 305, 318
dopamine systems 270
Dordogne (France) 107, *107*, 108, 126
Down's syndrome 139
Drake, C. 192
dreams 17
drumming 245
  Yupik 26
drum(s) 16, 17, 25, 31, 99
  -sticks 31
  see also beaters, percussion
Dryopithicus brancoi 173
ducks 93, 94
Dunbar, R. 221, 222, 223, 225

eagles 84, 90–2, 94–5
ear(s) 169–74, 179
  canal 104, 169
  -drums 171
  inner 319
  left *vs* right 187
  see also labyrinths
early hominins:
  brains 162
  infants 209
  minimal vocal capabilities 143
  see also anatomically modern humans;
    early modern humans; *under names of
    individual species*
early modern humans 293, 305
  cognitive abilities 36

earliest occupation of Europe 96
  see also anatomically modern humans
East Turkana (Kenya) 162
echolalia 185
ecological context in music 259–68, 273
Efe 19
efficaciousness 291–2, 303
Egypt, dancing representations 122
Ehrlé, N. 193
Ekman, P. 238
El Sidrón cave (Spain) 145, 169
elder pipes 51
electroencephalogram (EEG) 182, 191
Electron Spin Resonance (ESR) dating, Divje
  babe I pipe 39
elephants 20, 245
Elowson, M. 203, 204, 221
embodied
  gesture 309, 316, 317
  motor action 181
  music as embodied activity 8, 181, 247, 317
emotion(al) 181, 237–42, 279, 311, 321, 324
  and communication 215
  contagious 239, 240, 267, 315
  control of 165–8, 175, 197
  expression 183, 218
  information 198
  Motor System 238
  in music, biological correlates 268–73
  response to music 208, 252, 253, 255–74,
    295, 314, 317
  tone analysis 189
  vocalization 206, 209, 212–13, 217, 226–7,
    312–13
empathy 255, 261, 262, 273, 315
  lack of 265
emu 22
encephalization 159, 176, 211, 296
  and group size 221–3
endocasts, fossil 161–4, 175
England, pipes 38
English language:
  emotive vocalization 206
  infant-directed speech 205
engravings 89
  on artefacts 42
  on bullroarers 108
  images of 89
  and resonance 117
  stone plaques 53
  see also incised lines
enhanced working memory 301–3
enharmonic changes 269
entrainment 243–50, 253, 268, 314
enunciation ability 141
environment(al):

Index

information in songs 27
  and laryngeal lowering 156
  stress, and avian fauna 94–6
Eorola, T. 244
Epigravettian 94
Epi-Palaeolithic 92
episodic memory *see* memory
Erb, M. 239
d'Errico, F. 37–8, 41, 49, 52, 56, 60, 76, 77, 82–4, 86–7, 106, 110, 124
Escoural (Portugal) 117, 119
Eskimos 24–9
ESR, *see* Electron Spin Resonance (ESR)
Ethiopia 144
  early hominins 222
ethnographic research 278
ethnomusicology 6, 13, 248
Europe:
  anatomically modern humans 90–1
  Mediterranean, bevelled flutes 49
  Middle Palaeolithic 36
  Upper Palaeolithic 294, 302, 323
evolution:
  and sound perception 169–74
  of the vocal apparatus, fossil evidence for 131–53
exchange, long-distance 22, 53
expectation 257
expression 292
  theory 261, 265
expressiveness 273, 315
external symbolic representation 306
extrinsic emotion in music 256–9

facial:
  expression 207–8, 215, 218, 220, 225–6, 234, 238–9, 252, 259, 298, 313, 321
  muscles 163
  prognathism 141, 142, 159
Fages, G. 34
Fagg, M. C. 119
Falk, D. 209, 210
fallow deer, laryngeal lowering 156
feather plucking 26
Fernald, A. 153, 205, 213
fertility 157
Feyereisen, P. 232
Fidelholtz, J. 139
figurines 324
  Upper Palaeolithic 42
Fink, R. 39
Finland 104
fish-identification songs 27
fishing weights 106, 109
Fitch, W. T. 142, 146, 153, 154, 156, 157, 159

flint:
  acoustic properties 120
  blades 96
  knapping 242
  long-distance exchange 53, 126
flutes 17, 31, 112
  end-blown *18*, 21, 29
  bevelled 49, 89
  block and duct 88
  Germany 124
  modern 36
  nose 89
  see also pipes
Fontalès, *see* abri Fontalès (France)
Fontanet (France) 115
foraging 19, 20
foramen magnum 141, *147*
Forkhead Box Protein P2, *see* FoxP2 genetic mutations
formant frequencies 154, 156, 207
FoxP2 gene 193
  mutations 168–9
France 108
  avian faunal remains 91, 92
  bas-reliefs 112, 113, *113*
  bullroarers 107–8, *107*
  caves 117
  engraved images 89
  Neanderthals 138
  phalangeal whistles *101*
  pipes 85
  rasps *111*
  rock art 115–17
  whistles 88
Frayer, D. 151, 152
Freeman, W. 281
French emotive vocalization 206
frequencies 104, 108, 143, 169, 171, 225, 312
  control over 157
  and infant-directed speech 205, 206, 207
  infant perception of 211
Friesen, W. 238
frontal lobe *180*
fruits 23
functional magnetic resonance imaging (fMRI) 182, 183, 189, 192, 249
funerals 20

Gagnon, L. 259
gait control 248
Gallaher, M. 237, 240
game, large, limited resources 94
games, *see* throat games
Garfinkel, Y. 122, 123
Gebrian, M. 7
geese 90–5

Geissenklösterle (Germany) 53, 125, 128
  pipes 41–51, *48, 50*, 85, 86, 96
Geissman, T. 236, 286
Gelada baboons 221
gender roles:
  Aka 20, 21
  Blackfoot 16
  Pintupi 23
  Plains Indians 17
  Yupik 25
genes 290, 291
geological pianos 120
Georgia, avian faunal remains 91
German emotive vocalization 206
Germany:
  avian faunal remains 91, 92
  early hominins 138
  Palaeolithic art 324
  pipes 38, 41–51, *48, 50*, 85–6, 88, 96
gesture 10, 11, 208, 216, 218, 220, 228,
  229–37, 244, 246, 249, 251, 256, 264,
  298, 313, 314, 316, 317, 318, 321, 325
  corporeal (bodily) gesture 234, 236, 237,
    244, 250, 252, 253, 254, 293, 313, 314, 320
  facial gesture 253, 313, 316
  see also facial expression
  gesture and speech vocalization 229–37,
    250, 251
  intentional 232
  manual gesture 228, 229–37, 293, 313
  rhythmic gestures 244, 245, 251, 314, 320
  as sentences 233, 234
  vocal gesture 155, 234, 236, 237, 244, 247,
    252, 253, 293, 313, 316, 317, 320
  vocalization and meaning 233–7
  and vocalization in infants 231–2
ghost-game songs 27
Gibson Desert 22
Gilbert, W. H. 147, 148
Gintis, H. 290
Gironde (France) 85, 108
Glory, *see* Abbé Glory
gnawing action 40, 106
gods, of hunting 26
golden eagle 65
Goldin-Meadow, S. 232
Gorilla gorilla, *see* gorillas
gorillas 20, 147, 173, 245
  hyoid 144
Gosselin, N. 271
Gotfredsen, A. 94
gourds 99
  rattles 16, *18*
Goyet (Belgium) 88
grammar 224
Grande Sale, *see* Salle d'Isturitz

Grandin, Temple 266
Grass Dance 17
grasses 25
grasslands 15, 16
Grauer, V. 15
Gravettian:
  artefacts, Isturitz 54
  avian faunal remains 92
  instruments 112
  pipes 35, 42, 56–8, 63, 65–7, 69–71, 77–80,
    85, 88, 93
  technologies 110
  tools 86
great bustard 94
Greece:
  ancient, bullroarers 105
  avian faunal remains 91, 92
  dancing representations 122
  early hominins 138
Greenland, juggling-game songs 27
grey matter 149, 150
grey parrot 245
greylag geese 90, 95
griffon vulture 87
  bone pipes 42, 44, 46, *46*
grooming 221–3, 227, 283
Gros-Louis, J. 202
Grotta Romanelli (Italy) 94
Grotte des Trois Frères (France) 89
Grotte du Portel (France) 115
group cohesion 283, 290, 305–6, 309
  and music 280–3
group selection *see* selection
Grubgraben bei Kammern (Austria) 89
grubs 23
Gypaetus barbatus, *see* bearded lammergeier/
  vulture
Gypaetus sp., *see* vultures
Gyps fulvus, *see* Griffon vulture
Gyps sp., *see* vultures

Hadar (Ethiopia) 222
Hagan, E. 283, 284, 285
hand stencils 118
Harding, J. 108
hare bone pipe 38
harmony 6, 7, 11, 181, 188, 271, 272, 308
harpooning 26, 96
Harrison, R. 103, 104
harvesting 16
Haua Fteah (Libya) 37, 38, 93
Haute-Garonne (France) 108
hawks 94
hearing 117
  and the brain 161–76
  see also auditory, ear

hearths:
　Isturitz 53
　Upper Palaeolithic 42
Heim, J. 138
Heimbuch, R. C. 138
hemispheres, brain, *see* brain hemispheres
Heschl's gyrus *180*, 191
Hewes, G. 233, 236
Hewitt, G. 148, 149, 160
hierarchies, hunter-gatherer 123
Hinoki, M. 173
Hitch, G. 301
hmmmm 219
Hohle Fels (Germany), pipes 41–51, *46*, 85, 86, 128
Hohlenstein-stadel (Germany) 324
holistic tonal phrases 217
hollow logs, *see* logs, hollow
Holocene 89
hominin species 4
　evolution 293, 297
　speech capabilities 132
　vocal apparatus 130–60
Homo erectus 4, 138, 145, 159, 176, 293, 298, 299, 305, 320
Homo ergaster 4, 138, 149, 151–3, 155–7, 159, 160, 163, 172–5, 177, 293, 296, 298–9, 319, 320
Homo georgicus 4
Homo habilis, see *Homo rudolfensis* 163
Homo heidelbergensis 4, 31, 138–40, 144–7, 151–2, 159–60, 173, 176–7, 211–12, 222, 299, 305–6, 320–21
Homo neanderthalensis, *see* Neanderthals
Homo rudolfensis 4, 147, 156, 157, 162, 175, 293, 296, 319, 320
honey-ants 23
honey-collecting 20
hormones 281
horns, representations 121
horses 16
horses representations 118
Hull, D. 291
human *vs* primate vocal production 164–9
Hunter-gatherer(s) 11–31, 109, 121, 123, 282, 310, 324
　modern, sound-producers 98
　music 11–31, 224, 308
　subsistence 11, 14
　vs early agriculturalists 123
hunting:
　Aka 20
　coordination 104
　Mbuti 20
　methods, Plains Indians 16
　and muscular bonding 241
　and music 114

Pintupi 22
Pygmies 19
　representations 120, 121
　rituals 30
　songs 19, 26
　whistles for signalling 102
Hunting Shrine (Çatalhöyük) 121
Huron, D. 276, 282
Huyge, D. 110, 112, *113*
hyenas 40
　dens 106
Hylobates 286
hyoid bone 134, *136*, 137, *137*, 140, 141, 144–6, 152, 159
hypoglossal canal 146–8, *147*

ibex representations 118, 119
Ice Age 33, 92
iconic properties of music 258, 295
ID, *see* infant-directed speech
idiophones 16, 17, *18*, 99
　representations *113*, 114, 121
　scraped 109–14
　see also rasps
Igbo people 6
Igloolik 28
Ilsenhöhle (Germany) 38
imitation 168, 175, 239, 249, 298
improvisation and the brain 163
incised lines 45, 48, 53, 57, 62, 68, 74, 85–7, 106–8, *107*, 110, 111, 114
incus 104
indexical properties of music 258
India, ringing stones 119
Indre (France) *101*, 108
Indri 286
infant(s):
　-directed speech 204–14, 226, 312, 313, 322
　musical capabilities 293
　-parent bonding 214
　perceptual and vocalization abilities 202–14
infections, jaw 142
inferior frontal lobe 197
inflection 211
inner ear 172, *172*, 174, 176
innervation 146, 147, 158, 168, 227, 320
　of the thorax 149
　vertebral 148–53
innovation, use of avian fauna 94, 96
instruments *see* musical instruments
instruments *vs* voice 309
intensity 150, 152, 158, 159, 160, 269, 308
interactors 291
intercostal musculature 148–53
intermarriage 22
inter-onset interval (IOI) 192, 193

intonation 150, 211, 216, 227
intrinsic emotion in music 256–9
Intrinsic Motor Formation 218, 238, 247
Inuit 14, 24–9, 30, 31, 224
  throat-games 28
Inuktitut language 29
Inupiaq Eskimo 24
invention 168, 175
Ioannides, A. 8, 182, 188, 190, 192
Iran:
  dancing representations 122
  lack of bird bone 92
Iraq, lack of bird bone 92
Israel:
  hyoid 145
  lack of bird bone 92
Istállóskö (Hungary) 88
Isturitz (France) 52, 93, 94, 117, 126, 128
  communal gatherings 87
  pipes 41, 42, 47, 49, 51–88, 86
  stratigraphy 54
Italian emotive vocalization 206
Italy:
  avian faunal remains 91, 92
  rasp-like artefacts 110
  Upper Palaeolithic 94
Ito, S. 173
Ituri forest 19
ivory 31
  bullroarers 105
  decorated 53
  figurine 324
  mammoth 90
  pipes 35, 42–7, 46, 49, 50, 88
  rattles 99, 114

jaguars 143
Jaques, J. 229, 230, 231
jaws 134, 142, 167
Joffe, T. 293
Johnston, R. 239
Johnston, T. 24, 25
Jordan, avian faunal remains 92, 95
juggling-game songs 27
jump-rope songs 27
Jürgens, U. 148, 164, 166, 183, 266
Juslin, P. 257, 267

Kabwe (Zambia) 138, 147
Kalahari Desert 15
kangaroos 22
Kappas, A. 264
Karow, C. 188
Katajjait 28
Kay, R. 146, 147
Kebara 2 (hominin specimen) 145, 151

Kebara Cave (Israel) 145
Keil, C. 247
Kent's Cavern (England) 38
Kenya, early hominins 138, 162
Kenyanthropus platyops 4
kinaesthetic responses to music 240, 247
Kivy, P. 262
knapping 120
KNM-ER 406 (hominin specimen) 138
KNM-ER 1470 (hominin specimen) 162, 163
KNM-ER 3733 (hominin specimen) 138
KNM-WT 15000 (hominin specimen) 151, 152, 163
Koelsch, S. 199, 312
Kogan, N. 280
Koobi Fora (Kenya) 138
Krapina J and C (hominin specimens) 145
Krause, J. 169
Kraut, R. 239
Krumhansl, C. 240, 269
Kudo, H. 222
Kuhl, P. 139, 170
Kuhn, S. 110, 112
Kulna (Czech Republic) 106
Kunej, D. 39
!Kung San 15
Kupgal Hill (India) 120
kyphosis of the spine 141

La Chapelle (hominin specimen) 151
La Chapelle-aux-Saints (France) 138, 147
La Ferrassie (France) 147
La Riera cave (Spain) 110
La Roche de Birol (France) 107–8, 107
labyrinths 172, 173, 174
Lafaye Bruniquel, see Abri Lafaye Bruniquel (France)
Laitman, J. 136, 138, 140
Lalanne, J.-G. 112
Lalinde (France) 107–8, 107
land ownership 23
language 219, 234, 279, 294, 311
  abilities, early hominins 162
  and bipedalism 174
  and the brain 180–1
  cognition necessary for 236
  development 232
  emergence of 300, 301
  evolution 131, 205
  and infants 205
  lexical 235, 322
  and music 177–200
  perception and production 139, 196
  processing 181
  production and processing 162
  and singing 224, 225

and social networks 223
and speech dysfunction 168–9
syntactic 229, 233, 322
throat games as 29
vs rhythmic ability 244
vs vocalizations 163
see also linguistic
Larribau Gallery (France) 117
larynx/laryngeal 133–44, *134*, *136*, *137*
  activation 168, 175
  anatomy 152
  control 166–8, 175, 197–8, 223, 228, 320
  lowering 153–60
  physiology 140
  position 140–4
last glacial maximum 48, 92, 95
Laugerie Basse (France) 108
Laussel (France) 112, 113, *113*
Lavy, M. 256, 263, 264
Lawson, G. 41, 52, 56, 60, 76, 77, 82–3, 84, 86, 87, 115
Le Placard (France) 126
Le Portel (France) 116
Le Roc de Marcamps (France) 88
learning deficits 174
leaves 19
Lebanon, lack of bird bone 92
Lefèvre, C. 94
leopards 143
Les Fieux (France) 117
Les-Eyzies-De-Tayac (France) 108
Lespugue (France) 108
Levant, dancing representations 122
Levenson, R. 238
Lévi-Strauss, C. 1
Lewis, J. 13
lexical language 235, 322
lexical-tonal units 216
lexicon 311
Liberman, A. 170
Libya 37, 38, 93
  avian faunal remains 91, 92
Lieberman, P. 138, 139, 153
Liégois-Chauvel, C. 171
Liguria (Italy) 110
linguistic(s):
  capacities 205
  development, child 211
  emergence of 249
  meaning of words 190
  processing 169
  and rhythmic function 193
  speech 150; and melody 227
  see also language
lion-man figurine 324
lions 143

lip-reed sound production 36, 49
lips 133, 143, 167, 232
listener *vs* performer 248
lithophones 100, 115–21, 126
lizards 22, 23
Locke, D. 13
Locke, J. 231
logs, hollow 99, 114
Lone valley (Germany) 38, 42, 324
Lortet (France) 108
Lower Palaeolithic 34
Lucy's baby 144
lullabies 20, 211
lumbosacral cord 173
lungs 134, 150
lyrics 30
  dominance of 23

mabo song and dance 20
macaques 191
MacDonald, K. 93
MacLarnon, A. 148, 149, 160
Magdalenian 127
  artefacts 126
  bullroarers 107, *107*, 108, 109
  instruments 112
  pipes 35, 51, 73, 74, 81
  rasps 110, 111, *111*, 112
  techno-complexes 124
  tools 88
  whistles *101*, 102
magnetoencephalogram (MEG) 182, 192
Maisières Canal (Belgium) 88
Malaga (Spain) 117
Malaysia bullroarers 105
Malina, M. 42, 128
mallets 115
malleus 104
Maltzman, I. 187
mammoth(s) 100, 110, 114
  bone instruments 127
  ivory pipes 35, 42–7, *46*, 49, *50*, 88, 90
  ivory *vs* bird bone, use for pipes 50
  representations 118, 119
  skull, decorated 115
Mandarin infant-directed speech 205
mandibles 134, 144–6, *146*
Mang, E. 214
Maori 105
Marcamps (France) 126
Marin, O. 184, 185, 199, 312
marmoset 171
marriages 20
Martinez, M. J. 184
Mas d'Azil (France) 110, 111, *111*, 112, 126
Masataka, N. 231

maxillary width 140
Mayberry, R. 229, 230, 231
Mbuti 14, 19–22
  songs 29
McAllester, D. 13
McBrearty, S. 97
McClave, E. 230, 231
McNeill, W. 242, 243
medicine men 17
Mediterranean Europe, bevelled flutes 49
medulla 167, 168
melody 197, 225, 249
  in neonates 204
  perception 190, 202–14
  and speech production 183–5
memory 297, 298
  artificial systems (mnemonics) 87, 111
  declarative 302, 303
  episodic 297
  processes 181
  procedural 297, 302, 303
  semantic 297, 298
  short-term 249
  working 192, 297, 301–3
Menon, V. 188
Merker, B. 241, 242
Mesolithic:
  pipes 34
  whistles 102
Mesopotamia, dancing representations 122
Mexico, idiophones 114
Meyer, L. 257
Mezhirich (Ukraine) 115
Mezin (Ukraine) 114, 127
microconcavities 38
Middle Palaeolithic:
  bone for instrument manufacture 90–6
  Europe 36
  Isturitz 53
  reindeer phalanges 104
  use of bone 95
Middle Stone Age (MSA), Africa 36, 91, 92, 125
  tools 37
  use of bone 95
Miller, G. 286–9
mimesis 218, 264, 297–301, 316, 318
mimicry 24, 27, 105, 245, 246, 298
  of animal noises 28
  bovine noises 108
  Pygmies 21
mirror neurons 249
Mitchell, R. 237, 240
Mithen, S. 3, 5, 219, 294, 296, 300
mobandi ritual dance 20
Moldova, avian faunal remains 92

molimo music 21
monkeys 20, 165, 202
  brain anatomy 167
  grasping actions 249
  hypoglossal canal 147
  tonotopy 171
  vocal tract length 154
  vocalizations 218
monophonic music 16
monosyllabic utterances 235, 236, 251, 321
moon-phases, representations *113*, 114
Morais, J. 187
Moravia 88
Morin, see abri Morin (France)
Morocco, lack of bird bone 92
morphemes 28
Morton, E. 207
motherese, see infant-directed speech
motor:
  control 192, 231, 279
  cortex 164–5, 167–8
moulting 93
Mourer-Chauviré, C. 34
Mousterian 38–41, 91
  Isturitz 53
  rasps 110
  sound-producers 33
  technology 35
  use of avian fauna 93
  whistles 102
mouth 133, 208
MSA, see Middle Stone Age
muscular
  bonding 240, 241
  timing module 230
Museé des Antiquités Nationales, Paris 55–60, 63, 65– 70, 72–4, 82–3
music:
  anthropology 5, 11–31
  see also ethnomusicology
  as an art form 295
  biological reactions to 268–73
  biological *vs* cultural foundation 194–7
  and the brain 161–8, 177–200
  capability for 276, 278, 279
  as a coalition signalling system 283–5
  cognition 5, 177–200, 202–14, 237–54
  and cognitive evolution 294–303
  communal 20, 26, 29, 30, 122, 123, 127, 250, 266, 268, 280, 281, 282, 289, 308, 309, 315,
  coordination in 243
  and creative play 8
  and dance 309, 316, 317
  definition of 3–8

ecological and social context in 259–68
emotion and communication 208, 255–74
emotion in, biological correlates 268–73
in evolution, rationales for 275–306
experience of 266–8
first direct evidence for 32
foundations of 315–19
and group selection 288, 289–91
as having human-like properties 262, 315
about hereditary land 23
hunter-gatherer 11–31, 121, 224, 281, 310, 324
and hunting 114
and infant-directed speech 204–14
interdependence with dance 237
and language processing 168, 180, 185–90, 199, 214–20, 221–5,
in later prehistory 121–4
-makers, trained 192
meaning in 296
as a mnemonic aid 30
module 194–7
multiple meanings and cognitive development 291–4
nature of 308–15
neurological study of 161–8, 177–200
not essential for survival 275
origins of, summary: 307–325
non-adaptive models 277–9
perception 5, 247, 278, 322; as innate 278, 280
performance 303
phrase in 218
for political communication 13
production 5, 322
psychology 5
and ritual 324
and rock art 115
role of 13
and sexual selection 285–9
as a shared experience 268, 284, 310, 314, 315
and speech 185–90
as a 'technology' 279
in traditional societies 124, 282
use of, adaptive rationales 279–94
Western 5, 11, 12, 99, 130, 179, 261, 272, 322
musical behaviour 7, 14, 177
biological basis for 2
differences and similarities 11, 15
as an indicator of fitness 286
as a pro-social norm 290
roles of 11–12
selection for 276

musical instruments 12, 130, 310, 323
bone for 90–6
earliest evidence for 3, 33, 35–88
first in Europe 33
hunter-gatherer 98, 109
manufacture of 303
Native American 16, *18*
Pygmies 21
raw material 324
representations 89–90
in traditional societies 99
vs sound-producers 104
see also bullroarers, drums, flutes, idiophones, percussion instruments, pipes, rattles, rasps, sound producers, whistles
musicality 273, 306, 313, 318
biology of 5
emergence of 319–25
score reading 190
musicology 6, 13, 248
musilanguage 215, 219
mute swans, bone pipes 49
myelinization 191
mylohyoid groove 145–6, *146*
mythic stage 300
mythical creatures 28
mythologies 30
Pintupi 23

Nariokotome Boy 151
narrative thought 301
nasal cavity *137*
nasopharyngeal soundspace 140, 141
Native Americans 13, 15–19, 29, 31, 224, 282, 309
see also Blackfoot, Inuit, Sioux, Yupik
natural auditory categories 176
natural selection *see* selection
Neanderthals 4, 90, 97, 138–40, 144–7, 151–3, 158–9, 164, 173, 175–6, 286, 293, 298–9, 305, 320–4
abilities 41
basicranium 141
behavioural capabilities 33
genome 169
hyoid and larynx 146
information-processing 171
Isturitz 53
musical behaviour 38
symbolic thought 37
use of avian fauna 92–3
Near East, early modern humans 91
needle-holders, bone 97
needles, bone 97
Nelsilik 28

neocortical development 222
Neolithic:
  art 121
  cup-marks 120
  sites, Netherlands 95
Nerja (Spain) 117–18
Netherlands:
  avian faunal remains 91
  Neolithic sites 95
net(s) 19, 93
  -hunting dances 20
  -traps 93
Nettl, B. 6, 13
neural:
  activation 249
  canals 134
  pathologies 189
neuroanatomy:
  development of 161–76
  functional 181–3
neurobiology 153
neurochemicals 274, 315, 319
  see also dopamine, opioids, oxytocin
neurochemistry of emotion in music 268–73
neurology:
  of emotion in music 268–73
  of music and speech 177–200
  and physiology 139
  primate 221
  of vocal production 164–9
neurons 221
neuropathology 181–3, 271
neuropsychology 2
New Guinea bullroarers 105, 108
New World monkeys 147
New Zealand bullroarers 105
Ngarrindjeri 23
Niaux (France) 115, 116
Nicolay, C. 151, 152
Nigeria, Igbo people 6
Nobe, S. 230
nomadic people, 15–22
North America
  bullroarers 105
  whistles 102
  see also Native Americans
Norton Sound 24
Norway, avian faunal remains 92
nose flutes, see flutes, nose
notational systems 87
nucleus:
  ambiguus 166, 167, 168, 175, 320
  retro-ambiguus 166
nuts 19, 20
Nyklicek, I. 269

ochre 107, 114
OH7 (hominin specimen) 164
Ohalo II (Jordan) 92, 95
OIS, see Oxygen Isotope Stages (OIS)
Old World monkeys 147
Omo L.338-y-6 (hominin specimen) 147
operculum *180*
opiate receptors 166
opioids 270
  see also dopamine, oxytocin
oral:
  cavity 133, 134, 147
  movement 232
  /praxic ability 193, 198, 230, 236, 252
  tradition 23
orang-utan 173
organic materials 99, 105, 114, 309, 310
  for instruments 15, 31
Orgeldinger, M. 286
ornamentation 96
orofacial:
  anatomy 152
  control 163, 166–9, 175, 198, 223, 228, 232, 320
  musculature 143, 197, 207, 225, 238
Orrorin tugenensis 4
osseophones 114–5
ossicles 104
osteoarthritis 138
Otis tarda, see great bustard
Otis tetrax, see little bustard
owl monkey 171
owls 94
Oxygen Isotope Stages (OIS) 91
  OIS 1 92
  OIS 2 92
  OIS 3 91, 92
  OIS 4 91, 92
  OIS 5 91, 92
oxytocin 270, 281

painting(s):
  on mammoth bones 114
  Palaeolithic 115
  and resonance 117
pair bonding 286
Pair-non-Pair (France) 85
Palaeolithic:
  art 89–90
  painting 115
  pipes, introduction 34–5
  see also Lower Palaeolithic, Middle Palaeolithic, Middle Stone Age, Upper Palaeolithic
palaeontology 153
palate 133, 142, 143

Pan paniscus, see bonobo chimpanzees
pan pipes 36
Pan sp. 132
Pan troglodytes, see chimpanzees
Panksepp, J. 249, 272
Panthera sp. 143
Papousek, M. 211, 281
Papunya 22
parahippocampal cortex 271
Paranthropus sp.:
  labyrinths 172
  P. aethiopicus 4
  P. boisei 4, 138
  P. robustus 4, 222
parentese, *see* infant-directed speech
parent–infant:
  bonding 214
  vocalizations 316
parrots 245
Parsons, L. 184
partridge 94
parturition 149
Passemard, E. 51, 54–60, 63, 65, 67–71, 73–4, 77, 80, 82–3, 85
Patel, A. 186, 187, 190, 246
Patterson, N. 265
Pech de l'Azé (France) 106
Pech-Merle (France) 117, 119
Pekárna cave (Czech Republic) 110, 111, *111*, 112
pendants 106, 109
percussion 124, 242
  instruments 29, 99, 109–15, 323; Native American 16, 17, *18*; Pygmies 21; representations 121; Yupik 25
  non-instrumental 24, 31, 109
  struck 114–15
Peretz, I. 7, 177, 186, 187, 190, 194, 196, 259
perfect fifth interval 6, 7, 11, 308
performers 261, 265
  vs listeners 248
periaqueductal grey matter (PAG) *165*, 166–8, 175, *180*, 183, 320
Perry, D. 184, 185, 199, 312
Pessac-sur-Dordogne (France) 108
Peterson, S. 163
Petit Salle, *see* Salle de Saint-Martin
Petralona (Greece) 138, 140
Petralona 1 (hominin specimen) 177
Petsche, H. 191
phalangeal whistles 90, 100–5, *101*, 112
  vs pipes 36
  sound over long distances 104
phalanges:
  pierced 34; *see also* phalangeal whistles

reindeer *101*, 102–4
pharynx *137*
phoneme analysis 189
phonetic sounds 170, 171
physiology:
  animal, vocal 135
  of the brain 162
  of emotion in music 263–4, 274
  and neurology 139
  of the thorax 149
  for vocalizations 131–53
piercing, deliberate *vs* natural 102, 103, 106
Pinker, S. 2, 278, 279, 280
Pintupi-speaking Aborigines 14, 15, 22–4, 30, 31
pipes 32–98, 112, 323
  bird bone 35, 42, 44, 45
  debate about intentionality 37
  earliest reputed 35–41
  end-blown 29
  hare bone 38
  inventories: Swabian Alb 43–5; Isturitz 55–75
  mammoth ivory 35, 42–7, *46*, 49, *50*, 88
  for paint-blowing 97
  raw material 91–3
  reed-voiced 87
  Upper Palaeolithic 41–89
  vs phalangeal whistles 36
pitch 6, 7, 11, 120, 130, 150, 152, 159–60, 168, 170, 181, 188, 197, 205, 213, 226, 227, 308
  and aggression 207
  contour 154, 160, 186, 205, 206, 209, 211, 213, 217, 218, 226, 227, 264, 312, 319
  control 154, 158, 198, 319
  perception 186, 197
  in primate vocalizations 217, 218
  and rhythm 196
  variation 280
Plains Indians 13, 15–19, 29, 224, 282, 309
  percussion instruments 31
plaques, stone, engraved 53
plaquettes, stone 110
playsongs 211
Pleistocene reindeer phalanges 104
Plio-Pleistocene hominins 138
points:
  bone 96, 110
  leaf, as sound-producers 120
Poland, avian faunal remains 91, 92
polishing 106, 112
political communication 13
polyphonic music 21
pongids, *see* bonobos; chimpanzees
Popescu, M. 192
population pressure 94

Portugal:
    avian faunal remains 91, 92
    caves 117
positron emission tomography (PET) 182-4, 187, 189, 270
post-glacial period 92
posture 215, 237, 253
Potoçka Zijalka (Slovenia) 88
Predmosti 3 and 4 (hominin specimens) 140
prefrontal cortex 165, 221
prehistory, definition of 3
premotor area *180*
preservation conditions 95; *see also* organic materials
primates:
    basicranial flexion 141
    higher 132
    infants, vocal behaviours 202-4
    neurology 221
    singing behaviour 286
    social organization 204
    vocal behaviours 201-4
    vocalization 160, 220-1, 236, 284, 310, 319
    vs human vocal production 164-9
primatology 153
Prisoners' dilemma 290
procedural memory *see* memory
procreation rate 276
projectile weapons 93
prosimians 147
prosodic repetition 213
prosody 185, 186, 188, 190, 197, 198, 210, 229, 234, 251, 262, 273, 311
proto-language 209, 210, 214-20, 221-4
proto-music 214-20, 221-4, 293, 306
Psittacus erithacus, *see* grey parrot 245
puberty rituals 17, 20
puffin-bill rattles 25
punching tool 103
purification ritual 20
Pygmies 14, 15, 19-22, 29, 31, 224, 282, 309
    see also Aka, MButi
pygmy marmosets 203, 204
Pyrenees 94, 115
Pyrenées-Atlantiques 102, 104

Qafzeh (hominin specimen) 145
quails 94
Quebec, throat-games 28

radiocarbon dating:
    cave art 119
    Divje babe I pipe 39
    Grubgraben bei Kammern 89
    Isturitz 53
rainforest people 15, 19-22

raptor wing bones 54, 68; *see also* eagles; vultures
rasps 34, 89, 90, 99, 100, 109-14, *111*, 124
    Mousterian 110
    representations 110, 112
    see also idiophones
rattle, infant use of 231
rattles 16, 17, *18*, 99, 115
    ivory 114
    puffin-bill 25
ravens 94
Rebuschat, P. 177
reconstructions:
    bullroarer 109
    cranial 138
    Neanderthal vocal tract 145
    pipe 49, 84, *86*
    supralayrngeal 140
red deer, laryngeal lowering 156
register selection 218
Regourdou (hominin specimen) 145
regularity 193
Reidenberg, J. 140
reindeer 25, 90
    antler: bullroarer 107-8, *107*; mallet 115; rasp 110
    bone musical instruments 90, 112
    phalanges *101*, 102, 103, 104
    tibia sound-producer 89
    reaction to whistling 104-5
repetition 17, 213, 226, 230
replicas 108, 112; *see also* reconstructions
replicators 291
Repp, B. 278
reproduction whistles 104
resin 50, 88
resonance 135
rock art 115-17
respiration:
    rate 269
    muscles 150
Reznikoff, I. 115, 116, 119, 127
rhesus monkeys 165, 166, 168
rhythm/rhythmic 7, 8, 11, 189, 197, 199, 225, 237-42, 280, 316, 317
    behaviour, organized 242
    capability 230, 236, 252
    control 249, 251
    infant perception of 211
    information processing 190-4
    internal 308, 315
    linguistic function 193
    movement 313
    naturally occurring 243
    in neonates 204
    perception of 202-14, 245

Index

processing 181
production 130, 163, 198
rib:
  cages *149*
  muscles 148–53
Riparo Mochi (Italy) 110, 111, 112
rites of passage 20, 29, 123
ritual(s) 30
  Aka 20
  calendrical 123
  contexts and rock art 115
  and music 324
  shamanistic 102
river hogs 20
rock art 53
  and music 115
  instruments predate 124
  and resonance 115–17
rock gongs, *see* stones, ringing 119
Rodríguez, L. 145
Roederer, J. 281
Romania, avian faunal remains 91, 92
Roucador (France) 117
Rouffinac (France) 116
Rousseau, J.-J. 215
Rudabanya (Hungary) 173
Russia, avian faunal remains 91

Saccopastore 1 and 2 (hominin specimens) 140
Sacks, O. 247, 261, 265, 266
Sahelanthropus tchadensis 4
Saint-Avil 88
Saint-Germain-en-Laye, Paris 55– 60, 63, 65–70, 73–4, 80, 82–3
Saint-Jean-de-Verges (France) *101*
Saint-Marcel (France) *101*, 108
Saint-Périer, R. de 51, 54, 56–9, 63, 65, 67–71, 76–7, 80, 82
Saint-Périer, S. de 51, 54, 56–9, 63, 65, 67–71, 76–7, 80, 82
Sakai, K. 249
Salle d'Isturitz (Isturitz cave) 52, 53, 57–8, 60, 63, 65–7, 69–74, 77–9, 81, 82, *86*
Salle de Saint-Martin (Isturitz cave) 52, 53, 55, 85, *85*
Salle Nord, *see* Salle d'Isturitz
Salle Sud, *see* Salle de Saint-Martin
salmon-fishing 25
Sami people 105
Samson, S. 193
Sanskrit 6
Saudi Arabia, lack of bird bone 92
scale structures 6, 7
Scandinavia bullroarers 105
Schachner, Adena 246

Schenck, D. 240
Scherer, K. 206, 215
Schmidt, L. 208, 212
Schön, D. 8, 189
Schöningen spears 31
Schulen (Belgium) 110
Schulz, Geralyn M. 167, 183
Schweiger, A. 187
Scothern, P. 34, 35, 38, 41, 52, 71, 76, 88, 103, 115
  'lost' pipe inventory 77–81
scrapers 31
sea-mammals 25
seasonality 123
secondary auditory cortex 187
Seeberger, F. 49
seeds 23
selection 10, 195, 275–94, 303–5
  biological selection 195, 303
  cultural selection 277, 304
  group selection 288, 289–291, 305
  natural selection 275
  possible selection processes acting on music, summary 276–7
  selective (advantages, benefits, environments, forces, influences, mechanisms, processes) 1, 3, 10, 133, 141, 154–6, 158, 169–70, 172–3, 176, 178, 204–5, 212, 215, 222–3, 227, 240–2, 275–7, 279–82, 285–93, 300, 303–5, 317–8, 320–1
  sexual selection 157, 275, 277, 285–9, 304
self-awareness 298, 299
self-sacrifice 290
semantic(s) 227
  verbal meaning 198
semantic memory *see* memory
sentic modulation 264
separation anxiety 270, 282
Serjeantson, D. 94
sexual dimorphism 157
sexual selection *see* selection
Seyfarth, R. 220
shamans:
  rituals 102
  Yupik 26
Shanidar 2 and 3 (hominin specimens) 151
Shanon, B. 187
shell, worked 53
Shennan, S. 289
shepherd whistling language 102, 104
siamang 286
Siberia, juggling-game songs 27
Siebel, W. 199, 312
sign language 271

signalling 319
  behaviour 245
  in hunting 102
silex tool 89, 108
Sima de los Huesos (Spain) 145, 222
singing 198, 224, 286, 311, 312, 317
  infant-directed 211, 213, 214
  and language 224, 225
  styles 262
Sioux 13, 15–19
  songs 29
size exaggeration hypothesis 155–8
Skhul 5 (hominin specimen) 140, 145
Skoyles, J. 223, 224
slapping 24, 31
slate, Cumbrian 120
sledding 26
Slevc, L. R. 7
Sloboda, J. 30, 180, 257, 258, 265, 267, 269, 282
Slovenia, pipe 38, 39
smiling 239, 244
snails 19
Snow, D. 183
social:
  activities 23
  bonding 304–5
  context in music 267–8
  dancing 17
  -emotional communication 225
  events and music 17
  intelligence 296, 300
  interaction 255, 292; aversion to 265; music's effect on 281; primate 202
  network and group size 221, 222
  organization: and music 282; primate 204
  relationships 321; and music 22
  stratification 122
  vocalization 221–4; primate 220–1
sociality 273, 315
Solutrean:
  bullroarers 108, 109
  images 118, 119
  lack of bone pipes 95
  leaf points, as sound-producers 120
  lithophones 126
  pipes 35, 72
  techno-complexes 124
songs:
  hunting 16, 19
  as a legal mechanism 26
  Pintupi 23
  texts 18, 20, 23
  types, Yupik 25
sound:
  perception 169–74
  production 8; instrumental 12

window 88
sound-producers 89
  identification problems 127–8
  intentional *vs* natural 88
  modern hunter-gatherer 98
  most complete and oldest 42
  Mousterian 33
  from 19th and early 20th century excavations 37
  non-pipe 99–129
  see also bullroarers, drums, flutes, idiophones, musical instruments, percussion instruments, phalangeal whistles, pipes, rattles, rasps, whistles
South Africa:
  australopithecines 173
  early hominins 138, 222
South Baffin Island, throat-games 28
Soviet archaeologists 114, 115
Spain:
  avian faunal remains 91, 92
  caves 117
  early hominins 222
  Neanderthals 145, 169
  rasps 110
spear(s) 19, 20, 22, 96
  Schöningen 31
  -throwers 23
spectograms 108
speech 139, 197, 199, 311, 312
  acoustical properties of 170
  articulate 153
  capabilities, early hominins 132
  and language dysfunction 168–9
  and melody production 183–5
  and music 185–90
  see also infant-directed speech
Sperber, D. 280
sphagnum 25
spiders' nests 16, 99
spinal:
  cord 173
  injury 152
Spoor, F. 172, 173
Spy (Belgium) 88
squirrel monkeys, vocalizations 166
stalactites, impact marks 117, 118
stalagmites 118, 119
  decorated 53
stamping 24, 31
stapedius muscles 171, 176
stapes bones 104, 171
Steinheim (Germany) 138, 139, 140
Sterkfontein (South Africa) 138, 147, 222
sternohyoid 154
sternothyroid 154
sticks, representations 121, 122

Stiner, M. 94, 110, 112
Stockman, D. 121
stone:
  axes 22
  bullroarers 105
  plaques, engraved 53
  plaquettes 110
  rasp 109
  ringing 119
  tools 15, 164, 242; Isturitz 53; for whistle manufacture 102; Upper Palaeolithic 42
strap muscles 154
striations 112, 113
string-figure games 27
Sts 5 (hominin specimen) 138
stuttering 229, 230, 250
Stw 19, 53 and 187 (hominin specimens) 147
styloid process 137, 140
subcortical:
  areas 188, 198
  basal ganglia 188
  systems 269
subglottal:
  pressure 150
  system *134*
subsistence:
  Aka 19
  hunter-gatherer 14
  rituals 30
sulphur-crested cockatoo 245
Sun Dance 16, 18
superior temporal:
  cortex 190, 249
  gyrus *180*, 186, 187, 192, 197
supernatural:
  inspiration 17
  origins of music 20
supplementary motor area *180*, 192
supralaryngeal:
  reconstructions 140
  space 141, 142, 146, 160, 319, 320
  tract 133, *134*, 154
supramarginal gyrus *180*, 190
Swabian Alb 38, 42–51, 88, 90, 94, 124, 128, 324, 327, 328, 329, 330,
Swabian cave sites 47
Swabian Jura (Germany) 42
swans 90–4
  bone pipes 43, 47, *48*, 49, 50
Swanscombe (England) 139, 147
Swartkrans (South Africa) 222
Symbol
  in Peircean semiotics *21*, *257*
symbolic:
  behaviour 125, 294

properties of music 257
referential behaviour 220
thought 294–6; early evidence 36
symbolism 29, 96, 301, 309
  lack of, in music 19, 21, 28, 29, 30
  of language 190
  in Pintupi music 24
symmetrical entrainment 246, 247
sympathy 261
synchronization 7, 244, 245, 253, 314
  chorusing 241, 277
  vocalizations 241, 242
syncopation 269
syntax 216, 224, 227, 229, 233, 257, 311, 322
  sequences 198
  structure 223
Syria:
  lack of bird bone 92
  rasps 110

Tabun I and II (hominin specimens) 145
Taforalt 12 (hominin specimen) 140
Tagliacozzo, Antonio 93
tallies 111, 113
Tarn-et-Garonne (France) 108
Tarsius 286
teasing songs 26
technologies requiring enhanced working memory 302
teeth 133
  impacted 142
  pierced 53
tegmentum 166
tempo:
  and level of arousal 260
  modulation 218
temporo-parietal region *180*, 186, 197
territory defence 284
testosterone 282
Thai Buddhism 6
Thaut, M. 248
The Dreaming, *see* Tjukurrpa
theoretic culture 297
Thermoluminescence (TL) dating, Geissenklösterle 49
thoracic:
  innervation 158
  nerves 151, 152
  vertebral column 150, 151
throat-games 28, 29
tigers 143
timbre 181, 187, 189, 198, 199, 225, 312
  processing 311
  infant perception of 211
tipis 16
Tjukurrpa 23

TL, see Thermoluminescence (TL) dating
tonal:
  phrases, holistic 217
  processing 181, 185–90, 190–4, 197
  range, increased 153–8
  vocalization 7, 154, 155, 310, 311, 312
tonality of singing 198
tone 211
  of voice 218, 234
tongue 133, *137*, 143, 146–8, 167
  control of 134, 154, 155
  position 137
tonotopy 171
tortoises 19
Trainor, L. 207, 208, 212, 213
Tramo, M. 271, 272
travelling songs 26
tree bark 16
  containers 23
Trehub, S. 6
Trevarthen, C. 238, 239, 247, 249
trill vocalizations 202
trumpets, end-blown 87
Tübingen University 43, 45
tundra peoples 24
Tunisia, lack of bird bone 92
Tunnituartuit 28
Turk, I. 39
Turkey, Upper Palaeolithic 94
Turner, R. 8, 182, 188, 190, 192
Turner, S. P. 147, 148
turnips 16
turtle shells 16
Tuto de Camalhot (France) *101*
Tyrberg, T. 91

ungulate bones 96
United Kingdom, avian faunal remains 91
Upper Palaeolithic 33–4
  blades, as sound-producers 120
  bone for instrument manufacture 90–6
  bone instruments 90–6
  Europe 294, 302, 303, 323
  pipes 41–89
  reindeer phalanges 104
  revolution 36, 125
  sound-producing activities 99
  symbolic artefacts 42
upper respiratory tract 140, 150
urban societies 123

Valée d'Aas (France) 102, 104
valence 257, 269
Van Wijngaarden-Bakker, L. H. 95
Vaneechoutte, M. 223, 224
vegetables, wild 20

Venda peoples 224
ventral premotor cortex 165, 175
Venus of Laussel 89, 112, 113, *113*
verbal:
  vs manual tasks 232
  vs vocal 184
verbalizations 132
versatility, vocal 201–24
vertebral innervation 158
vestibulo-spinal tract 173
Villa, P. 37, 38, 106
visceromotor control 167, 183
vocables 17, 18, 19, 20, 26, 28, 29, 30
vocal:
  ability 135
  anatomy 321; preservation in fossil record 134
  apparatus, evolution of 130–60
  behaviours, primate 201
  communication 321
  content and manual gesture 229–31
  contour 217
  control 228–54, 299; evolution of 161–9
  patterns, complex 168
  perception 153
  production, neurology of 164–9
  range 299, 319
  signalling 312
  tonal production 199
  tract 159, 179, 227, 319; development 155, 223, 225; length 154
  versatility 153, 159, 234, 319, 320; and complexity 201–24
  vs verbal 184
vocalization(s) 7, 28, 29, 30, 109, 115
  abilities 202–14
  complexity of 221–4
  control of 197
  and dance 251
  emotive 206, 209, 212, 213, 217, 226, 227
  and gesture in infants 231–2
  gesture and meaning 233–7
  long 150, 160
  mammalian 135
  musical traits of 212, 213
  primate 217, 218, 236, 284, 310, 319
  prohibitive and calming 206
  quality 141
  synchronized 241, 242
  vs language 163
Vogelherd (Germany), pipes 41, 42–51
voice:
  recognition 188, 198
  vs instruments 309
  see also vocal; vocalization(s)
vowel sounds 135, 136, 138, 154, 166, 167

vultures 84, 87–8, 90–3
  bone pipes 42, 44, 46, *46*, 47, *47*,
    50, 55, 57–8, 63, 66–7, 69, 72–4, *85*
  bones 47

Walker, A. 151
wall paintings, Çatalhöyük 121
wallabies 22
war dances 17
Watt, R. 225, 262
weevils 19
weights, fishing 106, 109
Western music, *see* music, Western
whales 221
Wharton, T. 234
whistles 17, 34
  Le Roc de Marcamps 88
  phalangeal 100–5, *101*, 112
  see also flutes; phalangeal whistles; pipes
white matter 149
white-faced capuchin monkeys 202
white-fronted goose 94
white-tailed eagles 95
whooper swans 95
  bone pipes 47
Wild, B. 239

Williams syndrome 265–7, 273
willow 25
  pipes 51
Wind, J. 170
wolves 40
wood(en) 125
  artefacts 15, 19, 31
  bullroarers 105
  instruments 99
  pipes 51
  rasps 109
woodwind instruments, modern 36
working memory *see* memory
Wray, A. 217
Württembergisches Landesmuseum
  (Stuttgart) 43
Wynn, T. 297, 301, 302, 303

yams, wild 19
Yupik people 14, 24–9, 30, 31, 224, 282

Zambia, early hominins 138
Zatorre, R. J. 177, 187, 190, 191, 269, 270
Zentner, M. 244
Zonneveld, F. 173
zoology 153

Printed and bound by CPI Group (UK) Ltd, Croydon, CR0 4YY